I0120548

Daniel McAlpine

Systematic Arrangement of Australian Fungi

Daniel McAlpine

Systematic Arrangement of Australian Fungi

ISBN/EAN: 9783337312725

Printed in Europe, USA, Canada, Australia, Japan

Cover: Foto ©berggeist007 / pixelio.de

More available books at **www.hansebooks.com**

SYSTEMATIC ARRANGEMENT

OF

AUSTRALIAN FUNGI,

TOGETHER WITH

HOST-INDEX AND LIST OF WORKS ON THE SUBJECT,

BY

D. McALPINE,

GOVERNMENT VEGETABLE PATHOLOGIST,

MEMBER OF THE INTERNATIONAL PHYTO-PATHOLOGIC COMMISSION ;
MEMBER OF THE IMPERIAL LEOP.-CAROL. GERMAN ACADEMY OF NATURALISTS;
HONOURMAN OF THE SCIENCE AND ART DEPARTMENT, SOUTH KENSINGTON, LONDON, ETC., ETC.

AUTHOR OF A "BOTANICAL ATLAS" IN 2 VOLS.; A "ZOOLOGICAL ATLAS" IN 2 VOLS.;
A "BIOLOGICAL ATLAS"; "SHORT NOTES FOR BIOLOGICAL STUDENTS";
"ATLAS OF ELEMENTARY PHYSIOLOGY AND PHYSIOLOGICAL ANATOMY";
"LIFE-HISTORIES OF PLANTS," ETC. ETC.

By Authority
ROBT. S. BRAIN, GOVERNMENT PRINTER, MELBOURNE.

PREFACE.

In dealing with the diseases of plants due to Fungi, it is necessary to determine the name and nature of the Fungus causing the disease, in order to be able to cope with it and to take effectual measures for its prevention, palliation, or cure. Accordingly I considered it essential to have the various known Australian Fungi recorded for reference, just as the various higher forms of Australian vegetation are so ably set forth by the Government Botanist, Baron von Mueller, in his Systematic Census. The very useful *Handbook of Australian Fungi*, prepared by Dr. M. C. Cooke, the veteran mycologist, under the sanction and authority of the various colonial Governments, has been taken as a basis and prepared the way for the present publication. This *Systematic Arrangement of Australian Fungi* aims at giving in a compact and handy form a complete enumeration of all the known species up to date, systematically arranged so as to show their relationships, and briefly described, together with such additional information as may be of use in a future detailed and more directly useful account. The object being to bring together all the species recorded by the various workers in this field, to take stock, as it were, of what has been done, I had to consult the different works bearing on the subject previous to the addition of a large number of hitherto unrecorded Fungi to the list. I have accordingly prepared a " List of Works on Australian Fungi," the first of its kind. A complete list of Fungi having been compiled from the various publications, including several papers of my own read before the Royal Society of Victoria, together with the plants or parts of plants on which they occurred, in the case of parasitic forms, the material was supplied for a provisional Host-index ; the term "host" being applied to the plant on which the Fungus lives or preys, the Fungus being an unwelcome guest as a rule. The necessity for a Host-index became apparent from the time I was appointed Vegetable Pathologist.

There are thus three connected and interdependent divisions in this publication, which may now be briefly glanced at and explained.

I.—SYSTEMATIC ARRANGEMENT OF AUSTRALIAN FUNGI.

The plan pursued is the following :—

1st. A consecutive number is given to each species, for convenience of reference, and all future additions will be numbered consecutively. Varieties are distinguished by having a letter added to the number of the species.

2nd. The number in Dr. Cooke's *Handbook of Australian Fungi* is next given for ready reference to the description of any species in that work. This serves a double purpose, and shows not only the species recorded in that work, but also species omitted.

3rd. The volume and number is next quoted for every Australian species given in Saccardo's *Sylloge Fungorum*, consisting at present of ten thick volumes, which are in the Melbourne Public Library. This is the standard work on Fungi, and is the most complete and exhaustive at the present time. The references to Cooke and Saccardo will leave no doubt as to the particular Fungus meant.

4th. The scientific name adopted for each species of Fungus follows next. It is absolutely necessary, for purposes of accuracy, to have the scientific names as well as the common names, for otherwise serious mistakes may arise. Thus, the name of "Peach Yellows" (the dreaded American disease) is often applied to a disease of the Peach in this colony, but, fortunately, it is a very different and much more harmless disease, being none other than the Peach-leaf Rust (*Puccinia Pruni*). It may be noted that the sub-genera of *Agaricus* are raised to the rank of genera ; and, as the original generic name is thus set free, it is retained for the species to which the common Edible Mushroom belongs, and which were formerly included in the sub-genus *Psa''iota*.

5th. The authority for the name is next stated. With so many different names often applied to the same Fungus, and even the same name often applied by different authors to entirely different plants, it is necessary to give the authority for the particular name, in order to indicate the precise Fungus meant. The name of the authority is usually given in a contracted form, and it will be noticed that it is sometimes printed in italics. The reason of this is that it is customary in works on Fungi often to give two authorities, the first to indicate the original describer of the Fungus, and the second where some one has classified it differently on good scientific grounds. I simply give one authority, the name of the original describer being printed in Roman characters ; and, where the original name has been set aside, the correct classifier is given in italics. The year of publication is also stated.

As an illustration of the variety of naming, I may mention one kind of Rust of Wheat met with in the colony, and the following nine names have been given to it by the authors whose names are appended :—

Uredo rubigo-vera, De Candolle.
Uredo rubigo, Berkeley.
Cæoma rubigo, Link.
Trichobasis rubigo-vera, Leveillé.
Trichobasis glumarum, Leveillé.

Puccinia rubigo-vera, Winter.
Puccinia striæformis, Westendorp.
Puccinia straminis, Fuckel.
Æecidinm asperifolii, Persoon.

By a recognised principle the name of *Puccinia rubigo-vera* is adopted, although Dr. Cooke in his Handbook uses the name of *Puccinia straminis*.

6th. The English name follows. This is merely an attempt to give an English rendering to the specific name, and something of the kind is necessary in naming Fungus diseases to the average farmer or fruit-grower : but, as these diseases become better known as to their cause, some characteristic feature of the disease may be used as a distinguishing name, such as Leaf-curl, Shot-hole, Bitter-rot, Club-root, &c.

7th. The " Habitat " is next given, indicating the various colonies in which the species have been found being recorded. It has been thought advisable to add B. for British when it occurs there, as there may have been preventives or remedies applied in the old country which it would be profitable for us to know. I make no apology for dealing with Australian Fungi, including the five colonies of the Australian Continent and Tasmania, for Fungi do not respect our political boundaries and restrict themselves to artificial limits. There must be federation in the treatment of disease if it is to be thoroughly effectual, and this has been happily illustrated in dealing with the Rust in Wheat question, in which all the colonies are united for devising measures against a common enemy.

8th. The " Occurrence " follows, indicating on what plants or parts of plants the different kinds of Fungi may be looked for. This is afterwards collectively shown in the Host-index, each plant having all its known diseases due to Australian Fungi ranged under it.

9th. " General characters " conclude the whole, giving such superficial and easily-recognised characters as may serve as a guide in the rough discrimination of many species requiring immediate attention to check their spread.

From the very nature of this work and from our present very limited knowledge of the Fungi of Australia there will be constant additions made (in fact, I have quite a number of new species awaiting determination myself), and this will be met by the issue of supplements, when necessary, on the same lines. As Dr. Cooke truly says in his introduction to the Handbook—" It is quite probable that in the course of a few years, by working up the minute species, the total number contained in this volume would be more than doubled, even without the investigation of unexplored districts."

It ought also to be borne in mind that many of the more conspicuous Fungi—such as what are popularly called Mushrooms and Toadstools—work considerable mischief, although unseen and unnoticed. Thus, the Honey Agaric (*Armillaria mellea*), which is even considered edible, does a deal of damage, and by attacking the roots undermines the tree. It spreads from root to root in the soil by means of long purple-black cord-like strands, even in the absence of the tawny-yellow " Toadstools," which are simply the fructification of the Fungus, and I have seen orchard trees killed by this cause. In the soil and in the rotting roots or wood these strands are found, attacking the roots and bases of stems and often causing copious " gumming " there. The Vegetable Pathologist should therefore not only be more or less conversant with the Fungi of the different colonies, as they spread so readily by means of their spores, but he should be acquainted with Fungi as a whole, since even Mushrooms and Toadstools are not beyond his province.

II.—Provisional Host-index of Australian Fungi.

The list of Fungi, systematically arranged, enables us to classify them under their respective Host-plants. Strictly speaking, it is only those which are parasitic, or which prey upon living plants, that should be included; but it is so difficult with our present knowledge to distinguish between those which cause disease and those which attack decaying or decayed parts, that I have given all the Fungi found upon any particular plant. While special attention is paid to the Fungi occurring on the various vegetable products grown in the colony for commercial purposes, as given in the Government Statist's returns, the Fungi on so-called "weeds" are not neglected, because they may and often do pass over to the cultivated and therefore more delicate forms of vegetation. For example, the Fungus causing "Club-root" in Cabbages, Cauliflowers, Turnips, Radishes, Kale, &c., also infests two of our common weeds, viz., Shepherd's Purse (*Capsella Bursa-pastoris*, Mœnch) and Hedge Mustard (*Sisymbrium officinale*, Scop), and many similar instances could be given. This fact is strikingly put by Mr. Bailey, who says—"As we find in the animal kingdom the wild man preferring sheep to kangaroo, the flying fox peaches to quandong, the grasshopper the more succulent vegetation of our gardens to the dry herbage of the plains, so in like manner we shall doubtless find from time to time blight-fungi, at present unknown, will come from the indigenous plants to exotic ones which may be more congenial to their development."

There can be no doubt that many of the Fungi on our native vegetation will attack introduced plants, and it would be very desirable, both in the interests of science and of practical utility, to have a record of the Fungi preying upon our native plants. I have seen some of our richest soils with the decaying roots of Eucalypts and the mycelium of Fungi passing from them to the roots of orchard trees and causing their decay.

The Host-index should serve various useful purposes. First of all, it will enable the intelligent grower to determine with some degree of certainty the cause of the disease when it is due to a Fungus, and that is often the first step towards its eradication. Thus, if his Peach trees are affected with some Fungus disease on the leaves, he turns up the Index and finds two Fungi recorded there. He then turns to the General Characters in the "Systematic Arrangement" and can easily tell whether it is the "Peach-leaf Rust" or the "Leaf-curl." Or if his Cabbages and Cauliflowers begin to turn yellow and the roots become distorted, he finds from the Index that it is due to a Fungus, a knowledge of which enables him to battle with the disease. Having traced the disease to its source, he may find treatment already prescribed in some of the Government publications, or can apply to the Department for advice. If there is no record of the disease in the Index, then the grower knows it is a subject requiring investigation.

Further, the Host-index may be used in assisting growers to "spot" diseases due to Fungi before they have spread too far and become established. A great many Fungus diseases are overlooked for a number of years and allowed to spread freely before active measures are taken for their suppression, and thus what might have been easily nipped in the bud is now difficult to eradicate; so that another important use of this publication will be to enable Fungus diseases to be recognised at the earliest possible moment and action taken accordingly.

Onion Mould, Ergot in Rye and other Grasses, Powdery Mildew in Apple, and various other diseases, are not recorded in Cooke's Handbook, and, presumably, have been neglected.

A third use will be to assist in the carrying out of any legislation which may be passed for the suppression of Insect and Fungus pests. Many growers err in ignorance, because they are not aware of the disease being present until it has got a firm hold, but now a record of the various Fungus pests is available.

And there is a final purpose to be served which is not the least important. New diseases are continually cropping up, and the sooner they are recognised the better. If the disease is not recorded in the Index there is a strong probability of its being some new one, and then it can be traced to its source without delay.

The names of the Host-plants are given according to Baron von Mueller's *Second Systematic Census of Australian Plants* or Hooker and Jackson's *Index Kewensis*, as far as published. The Fungi belonging to Victoria are indicated by the letter V.

III.—List of Works on Australian Fungi.

It was necessary, as already stated, to draw up a list of works in order to have the list of Fungi as complete as possible. I have only included those publications in which there is special reference to Australian forms, and no doubt several have been overlooked. To Dr. Alexr. Morrison I am much indebted for bringing under my notice some references to the subject in scattered publications. The "List of Works," the "Systematic Arrangement," and the "Host-index" should serve to focus our present knowledge and prepare the way for further additions to it.

In giving the general characters of the various Fungi, I have endeavoured to use as simple terms as possible, but it was difficult to avoid the employment of technical terms occasionally. For those who wish to enter into the subject more fully and to study in an elementary way the disease-causing Fungi, the following works among others may be mentioned : —

Diseases of Plants, by Professor Marshall Ward, and published by the Society for promoting Christian Knowledge (2s. 6d.). This is a readable little book, and treats in a popular manner such diseases as Rust in Wheat, Smut of Corn, Ergot of Rye, Hop disease, Potato disease, &c.

Diseases of Field and Garden Crops, by Worthington G. Smith, and published by MacMillan and Co. (1s. 6d.). This work is beautifully illustrated, and treats of Onion, Pea, Parsnip, Lettuce, Potato, and other diseases, in addition to those of Wheat and Oats.

Diseases of Crops and their Remedies, by Dr. A. B. Griffiths, and published in Bell's Agricultural Series (2s. 6d.). The diseases of leguminous, gramineous, root, and miscellaneous crops are considered, together with the Fungi or insects causing them, and the best methods of prevention.

Fungus Diseases of the Grape and other Plants, and their Treatment, by F. Lamson-Scribner, and published in America (5s.). This is a thoroughly practical work, and deals with the principal Fungus diseases of Fruit trees as well as of the Vine.

Fungi and Fungicides, by Dr. C. M. Weed, and published in New York (5s.). It is divided into five parts—Fungi affecting the larger fruits, the small fruits, shade trees, &c.. vegetables, cereals, and forage crops ; and practical remedies, as a rule, are given.

There is still a want of proper works dealing with the subject of Fungus disease from an Australian standpoint and suited to the wants of our orchardists and vignerons especially, but the strong necessity which exists for such information will probably soon lead to its being supplied.

The preparation of this work has entailed a vast amount of labour, done single handed and in my spare time, but it was absolutely necessary as a preliminary for the proper carrying out of my duties. To all those who have supplied me with information my best thanks are due and are hereby tendered. It is hardly necessary to mention special names, since the "List of Works" will afford the best evidence of work done. The Government Botanist, Baron von Mueller, has always aided me with the free use of his library and the benefit of his rare and critical knowledge in connexion with some of the Host-plants. Mr. F. M. Bailey, F.L.S., Colonial Botanist of Queensland, has given me every assistance in his power in connexion with Queensland Fungi, and Mrs. Flora Martin is well known for her indefatigable labours in extending our knowledge of Australian species. I am indebted to A. de Bavay for a list of the Yeasts identified by him in Australia, and he adds that they will be largely increased from time to time. Wine Yeasts especially will yet play an important part in connexion with that industry, and there are kinds of Yeast causing decomposition and disease in Onions, &c. The officers in the neighbouring colonies have also willingly given me the benefit of their advice when asked. Amid such a mass of detail some important points may have been overlooked, and I shall be pleased to have any errors or omissions pointed out, such additions and corrections to be subsequently issued as a supplement.

It must not be imagined that because we have tabulated and briefly described a number of Fungi we therefore know all that is necessary about them. The most fascinating branch is the life-history—the story of their lives from year to year ; and it is this knowledge as to their various and often disguised phases, how they spread, and where they winter, which will help us to cope with them successfully. There is room for plenty of workers, and it is hoped that some of our young and rising fruit-growers and farmers may be induced to attend to this subject, on account of its great interest and practical importance.

1st January, 1895.

CONTENTS.

I. SYSTEMATIC ARRANGEMENT OF AUSTRALIAN FUNGI.

I.—SYSTEMATIC ARRANGEMENT OF AUSTRALIAN FUNGI.

SYNOPSIS OF GROUPS.

IN selecting a system of classification I have adopted that which best expresses the present state of our knowledge as regards the life-histories of the various forms, which after all constitute the ultimate court of appeal in settling affinity. But unfortunately there are numerous cases where the life-history has not been wrought out, and so certain groups have to be provisionally placed along with those to which they seem to be most nearly related. Saccardo's *Sylloge Fungorum* has been mainly followed, while Dr. Cooke's *Handbook of Australian Fungi*, G. Massee's *British Fungus-Flora*, De Bary's *Fungi*, and Brefeld's works have all been consulted.

GROUPS OF AUSTRALIAN FUNGI.

MYCOMYCETES ...	I.—HYMENOMYCETES ...	} Basidiomycetes.
	II.—GASTROMYCETES ...	
	III.—UREDINES	Æcidiomycetes.
	IV.—PYRENOMYCETES ...	
	V.—DISCOMYCETES	} Ascomycetes.
	VI.—TUBEROIDES	
	VII.—HYPHOMYCETES ...	
	VIII.—SPHÆROPSIDES ...	} Imperfect forms of Ascomycetes ?
	IX.—SACCHAROMYCETES ...	
	X.—USTILAGINES	Transitional forms.
	XI.—PHYCOMYCETES ...	
	XII.—MYXOMYCETES ...	Transitional to animals.

The systematic sequence of the groups is at present a matter of individual opinion, but they are arranged in the order in which they will be treated, and are reduced within the smallest limits consistent with clearness. The two main divisions are *Mycomycetes*, in which there are no sexually produced reproductive bodies, and *Phycomycetes*, or those approximating to sea-weeds, in which reproduction is sexual as well as asexual.

The *Mycomycetes* are divided into two chief classes—*Basidiomycetes* and *Ascomycetes*—the former producing naked spores at the ends of large terminal cells called *basidia*, and the latter producing spores in an *ascus* or bag. The *Ustilagines* are regarded as transition forms to the *Phycomycetes*. The *Mycomycetes* differ in important points from fungi, and are regarded as more nearly related to animals, but they are conveniently retained here for the present. The *Schizomycetes* or Bacteria also differ from fungi in chlorophyll being sometimes present, and the hyphæ or the threads of the ordinary fungus absent. To this group belong some of the organisms causing disease in plants, but as the greater part of the forms belong to medicine, I have finally decided to omit them. The *Uredines* are doubtful in their affinities, and they are placed in a class, *Æcidiomycetes*, between the other two until their position is properly settled.

B

The imperfect forms are those which are assumed to be genetically related to other fungi, probably *Ascomycetes*, in contradistinction to the perfect fungi, which have an independent life-history.

The characteristic features of each of the twelve groups are here shown, then the general classification of each group is given in its proper connexion.

BASIDIOMYCETES.—Naked spores borne on basidia. Receptacle distinct.

    I.—HYMENOMYCETES,—Hymenium external.

    II.—GASTROMYCETES.—Hymenium internal.

ÆCIDIOMYCETES.—Æcidium or cluster-cup forms a feature of the life-history.

    III.—UREDINES.—Receptacle none or obsolete.

ASCOMYCETES.—Spores produced in asci or spore-sacs.

    IV.—PYRENOMYCETES.—Receptacles (*Perithecia*) flask-shaped or spherical, opening at apex.

    V.—DISCOMYCETES.—Receptacles (*Apothecia*) disc- or cup-shaped.

    VI.—TUBEROIDES.—Subterranean, sub-globose, indehiscent.

IMPERFECT FORMS OF ASCOMYCETES ?—

    VII.—HYPHOMYCETES.—Perithecia absent.

    VIII.—SPHÆROPSIDES.—Perithecia present.

    IX.—SACCHAROMYCETES.—Multiplication by gemmation and ascospores.

TRANSITIONAL FORMS.—

    X.—USTILAGINES.—Minute, parasitic, usually spores of one kind only.

ALGA-LIKE FORMS.—

    XI.—PHYCOMYCETES.—Mycelium without septa. Sexual and asexual reproduction.

ANIMAL-LIKE FORMS.—

    XII.—MYXOMYCETES.—Plasmodium or naked mass of motile protoplasm formed and hyphæ absent.

GENERAL CLASSIFICATION OF HYMENOMYCETES.

GROUP I.—HYMENOMYCETES, FRIES.

ARRANGEMENT OF ORDERS (6).

Hymenium or spore-bearing surface normally inferior—

1. AGARICACEÆ—Hymenium spread over gills.
2. POLYPORACEÆ—Hymenium spread over tubes or pores.
3. HYDNACEÆ—Hymenium spread over prickles.
4. THELEPHORACEÆ—Hymenium spread over an even surface.

Hymenium superior or encircling—

5. CLAVARIACEÆ—Plants club shaped or branched, rarely lobed.
6. TREMELLACEÆ—Plants lobed, convolute, or disc-like : gelatinous.

ORDER I.—AGARICACEÆ, FRIES.

ARRANGEMENT OF GENERA (55).

Section I.—Leucosporæ—Spores white, or nearly so.

Series 1. Haplophyllæ—Gills entire at edge.

Sub-section 1. Molles—Plants fleshy, more or less firm, putrescent, not reviving when once dried.

Genera (16)—

1. Amanita, Pers.	5. Armillaria, Fries.	9. Mycena, Pers.	13. Hygrophorus, Fries.
2. Amanitopsis, Roze.	6. Tricholoma, Fries.	10. Hiatula, Fries.	14. Lactarius, D.C.
3. Lepiota, Pers.	7. Clitocybe, Fries.	11. Omphalia, Fries.	15. Russula, Pers.
4. Schulzeria, Bres.	8. Collybia, Fries.	12. Pleurotus, Fries.	16. Cantharellus, Pers.

Sub-section 2. Tenaces—Plants tough and leathery, or hard, reviving when moistened.

Genera (6)—

17. Marasmius, Fries.	19. Panus, Fries.	21. Trogia, Fries.	22. Lenzites, Fries.
18. Lentinus, Fries.	20. Xerotus, Fries.		

Series 2. Schizophyllæ—Gills split at edge.

Genus (1)—
23. Schizophyllum, Fries.

Section 2. Rhodosporæ—Spores rosy or salmon pink.

Genera (10)—

24. Metraria, C. and M.	27. Pluteus, Fries.	30. Leptonia, Fries.	32. Eccilia, Fries.
25. Volvaria, Fries.	28. Entoloma, Fries.	31. Nolanea, Fries.	33. Claudopus, Smith.
26. Annularia, Schulz.	29. Clitopilus, Fries.		

Section 3. Ochrosporæ—Spores ochrey brown or red brown.

Genera (11)—

34. Pholiota, Fries.	37. Hebeloma, Fries.	40. Galera, Fries.	43. Cortinarius, Pers.
35. Locellinia, Gill.	38. Flammula, Fries.	41. Tubaria, Smith.	44. Paxillus, Fries.
36. Inocybe, Fries.	39. Naucoria, Fries.	42. Crepidotus, Fries.	

Section 4. Melanosporæ—Spores blackish purple, purplish brown, black, or nearly black.

Genera (11)—

45. Agaricus, Linn.	48. Psilocybe, Fries.	51. Bolbitius, Fries.	54. Anellaria, Karst.
46. Stropharia, Fries.	49. Deconica, Smith.	52. Coprinus, Pers.	55. Psathyrella, Fries.
47. Hypholoma, Fries.	50. Psathyra, Fries.	53. Panæolus, Fries.	

Total number of species = 552.

4

SYSTEMATIC ARRANGEMENT

Number.	Cooke's Number.	Saccardo's Number.	Scientific Name.	Authority for Name.	English Name.

GROUP I.—HYMENOMYCETES.—FRIES, SYST. MYC. I. 53 (1821).

1. AMANITA.—Pers. Syn. 246 (1801).

1	8	V. 36	A. ananæceps	... Berk., Hook., Lond. Journ. VII. 572 (1848)	... Pine-apple-headed amanita ...
2	16	„ 23	A. grossa	... Berk., Fl. Tasm. II. 242 (1860)...	... Large amanita ...
3	4	„ 8	A. mappa	... Fries, Epicr. 6 (1838) ...	... Napkin amanita ...
4	6	IX. 5	A. murina	..., Cooke and Mass., Grev. XVIII. 1 (1889)	... Mouse-coloured amanita ...
5	7	V. 18	A. muscaria	... Linn., in Fries S.M. I. 16 (1821)	... Fly amanita ...
6	1	„ 3	A. ovoidea	... Fries, Hym. Eur. 18 (1874)	... Ovoid amanita ...
7	2	„ 4	A. Preissii	... Fries, Pl. Preiss. II. 131 (1846)	... Preiss's amanita ...
8	9	„ 31	A. spissa ...	... Fries, Epicr. 9 (1838)	... Clammy amanita ...
9	5	...	A. strobilacea	... Cooke, Grev. XIX. 82 (1891)	... Cone-like amanita ...
10	3	V 7*	A. verna ...	... Fries, Hym. Eur. 18 (1874)	... Spring amanita ...

2. AMANITOPSIS.—Roze, in Karst.

11	14	IX. 9	A. curta ...	... Cooke and Mass., Grev. XVI. 72 (1888)...	... Short-stalked amanitopsis ...
12	13	„ 7	A. farinacea	... Cooke and Mass., Grev. XVIII. 1 (1889)	... Mealy amanitopsis ...
13	12	„ 4	A. illudens	... Cooke and Mass., Grev. XVI. 30 (1887)	... Illusive amanitopsis ...
14	15	„ 6	A. pulchella	... Cooke and Mass., Grev. XVIII. 1 (1889)	... Beautiful amanitopsis ...
15	11	V. 47	A. vaginata	... Roze, Karst. Hattsv. I. 6 (1879)	... Sheathed amanitopsis ...

3. LEPIOTA.—Pers.,

16	39	V. 150	L. asprata	... Berk., Hook., Lond. Journ. VI. 481 (1847)	... Warty lepiota ...
17	43	„ 258	L. australiana	... Fries, Pl. Preiss. II. 131 (1846)	... Australian lepiota ...
18	25	„ 185	L. Beckleri	... Berk., Linn. Journ. XIII. 156 (1873)	... Beckler's lepiota ...
19	32	„ 243	L. bubalina	... Berk., Linn. Journ. XIII. 156 (1873)	... Gazelle lepiota ...
20	37	„ 130	L. cepæstipes	... Fries, Hym. Eur. 35 (1874)	... Onion-stalked lepiota ...
20A	„	„ ..	L. cepæstipes, var. cretacea	... Bull. Champ. 374 (1798)	... Chalky lepiota ...
21	33	„ 236	L. cheimonoceps	... Berk. and Curt., Linn. Journ. X. 283 (1869)	... Winter-capped lepiota ...
22	24	„ 101	L. clypeolaria ...	... Fries, S.M. I. 21 (1821)	... Shield-like lepiota ...
23	29	„ 111	L. cristata	... Fries, S.M. I. 22 (1821)	... Crested lepiota ...
24	20	„ 84	L. dolichaula	... Berk. and Br., Linn. Trans. XXVII. 150 (1869)	... Long-tubed lepiota ...
25	18	„ 83	L. excoriata	... Fries, Hym. Eur. 30 (1874)	... Flaky lepiota ...
26	28	IX. 29	L. fimetaria	... Cooke and Mass., Grev. XVIII. 1 (1889)	... Dung lepiota ...
27	38	V. 145	L. granulosa	... Fries, Hym. Eur. 36 (1874)	... Granular lepiota ...
28	42	IX. 36	L. lavendulæ	... Cooke and Mass., Grev. XVI. 72 (1888) ...	... Lavender lepiota ...
29	34	V. 234	L. leontoderes	... Berk. and Br., Linn. Journ. XI. 499 (1871)	Tawny lepiota ...
30	21	„ 170	L. lepidophora	... Berk. and Br., Linn. Journ. XI. 498 (1871)	... Scaly lepiota ...
31	36	„ 152	L. licmophora	... Berk. and Br., Linn. Journ. XI. 500 (1871)	... Shovel-bearing lepiota ...
32	19	„ 88	L. mastoidea	... Fries, S.M. I. 20 (1821)	... Bossed lepiota ...
33	...	...	L. membranacea ...	... Cooke and Mass., Grev. XXI. 36 (1892)	... Membranous lepiota ...
34	40	V. 165	L. mesomorpha	... Fries, Fl. I. 2 (1828) ...	... Intermediate lepiota ...
35	31	„ 127	L. naucina	... Fries, Epicr. 17 (1838) ...	... Short-stalked lepiota ...
35A	31	„ 127	L. naucina, var. sphærospora	Cooke and Mass., Grev. XVIII. 5 (1890)	... Globose spored lepiota ...
36	35	IX. 37	L. obclavata	... Cooke and Mass., Grev. XVI. 30 (1887) ...	... Obclavate lepiota ...
37	23	IX. 13	L. ochrophylla	... Cooke and Mass., Grev. XVIII. 2 (1889)	... Ochre-gilled lepiota ...
38	16	V. 70	L. procera	... Fries, S.M. I. 20 (1821)	... Tall lepiota ...
39	17	„ 74	L. rhacodes	... Fries, Hym. Eur. 29 (1874)	... Stripped lepiota

OF AUSTRALIAN FUNGI.

Number	W.A.	S.A.	T.	V.	N.S.W.	Q.	B.	Occurrence.	General Characters.

ORDER I.—AGARICACEÆ.—FRIES, PL. HOMON. 65 (1825).

Agaricus, Amanitopsis.

Number	W.A.	S.A.	T.	V.	N.S.W.	Q.	B.	Occurrence.	General Characters.
1	...	...	T.	...		Q.	...	Ground	Broad, smooth, shining, breaking into distinct angular spaces or areola at the centre, each bearing a conical wart; pine-apple-like.
2	...	...	T.	...		...	...	Ground ...	White. Thick, fleshy, warted. Stem bulbous; ring obsolete.
3	...	...	...	V.		...	B	Ground ...	Delicate. Without separable cuticle, dry, primrose, white, or buff.
4	...	...	...	V.		Q.	...	Sandy soil	Bell shaped, shining, mouse coloured. Stem thin, straight, whitish.
5	...	S.A.	...	V.		...	B	Woods ...	Large. Orange scarlet, clothed with scattered warts. Very common.
6	...	...	...	V.		...	...	Ground ...	White. Hemispherical, margin inflexed. *Edible.*
7	W.A.	...	...	...		...	...	Sandy soil, woods, &c.	Fleshy, viscid. Stem stuffed, mealy, rooting.
8	...	S.A.	...	...		...	B	Woods ...	Rough, with minute crowded mealy warts, amber grey.
9	...	...	...	V.		...	...	Ground ...	Hemispherical, covered with large persistent obtusely-conical warts, arranged after the manner of a fir-cone, ochrey yellow.
10	...	...	...	V.		Q.	B	Moist woods	Snow white, beautiful, viscid. Appearing in spring and summer.

Hattsv. l. 6 (1879).—Agaricus, Amanita.

Number	W.A.	S.A.	T.	V.	N.S.W.	Q.	B.	Occurrence.	General Characters.
11	...	...	...	V.	...	...	...	Ground ...	Ochrey white. Stem short, bulbous, brick red.
12	...	...	...	...	...	Q.	...	Ground ...	White and mealy. Fleshy, sprinkled with erect prominent warts, chiefly at disc.
13	...	...	...	V.	...	...	...	Ground	Ochrey yellow, clad with scattered broad unequal warts, which soon fall away.
14	...	...	...	V.	...	...	...	Ground ...	Vermilion, clad with irregular deciduous whitish warts. Stem hollow, white.
15	...	...	...	V.	N.S.W.	Q.	B	Woods and under trees	Size and colour very variable, grey, brown, &c. Thin, margin membranaceous, deeply furrowed. Common. *Edible.*

Tent. Disp. 68 (1797).—Agaricus.

Number	W.A.	S.A.	T.	V.	N.S.W.	Q.	B.	Occurrence.	General Characters.
16	...	...	...	...	N.S.W.	Q.	...	Ground, trunks ...	Hemispherical, pallid yellow, often deep orange, rough with warts.
17	W.A.	...	...	...	...	...	...	Sandy soil ...	Large. Slightly fleshy, viscid. Stem long, club shaped downwards.
18	...	...	...	...	N.S.W.	...	...	Ground, in scorched places	Spongy, rough about apex with little scales. Stem long, minutely warted.
19	W.A.	...	...	V		...	...	Cow-dung, &c.	Snow white. Cap ovate to hemispherical. Stem thickened downwards.
20	...	...	...	...		Q.	B	Ground ...	Sub-membranaceous, mealy and scaly, yellow. Stem hollow and bulbous.
20A	...	...	...	...		Q.	...	Ground ...	Chalky white, with darker scales.
21	...	...	...	...		Q.	...	Trunks ...	Snow white. Thin, powdery. Stem thickened downwards; mealy.
22	...	...	...	V.		Q.	B	Woods and hot-houses	Slender. Fleshy, variable in colour, yellowish or pinkish. Common.
23	...	...	T.	V.	...		B	Fields, lawns, &c. ...	Small and delicate. Slightly fleshy, whitish yellow, red scales.
24	...	...	...	...		Q.	...	Ground ...	Fleshy, centre smooth, otherwise with point-like scales, margin torn.
25	W.A.	...	...	V.	N.S.W.	Q.	B	Pastures ...	Small and delicate. Fleshy, pale-fawn colour. Cuticle thin, breaking up into scattered papillæ. *Edible.*
26	...	...	...	...	...	Q.	...	Dung ...	Fleshy, thin, pallid, ornamented with darker scales. Stem slender, scaly below.
27	...	...	...	V.		Q.	B	Woods and heaths ...	Small. Fleshy, tawny or dull reddish yellow, mealy and granular.
28	...	S.A.	...	V.		...	...	Ground ...	Rather fleshy, mealy, greyish-blue or dove colour. Stem cylindrical, whitish.
29	...	...	...	...		Q.	...	Ground ...	Tawny, with a few pallid warts, minutely tomentose or downy.
30	...	...	...	...	N.S.W.	...	...	Ground ...	Rather fleshy, white, sprinkled with minute reddish scales, yellow on drying.
31	...	...	...	V.	...		B	Ground ...	Membranaceous, lemon coloured, folded and furrowed, margin notched.
32	...	S.A.	...	V.	N.S.W.	Q.	B	Ground ...	Whitish brown. Rather fleshy, umbo or boss acute, scales papillate. *Edible.*
33	...	...	...	...		Q.	...	Chips of wood buried in ground	Membranaceous, thin, pale-cream colour. Stem slender, hollow.
34	...	...	...	V.	...		B	Ground ...	Slender. Rather fleshy, tawny. Stem even and smooth, as well as cap.
35	...	...	...	V.	...		B	Fields ...	Whitish. Fleshy, silky. Stem short, almost hollow, thickened at base.
35A	...	...	...	...		Q.	...	Ground ...	Spores globose.
36	...	S.A.	...	V.		...	...	Charred ground, under *Eucalyptus*	Rather fleshy, mealy, rufous with a tawny tinge. Stem slender, cylindrical
37	...	...	...	...		Q.	...	Sandy ground	Fleshy, pale ochre, variegated with darker concentric scales.
38	...	...	T.	V.	N.S.W.	Q.	B	Pastures ...	Large and tall, very shaggy, brownish. Fleshy, cuticle thick and torn up into broad evanescent scales. *Edible.*
39	...	...	...	V.		...	B	Shady pastures ...	Very large. Fleshy, grey, cuticle thin, broken into persistent brown scales. *Edible.*

Number.	Cooke's Number.	Saccardo's Number.	Scientific Name.	Authority for Name.	English Name.
					3. LEPIOTA.—Pers.,
40	30	V. 119	L. rhizobola ...	Berk., Hook., Lond. Journ. IV. 42 (1845)	... Bulbous-stalked lepiota ...
41	41	,, 204	L. rhyparophora ...	Berk. and Br., Linn. Journ. XI. 500 (1871)	... Spot-bearing lepiota ...
42	22	IX. 28	L. rhytipelta ...	F. v. M., Linn. Soc. N.S.W. 104 (1882) ...	... Wrinkle-shield lepiota ...
43	26	,, 16	L. stenophylla ...	Cooke and Mass., Grev. XV. 98 (1887) ...	... Narrow-gilled lepiota ...
44	27	V. 237	L. subclypeolaria ...	Berk. and Curt., Linn. Journ. X. 283 (1869)	... Sub-clypeolate lepiota ...
					4. SCHULZERIA.—Bres,
45	45	IX. 45	S. revocans	... / Cooke and Mass., Grev. XVIII, 2 (1889)...	... / Recalling schulzeria ... ... /
					5. ARMILLARIA.—Fries,
46	48	IX. 49	A. fulgens	Cooke and Mass., Grev. XVIII. 2 (1889)	... Shining armillaria ... ...
47	47	V. 289	A. mellea	Fries, Hym. Eur. 44 (1874)	... Honey-coloured armillaria ...
48	46	,, 265	A. robusta ...	Fries, S.M. I. 26 (1821)	... Robust armillaria ... ...
48A	,,	,, ,,	A. robusta, var. subannulata	Batsch, Consp. f. 17 (1783)	... Smaller-ringed armillaria
					6. TRICHOLOMA.—Fries,
49	55	V. 415	T. cerinum	... Fries, S.M. I. 80 (1821)	... Wax-coloured tricholome ...
50	56	,, 466	T. civile ...	Fries, Icon. t. 42, f. 1 (1867)	... Civil tricholome ... ...
51	51	IX. 52	T. coarctatum	Cooke and Mass., Grev. XVIII. 2 (1889)	... Compressed tricholome ...
52	53	V. 387	T. cuneifolium	... Fries, S.M. I 99 (1821)	... Wedge-gill tricholome
53	59	,, 485	T. humile...	Fries, S.M. I. 51 (1821)	... Humble tricholome ...
54	58	,, 480	T. melaleucum	Fries, S.M. I. 114 (1821)	... Black and white tricholome ...
55	50	,, 329	T. muculentum	Berk., Hook., Lond. Journ. IV. 43 (1845)	... Glutinous tricholome ... ...
56	57	,, 470	T. nudum...	Fries, Hym. Eur. 72 (1874)	... Naked-margined tricholome ...
57	60	,, 488	T. persicinum	Fries, S.M. I. 52 (1821)	... Peach-coloured tricholome ...
58	61	,, 501	T. putidum	Fries, Epicr. 54 (1838)...	... Fœtid tricholome ... ...
59	49	,, 326	T. resplendens	Fries, Mon. I 55 (1857)	... Resplendent tricholome ...
60	52	,, 344	T. rutilans	Fries, S.M. I. 41 (1821)	... Red-haired tricholome ...
61	54	,, 401	T. sulphureum	Fries, S.M. I. 110 (1821)	... Sulphur-coloured tricholome ...
					7. CLITOCYBE.—Fries,
62	76	IX. 106	C. canaliculata	... / Cooke and Mass., Grev. XVIII. 2 (1889)	... Channelled clitocybe ...
63	63	V. 553	C. cerussata	Fries, S.M. I. 92 (1821)	... White clitocybe ... ...
64	62	,, 517	C. curtipes	... Fries, S.M. I. 88 (1821)	... Short-stalked clitocybe ...
65	73	,, 637	C. expallens	Fries, Mon. I. 129 (1857)	... Bleaching clitocybe ...
66	71	,, 621	C. flaccida ...	Fries, S.M. I. 81 (1821)	... Flaccid clitocybe ...
67	64	,, 580	C. fumosa ...	Fries, Hym. Eur. 91 (1874)	... Smoky clitocybe ...
68	68	,, 612	C. gilva	Fries, Hym. Eur. 95 (1874)	... Yellowish-tan clitocybe ...
69	66	,, 595	C. infundibuliformis	Fries, Hym. Eur. 93 (1874)	... Funnel-shaped clitocybe
70	70	,, 619	C. inversa	Fries, Hym. Eur. 96 (1874)	... Inverted clitocybe ...
71	75	,, 720	C. laccata	Fries, S.M. I. 106 (1821)	... Sealing-wax clitocybe...
72	67	IX. 75	C. myriophylla	Cooke and Mass., Grev. XVI. 113 (1888)	... Myriad-gilled clitocybe ...
73	74	V. 643	C. pruinosa	Lasch, in Fries, Epicr. 75 (1836) ...	... Pruinose clitocybe ...
74	65	,, 572	C. schizophylla	Berk., Fl. Tasm. II. 242 (1860) ...	... Split-gilled clitocybe ...
75	69	IX. 96	C. subsplendens	Cooke and Mass., Grev. XVIII. 2 (1889)	... Shining clitocybe ...
76	72	V. 632	C. tuba ...	... Fries, Epicr. 72 (1838) ...	... Trumpet clitocybe ...
					8. COLLYBIA.—Fries,
77	85	V. 761	C. butyracea	Fries, Hym. Eur. 113 (1874)	... Buttery collybia ... ...
78	92	,, 865	C. coagulata	Berk. and Br., Linn. Trans. II. 53 (1883)	... Coagulated collybia ...
79	93	,, 871	C. dryophila	... Fries, Hym. Eur. 122 (1874)	... Wood-loving collybia ... ...

OF AUSTRALIAN FUNGI—*continued.*

Number	W.A.	S.A.	T.	V	N.S.W.	Q.	B.	Occurrence	General Characters.
			Habitat.						

Tent. Disp. 68 (1797).—Agaricus—*continued.*

40	W.A.	...	...	...	...	...	...	Ground ...	Fleshy, shining, white, centre ornamented with pyramidal wart-like scales. Eaten largely by the smaller marsupials.
41	...	...	...	...	N.S.W.	...	...	Ground ...	Small, white, marked with brownish spots. Stem club shaped.
42	W.A.	...	...	V.	...	Q.	...	Ground ...	Fleshy, at first umber, then white. Stem rather bulbous at base.
43	...	...	...	V.	...	Q.	...	Ground ...	Fleshy, brownish, with depressed persistent scales. Stem long, bulbous.
44	...	...	...	V.	...	...	...	Ground, roots of trees, or dead wood	Thin, white, umbo or boss dusky. Stem smooth, white.

Trid. 7 (1881).—Agaricus.

45	...	...	...	...	...	Q.	...	Gardens ...	Somewhat fleshy, pallid, spotted chiefly about disc, with darker scales.

S.M. I. 26 (1821).—Agaricus.

46	...	...	...	...	...	Q.	...	Sandy soil	Bright golden yellow, smooth, shining. Stem erect, slender, hollow.
47	...	S.A.	...	...	N.S.W.	Q.	B	Dead stumps	Fleshy, honey brown, scaly fibrous. In tufts on stumps. Very common. *Edible.*
48	...	...	...	V.	...	...	B	Woods, &c.	Robust. Fleshy, compact, brown. Stem solid, short, tapering downwards.
48ᴀ	...	...	...	V.	...	...	...	Woods, &c.	Smaller than typical form, with smaller ring.

S.M. I. 36 (1821).—Agaricus.

49	...	...	...	V.	...	...	B	Lawns, &c.	Fleshy, brown or yellow. Stem stuffed, grooved, with fibrils. Rare.
50	...	...	...	V.	...	Q.	B	Pine woods	Fleshy, soft, moist, ash coloured, becoming pallid. Rare.
51	...	...	...	V.	...	...	...	Sandy soil	Pressed together and deformed. Fleshy, viscid, tan coloured, cracked when dry.
52	...	...	...	V.	...	...	B	Pastures ...	Small, very brittle. Rather fleshy, buff. Stem hollow, tapering downwards. Common.
53	...	...	...	V.	...	...	B	Ground, among grass	Fleshy, blackish brown to ash grey. Stem stuffed, powdery, and shaggy. Common.
54	...	S.A.	...	V.	...	...	B	Ground ...	Fleshy, moist, changing colour dingy black then livid brown. Gills white. Very common.
55	W.A.	...	...	...	...	...	...	Among moss	Rather fleshy, glutinous, whitish. Stem solid, viscid.
56	W.A.	...	T.	V.	...	...	B	Woods, &c., among dead leaves	Fleshy, rather thin, moist, changing colour. Rare. *Edible.*
57	...	...	...	V.	...	...	...	Grassy places	Fleshy, moist, thin at the naked margin. Stem stuffed, cartilaginous.
58	...	...	...	V.	...	...	B	Firwoods...	Somewhat fleshy, olive grey, hoary when dry. Odour mealy, rancid.
59	...	...	...	...	...	Q.	B	Shady places	White. Fleshy, shining when dry. Odour agreeable.
60	W.A.	S.A.	...	...	...	...	B	Pine stumps	Fleshy, with red or purplish down. Odour strong. Common.
61	W.A.	...	...	...	...	...	B	Woods ...	Fleshy, more or less sulphur coloured. Odour disagreeable.

S.M. I. 78 (1821).—Agaricus, Laccaria.

62	...	...	...	...	...	Q.	...	Under Casuarina (Sheoak) trees	Somewhat membranaceous, velvety, bright tawny, with radiating channels.
63	...	...	...	V.	...	Q.	B	Woods ...	White. Fleshy, moist. Stem spongy, elastic. Common. *Edible.*
64	...	...	T.	...	...	...	...	Grassy places	Rather fleshy, brown to livid. Stem solid, short, rigid.
65	...	...	...	V.	...	...	B	Ground ...	Fleshy to membranaceous, becoming tawny, when dry clay coloured. *Edible.*
66	...	S.A.	...	V.	...	...	B	Firwoods...	Rather fleshy, flaccid, funnel shaped, bright brown, becoming pale.
67	W.A.	...	...	...	...	...	B	Woods, waste ground	Rigid. Fleshy, smoky, turning pale. *Edible.*
68	W.A.	...	...	...	...	...	B	Pine woods	Fleshy, moist, yellowish tan. Stem fleshy, solid, stout. *Edible.*
69	...	S.A.	...	V.	...	...	B	Fields and woods, among moss	Fleshy, downy, funnel shaped, flaccid, pale-tan colour or cinnamon. Common. *Edible.*
70	...	...	T.	V.	...	...	B	Woods ...	Fleshy, fragile, brownish red at first, then tan coloured. Margin inverted.
71	...	...	T.	V.	N.S.W.	Q.	B	Woods ...	Tall and slender. Membranaceous, red, brown, or amethyst, mealy. Very common. *Edible.*
72	...	...	...	V.	...	...	...	Grassy places	Fleshy, shining, tawny, grey or ochrey, white. Stem solid.
73	...	...	...	...	...	Q.	B	Pine woods and on trunks	Slender, rigid, inodorous. Fleshy to membranaceous, brownish or ash coloured, sprinkled with a greyish bloom. *Edible.*
74	...	...	T.	...	...	...	...	Rotten wool	In tufts. Gills splitting at the edge. Stem stringy.
75	...	...	...	...	...	Q.	...	Among grass in garden	Somewhat fleshy, shining, rufous or yellowish. Stem solid.
76	...	...	...	V.	...	...	B	Among leaves, chiefly of Pines	White. Fleshy, moist, shining with a whitish silky lustre. Stem soon hollow.

S.M. I. 129 (1821).—Agaricus, Amanita, Marasmius.

77	...	...	...	V.	...	...	B	Woods ...	Small. Fleshy, changing colour, flesh becoming white. Very common.
78	...	...	...	...	...	Q.	...	Ground ...	Cream colour, yellow when dry. Stem slender, twisted, yellow.
79	...	...	...	...	...	Q.	B	Among leaves in woods	Somewhat fleshy, turning pale bay red, yellowish, clay coloured, white. Very common.

Number	Cooke's Number	Saccardo's Number	Scientific Name	Authority for Name	English Name
					8. COLLYBIA.—Fries,
80	79	V. 734	C. cradicata	Kalch., Grev. VIII. 151 (1880) ...	Non-rooting collybia ...
81	91	„ 810	C. esculenta	Fries, Hym. Eur. 121 (1874) ...	Esculent collybia ...
82	81	„ 748	C. fusipes	Fries, Hym. Eur. 111 (1874) ...	Spindle-stalked collybia
83	77	„ 807	C. laccatina	Berk., Linn. Journ. XVII. 383 (1881) ...	Scaling-wax collybia ...
84	99	„ 918	C. lacerata	Lasch, in Fries, Hym. Eur. 127 (1874) ...	Torn collybia ...
85	86	„ 831	C. lepidopoda	Fries, in Pl. Preiss. II. 131 (1846)	Scaly-stalked collybia ...
86	80	„ 735	C. longipes	Fries, Hym. Eur. 110 (1874) ...	Long-staked collybia ...
87	87	„ 808	C. morula	Berk., Fl. Tasm. II. 243 (1860) ...	Mulberry-brown collybia
88	95	...	C. nivosula	Berk., Cuban Fungi III.	Snowy collybia
89	90	V. 839	C. nummularia	Fries, Epicr. 91 (1838)	Coin-like collybia ...
90	82	IX. 116	C. olivaceo-alba	Cooke and Mass., Grev. XV. 93 (1887) ...	Olive-white collybia ...
91	97	V. 904	C. ozes ...	Fries, Epicr. 95 (1838) ...	Smelling collybia ...
91A	„	IX. 139	C. ozes, var. crassipes	Cooke and Mass., Grev. XV. 93 (1887) ...	Thick-stalked collybia...
92	98	V. 907	C. plexipes	Fries, S.M. I. 146 (1821)	Twisted-stalked collybia
93	78	„ 728	C. radicata	Fries, Hym. Eur. 109 (1874) ...	Rooting collybia
93A	„	„ „	C. radicata, var. superbiens	Berk., Hook., Lond. Journ. IV. 43 (1845)	Superb collybia
94	96	„ 899	C. rancida	Fries, S.M. I. 141 (1821)	Rancid collybia ...
95	100	„ 784	C. rheicolor	Sacc. Syll. I. 214 (1887)	Rhubarb-coloured collybia
96	88	„ 826	C. tuberosa	Fries, Hym. Eur. 119 (1874) ...	Tuberous collybia ...
97	100	„ 929	C. tylicolor	Fries, S.M. I. 132 (1821)	Greyish collybia ...
98	91	IX. 115	C. veluticeps	Cooke and Mass., Grev. XVI. 30 (1887) ...	Velvet-head collybia ...
99	85	V. 773	C. velutipes	Fries, Hym. Eur. 115 (1874) ...	Velvet-stem collybia ...
100	89	„ 836	C. xanthopoda	Fries, Hym. Eur 120 (1874) ...	Yellow-stalked collybia
101	84	„ 768	C. xylophila	Weinm., in Linn. X. 54 (1836) ...	Timber-loving collybia
					9. MYCENA.—Pers.,
102	109	V. 1050	M. actites	Fries. Epicr. 110 (1838)	Beaked mycena ...
103	107	„ 1037	M. atro-cyanea	Fries, S.M. I. 147 (1821)	Dark-blue mycena ...
104	123	„ 1152	M. capillaris	Fries, S.M. I. 154 (1821)	Thread-stalked mycena
105	432	„ 991	M. coharens	Fries, Epicr. 105 (1838)	Cohering mycena
106	120	„ 147	M. corticola	Fries, S.M. I. 150 (1821)	Bark-growing mycena
107	113	„ 1088	M. crinalis	Berk., Hook., Lond. Journ. IV. 44 (1845)	Hair-like mycena
108	111	„ 1071	M. debilis	Fries, Epicr. 112 (1838)	Tender mycena
109A	110	...	M. filipes, var. acutata	Kalch., Linn. Soc. N.S.W. 104 (1882)	Acutely conical mycena
110	101	IX. 45	M. flavovirens	Cooke and Mass., Grev. XIX. 45 (1890) ...	Yellowish-green mycena
111	105	V. 1092	M. galericulata	Fries, Hym. Eur. 138 (1874)	Little-cap mycena
112	115	„ 1097	M. hæmatopoda	Fries, S.M. I. 149 (1821)	Dark-red juiced mycena
113	122	„ 1118	M. hiemalis	Osbeck, in Retz. Supp. 19 (1805)	Winter mycena
114	119	„ 1135	M. interrupta	Berk., Fl. Tasm. II. 243 (1860) ...	Interrupted mycena ...
115	124	„ 1154	M. juncicola	Fries, Hym. Eur. 154 (1874)	Rush-growing mycena
116	108	„ 1041	M. leptocephala	Fries, Hym. Eur. 141 (1874)	Delicate-head mycena ...
117	102	„ 952	M. pura ...	Fries, Hym. Eur. 133 (1874)	Pure mauve-cap mycena
118	101	„ 944	M. rosella	Fries, S.M. I. 151 (1821)	Rose-coloured mycena
119	116	„ 1100	M. sanguinolenta ...	Fries, Hym. Eur. 148 (1874) ...	Light red juiced mycena
120	103	„ 983	M. Silenus	Berk. and Br., Linn. Journ. XI. 524 (1871)	Bacchanalian mycena
121	112	„ 1080	M. spirea	Fries, S.M. I. 159 (1821)	Twisted mycena
122	117	„ 1124	M. stylobates	Fries, Hym. Eur. 150 (1874) ...	Pillar-shaped mycena ...

OF AUSTRALIAN FUNGI—continued.

Number	W.A.	S.A.	T.	V.	N.S.W.	Q.	R.	Occurrence	General Characters.

S.M. I. 129 (1821).—Agaricus, Amanita, Marasmius—continued.

Number	W.A.	S.A.	T.	V.	N.S.W.	Q.	R.	Occurrence	General Characters.
80				V.	N.S.W.			Ground ...	Resembling C. radicata, but not rooting.
81				V.			B	Pastures ...	Small, in clusters, buffish. Somewhat fleshy. Edible.
82							B	Stumps ...	Fleshy, dull vinous brown or chestnut. Edible.
83						Q.		Dead wood among leaves	Pale fleshy red, margin grooved. Stem paler, fibrillose.
84				V.			B	Pine woods	Fleshy to membranaceous, moist, streaked brown, lacerated when old.
85	W.A.							Ground ...	Rather fleshy, orange, stem rough with scales.
86				V.		Q.	B	Old stumps, &c.	Fleshy, dry, slightly velvety, tan brown. Stem stuffed, tall.
87			T.	V.				Dead wood	Purple red or dark mulberry brown. Fleshy. Stem horizontal, rough.
88				V.				Logs	Thin, whitish. Stem slender, smooth, solid, white, rooting copiously.
89				V.		Q.	B	Among leaves, on wood, &c.	Beautiful, white becoming pale, variegated with light yellow and red. Rather fleshy.
90		S.A.		V.				Ground under Casuarina (Sheoak)	Fleshy, shining, dark sooty olive. Stem smooth, whitish, black below.
91				V.			D	Pine leaves	Fleshy, ashy brown when moist, pale clayey brown when dry. Strong scented, colour of meal.
91A		S.A.						Low damp ground	Conical to bell shaped. Stem tapering upwards, umber.
92				V.			D	Trunks ...	Sub-membranaceous, grey. Stem hollow, silky fibrous. Inodorous.
93	W.A.		T.	V.		Q.	B	Ground around bases of stems	Fleshy, glutinous, with long, pale, slender, twisted, rooting stems.
93A	W.A.							Ground ...	Brown, and stem clad at base with velvety scurf.
94				V.			B	Under trees	Strong scented. Rather fleshy, grey, whitish, silky. Stem rooting, smooth.
95						Q.		Trunks ...	Rhubarb colour. Thin, and stem clothed with a velvety down.
96						Q.	B	Putrid Agarics, such as Russula, &c., and on ground	Slightly fleshy, white. Root springing from sclerotioid tuber. Common.
97		S.A.		V.			B	Woods ...	Rather fleshy, ash colour. Stem hollow, powdery.
98				V.				Fern-tree Gully	Fleshy, velvety, liver coloured. Stem short, pale upwards.
99				V			B	Logs and trunks of trees—Willow, Beech, &c.	Fleshy, viscid, tawny yellow or fawn. Stem stuffed, velvety, dark bay. Common.
100				V.			B	About stumps of trees	Rather fleshy, becoming pale. Stem yellow, and rooting at base.
101				V.			B	Stumps ...	Rather fleshy, whitish or clay coloured, bell shaped. Stem hollow.

Tent. Disp. 69 (1797).—Agaricus, Marasmius.

Number	W.A.	S.A.	T.	V.	N.S.W.	Q.	R.	Occurrence	General Characters.
102		S.A.		V.			B	Among mosses	Fragile. Membranaceous, brownish, growing pale, with broad obtuse prominent umbo.
103			T.				B	Ground ...	Fragile, inodorous. Membranaceous, brownish, then grey, becoming bluish.
104			T.	V.			B	Dead leaves in woods	Very delicate, white. Bell shaped, smooth. Stem thread-like, smooth.
105			T.				B	Ground ...	Rather fleshy, velvety, cinnamon brown, growing pale. Stem horny, rigid.
106				V.	N.S.W.	Q.	B	Mossy bark	Colour various—reddish brown, blue, or ash coloured. Stem incurved, scurfy.
107	W.A.							Decayed wood	Very delicate. Membranaceous, white. Stem thread-like, brown.
108				V.	N.S.W.		B	Woods ...	Tender. Membranaceous, brownish. Stem thread-like.
109A				V.	N.S.W.			Ground ...	Dark, ash coloured, and acutely conical. Stem thread-like, rooting.
110				V.				Tree ferns	Membranaceous, yellowish green. Stem slender, erect, hollow.
111			T.	V.			B	Trunks of trees	Sub-membranaceous, flesh coloured, drab, or various. Densely clustered. Common.
112				V.			B	Old dead trunks	Fleshy. Stem yielding a dark-red juice, rigid, powdery.
113		S.A.		V.			B	Trunks of trees	Thin, brighter coloured than M. corticola, hardly ashy brown.
114			T.					Bark ...	Rather thick, livid, gelatinously fleshy. Gills descending interruptedly into flesh of cap.
115				V.			B	Dead rushes in bogs	Very delicate. Somewhat red. Stem thread-like, smooth, brownish.
116				V.			B	Trunks ...	Fragile, with nitrous odour. Sub-membranaceous, furrowed, frosted.
117				V.			B	Ground in woods ...	Strong smelling, odour of radish. Rather fleshy, violet or roseate, becoming pallid and variously coloured.
118				V.			B	Among fir leaves ...	Rose coloured. Membranaceous, boss obtuse. Stem thin, juiceless.
119		S.A.		V.			B	Among leaves and damp moss	Delicate pale red, becoming brown, membranaceous. Stem yielding pale-red juice.
120						Q.		Dead wood	Small. Fleshy, red to vinous brown. Stem short, hollow.
121		S.A.		V.	N.S.W.		B	Mossy trunks	Membranaceous, greyish brown, disc darker. Stem thread-like.
122				V.		Q.	B	Fern, twigs, &c. ...	Membranaceous, white, somewhat hairy. Stem thread-like.

Number.	Cooke's Number	Saccardo's Number	Scientific Name.	Authority for Name.	English Name.
					9. MYCENA.—Pers.,
123	121	IX. 143	M. subcorticalis ...	Cooke and Mass., Grev. XV. 93 (1887) ...	Subcortical mycena ...
124	118	V. 1129	M. tenerrima ...	Fries, Hym. Eur. 151 (1874) ...	Very tender mycena ...
125	106	., 1025	M. trachycephala ...	F. v. M. and Kalch., Grev. VIII. 151 (1880)	Rough-headed mycena
126	114	„ 1090	M. tuberigena	Berk., Linn. Journ. XIII. 156 (1873) ...	Tuber-bearing mycena
					10. HIATULA.—Fries,
127	359	V. 1168	H. Wynniæ	Berk. and Br., App. Nat. Hist., III., 5. 206 (1879)...	Wynne's hiatula, Green-light fungus
					11. OMPHALIA.—Fries.
128	136	V. 1247	O. carneo-rufula ...	Berk., Fl. Tasm. II. 243 (1860) ...	Fleshy-red omphalia ...
129	126	„ 1181	O. damosa ...	Fries, Hym. Eur. 155 (1874) ...	Thicket-loving omphalia ...
130	129	„ 1205	O. epichysium ...	Fries, S.M. I. 169 (1821) ...	Watery omphalia ...
131	140	„ 1283	O. fibula ...	Fries, Hym. Eur. 164 (1874)	Pin-like omphalia ...
132	137	„ 1265	O. flavo-crocea ...	Berk., Fl. Tasm. II. 244 (1860) ...	Bright-yellow omphalia
133	133	IX. 179	O. glaucescens ...	Kalch., Linn. Soc. N.S.W. 105 (1882) ..	Sage-green omphalia ..
134	141	V. 1286	O. gomphomorpha	Berk., Linn. Journ. XVIII. 383 (1881) ...	Club-shaped omphalia
135	142	„ 1289	O. gracillima	Weinm., Ross 121 (1836) ...	Slender omphalia ...
136	128	„ 1146	O. hohochlora	Berk. and Br., Linn. Journ. XI. 525 (1871)	Green omphalia ...
137	125	„ 1179	O. hydrogramma ...	Fries, S.M. I. 169 (1821) ...	Water-line omphalia ...
138	143	„ 1313	O. integrella	Fries, Hym. Eur. 165 (1874) ...	Perfect omphalia ...
139	144	„ 1321	O. Muelleriana	Berk., in Cooke's Handb. Aust. Fungi 30 (1892) ...	Mueller's omphalia ...
140	134	„ 1239	O. muralis	Fries, Hym. Eur. 160 (1874) ...	Wall omphalia ...
141	130	„ 1208	O. oniscus	Fries, S.M. I. 172 (1821) ...	Grey omphalia ...
142	131	„ 1215	O. pumilio	Kalch., Grev. VIII. 151 (1880) ...	Dwarf omphalia ...
143	127	„ 1199	O. pyxidata	Fries, S.M. I. 164 (1821) ...	Box-like omphalia ...
144	132	„ 1216	O. scyphiformis ...	Fries, Hym. Eur. 159 (1874) ...	Goblet-shaped omphalia ...
145	139	„ 1282	O. setipes ...	Fries, Hym. Eur. 164 (1874) ...	Hairy-stalked omphalia
146	133	„ 1241	O. um ellifera ...	Fries, Hym. Eur. 160 (1874) ...	Umbrella-like omphalia
147	138	„ 1279	O. umbratilis	Fries, Epicr. 127 (1838)	Shade-loving omphalia ...
					12. PLEUROTUS.—Fries,
148	147	V. 1346	P. abbreviatus	Kalch., Grev. VIII. 152 (1880) ...	Abbreviated pleurote ...
149	170	„ 1444	P. affixus	Berk., Hook, Lond. Journ. VII., 573 (1848)	Affixed pleurote ...
150	180	„ 1504	P. applicatus	Fries, Hym. Eur. 180 (1874)	Sessile pleurote
151	179	„ 1492	P. atro-cœruleus ...	Fries, S.M. I. 190 (1821) ...	Dark-blue pleurote ...
152	188	IX. 187	P. australis ...	Cooke and Mass., Grev. XV. 93 (1887) ...	Southern pleurote ...
153	177	V. 1487	P. bir æformis ...	Berk., Fl. Tasm. II. 245 (1860) ...	Pouch-like pleurote ...
154	156	„ 1400	P. candescens ...	F. v. M., Linn. Journ XIII. 157 (1873) ...	Glowing pleurote ...
155	169	„ 1442	P. caryophylleus ...	Berk., Linn. Journ. XIII. 157 (1873) ...	Clove-like pleurote ...
156	167	IX. 200	P. chætophylleus ...	Sacc. Hedw. 126 (1889) ...	Hairy-gill pleurote ...
157	187	V. 1527	P. chion us ...	Pers. M. Eur. 3 (1828)..	Snow-white pleurote ...
158	159	IX. 199	P. clitocyboides ...	Cooke and Mass., Grev. XV. 98 (1887) ...	Clitocybe-like pleurote ...
159	143	V. 1322	P. corticatus ...	Fries, S.M. I. 179 (1821) ...	Corticated pleurote ...
160	183	„ 1511	P. diversipes ...	Berk. Fl. Tasm. II. 244 (1860) ...	Variable-stalked pleurote
161	164	„ 1423	P. Eucalyptorum ...	Fries, Pl. Preiss. II. 131 (1846)...	Eucalyptus pleurote ...
162	189	..p.387	P. euphyllus ...	Berk., in Handb. N.Z. Flora 755 (1867) ...	Broad-gilled pleurote ...
163	172	„ 1449	P. flabellatus ...	Berk. and Br., Linn. Journ. XI. 528 (1871)	Fan-shaped pleurote ...
164	149	„ 1369	P. Gardneri ...	Berk., Hook., Journ. II., 427 (1840)	Gardner's pleurote ...
165	161	„ 1409	P. Guilfoylei ...	Berk., Linn. Journ. XIII. 158 (1873) ...	Guilfoyle's pleurote ...
166	150	„ 1379	P. illuminans ...	F. v. M., Linn. Journ. XIII. 157 (1873)...	Luminous pleurote ...
167	168	„ 1429	P. imberbis ...	Kalch., Grev. VIII. 152 (1880) ...	Beardless pleurote ...
168	148	„ 1347	P. lacticolor	Kalch., Grev. VIII. 151 (1880) ...	Bright-coloured pleurote

OF AUSTRALIAN FUNGI—*continued.*

Number.	Habitat.						B.	Occurrence.	General Characters.
	W.A.	S.A.	T.	V.	N.S.W.	Q.			

Tont. Disp. 69 (1797).—Agaricus, Marasmius—*continued.*

123	...	S.A.	...	...	...	...	...	Log of *Banksia* (Native Honeysuckle)	Thin, lilac, disc brick red. Stem ascending, thin, hollow.
124	...	...	...	...	Q.	B	Fir cones, sticks, &c.	White, very delicate. Cap frosted with scurfy granules. Stem hairy.
125	...	...	V.				...	Rotten trunks	Membranaceous, ashy brown, covered with papillæ. Stem thread-like.
126	...	...	V.				...	Ground	Tender and small, white. Stem thread-like, arising from sclerotium.

Nov. Symb. 27 (1851).

| 127 | ... | ... | ... | V. | | Q. | B | Ground | White. Tender, luminous, emitting a greenish light. Stem slender. |

S.M. I. 162 (1821).—Agaricus.

128	...	...	T.				...	Rotten wood	Rather fleshy, pale red. Stem flexuous, stuffed.
129	...	...	...	V.			...	Woods	Rather membranaceous, brick red. Stem hollow, smooth. Rare.
130	...	...	T.				...	Rotten wood	Soft. Membranaceous, sooty, ash coloured. silky or scaly.
131	W.A.	S.A.	...	V.		Q.	B	Among moss in moist places	Tiny. Membranaceous, nearly orange colour or orange fawn. Common.
132	...	...	T.				...	Branches	Yellow, and gills saffron yellow. Stem elongated, solid.
133	...	...	...	V.	N.S.W.		...	Ground	Small. Grey, sage green. Stem thread-like, greenish yellow.
134	...	...	...	...		Q.	...	Ground in tufts	Club-shaped, lurid. Stem thickened upwards, reddish brown.
135	...	...	...	V.			B	Marshy ground	Snow white. Membranaceous, turrowed. Stem thread-like.
136	...	...	...	...		Q.	...	Dead wood	Membranaceous, yellow brown, reddish brown when dry.
137	...	...	...	V.	N.S.W.		B	Dead leaves and moist places	Sub-membranaceous, livid; margin spreading, streaked. Stem hollow.
138	...	...	T.				B	Decayed wood	White, fragile. Membranaceous, pellucid. Stem very slender, short.
139	...	...	...	V.			...	Ground	Smooth, tawny. Stem elongated, smooth, or streaked lengthwise.
140	...	S.A.	...	V.			B	Ground, banks, and walls	Sub-membranaceous, reddish brown, radiately striate.
141	...	...	...	...		Q.	B	Swampy ground	Sub-membranaceous, dark ash coloured. Stem firm, partially hollow.
142	...	...	...	...	N.S.W.	Q.	...	Wood	Membranaceous, fawn coloured. Stem hollow, thin, curved.
143	...	S.A.	...	V.		...	B.	Among short grass, on lawns, &c.	Sub-membranaceous, brick red, funnel shaped. Stem stuffed at first. Common.
144	...	...	...	V.		Q.	...	Bare ground	White. Membranaceous. Stem rather hollow, short, thin.
145	...	...	...	V.	N.S.W.		...	Shady places	Membranaceous, brownish grey. Stem thread-like, downy at base.
146	W.A.	...	T.	V.		Q.	B	Swamps, exposed pastures, &c.	Fleshy to membranaceous, buff or variable in colour. Very common.
147	...	...	...	V.			B	Damp places	Sub-membranaceous, umber brown. Stem stuffed, tough.

S.M. I. 178 (1821).—Agaricus.

148	...	...	...	...	N.S.W.	...	...	Wood	Entirely reddish brown. Stem shorter than diameter of cap.
149	...	...	T.	...		...	...	Bark of young *Eucalyptus amygdalina*	White. Cap at length attached by the side, cup shaped. Stem short.
150	W.A.	...	T.	V.		Q.	B	Dead fallen branches, and on *Eucalyptus viminalis*	Very small. Dark-ash colour. Sub-membranaceous, cup shaped.
151	W.A.	...	...	V.			B	Trunks	Fleshy, dark blue, rarely brown, downy.
152	...	S.A.	...	...			...	Roots of *Leptospermum* (Tea-tree)	Fleshy, umber. Stem rather lateral, thick, clad with white down.
153	...	...	T.	...			...	Rotting bark	Cap affixed behind, pouch-like, whitish. Stem short, almost smooth.
154	...	...	...	V.			...	Dead wood	Strongly phosphorescent. White, becoming dingy. Stem dilated above.
155	...	...	...	V.	N.S.W.		...	Wood	Pale-tawny colour. Fan shaped. and much lobed.
156	...	S.A.	...	...			...	Branches	Thin, white, tomentose, spoon shaped. Stem thick, wrinkled.
157	W.A.	...	...	V.			B	Wood or dung	Snow white. Very thin, woolly. Stem very short, hairy.
158	...	...	...	V.			...	Old fern stems	Thin, ochrey, becoming reddish.
159	...	...	...	...		Q.	B	Trunks, living and dead	Beautiful large species. Fleshy, greyish white. Stem firm.
160	...	...	T.	...		Q.	...	Rotten wood	Pellucid, covered with a gelatinous layer. Stem very variable.
161	W.A.	...	...	...		Q.	...	*Eucalyptus* bark	Fleshy, bay brown, clad with a rough wool, sessile, kidney shaped.
162	...	...	...	...		Q.	...	Wood	Pale chestnut. Stem none or obsolete. Gills broad.
163	...	...	...	...		Q.	...	Dead wood	Fan shaped, thin, white, becoming reddish. Adhering to wood by spongy base.
164	...	...	...	...		Q.	...	Petioles and half-rotten fronds of palms	Fleshy to leathery, yellow, funnel shaped. Stem short. Phosphorescent.
165	...	...	...	...	N.S.W.	Q.	...	Trunks	Whitish, very much wrinkled when dry, cap kidney shaped.
166	...	...	...	V.	N.S.W.	Q.	...	Dead wood	Phosphorescent. Tawny, smooth. Stem thick.
167	...	...	...	...	N.S.W.	...	...	Wood	Membranaceous, kidney shaped, sessile, horizontal.
168	...	...	...	...	N.S.W.	...	...	Wood (?)	Rather fleshy, golden yellow. Stem stuffed, naked.

Number.	Cooke's Number.	Saccardo's Number.	Scientific Name.	Authority for Name.	English Name.

12. PLEUROTUS.—Fries,

169	155	V. 1397	P. lampas	...	Berk., Hook., Lond. Journ. IV. 44 (1845)	...	Shining pleurote	...	...
170	181	„ 1506	P. lenticula	...	Kalch., Grev. VIII. 151 (1880) ...	...	Freckled pleurote	...	...
171	166	„ 1427	P. limpidus	...	Fries, Epicr. 135 (1838)	...	Transparent pleurote	...	
172	178	„ 1488	P. lividulus	...	Berk. and Curt. Exp. No. 33 (1859)	...	Livid pleurote	...	
173	151	„ 1386	P. luteo-aurantius...	Kalch., Grev. VIII. 151 (1880) ...	...	Orange-yellow pleurote		
174	165	„ 1425	P. mitis ...	Fries, S.M. I. 188 (1821)	...	Mild pleurote...		
175	154	„ 1396	P. nidiformis	Berk., Hook., Lond. Journ. III. 185 (1844)	...	Nest-shaped pleurote ...		
176	152	„ 1390	P. ostreatus	Fries, Hym. Eur. 173 (1874)	...	...	Oyster-like pleurote	...	
177	186	„ 1523	P. perpusillus	Fries, S.M. I. 195 (1821)	...	...	Very small pleurote	...
178	162	„ 1412	P. petaloides	Fries, S.M. I. 183 (1821)	...	...	Petal-like pleurote	...	
179	157	„ 1401	P. phosphorus	Berk., Hook., Lond. Journ. VII. 572 (1848)	...	Phosphorescent pleurote	...
180	153	IX. 196	P. polyphemus	Cooke and Mass., Grev. XVI. 72 (1888) ...	...	Variegated pleurote	...
181	163	V. 1416	P. pulmonarius	...	Fries, S.M. I. 187 (1821)	...	Lung-like pleurote
182	158	„ 1405	P. saliguus	Fries, Hym. Eur. 174 (1874)	...	Willow-sprout pleurote
183	176	„ 1475	P. scabriusculus	...	Berk., Linn. Journ. XIII. 157 (1873)	...	Rough pleurote	...
184	174	„ 1470	P. semiliber	...	Berk. and Br., Linn. Trans. II. 54 (1885)	...	Half-free pleurote	...
185	173	„ 1469	P. semisupinus	...	Berk. and Br., Linn. Journ. XI. 529 (1871)	...	Semisupine pleurote	...
186	175	„ 1473	P. sordulentus	...	Berk. and Br., Linn. Trans. II. 54 (1885)	...	Dirty-white pleurote	...
187	184	„ 1518	P. striatulus	...	Fries, S.M. I. 193 (1821)	...	Striate pleurote	...
188	185	„ 1522	P. subbarbatus	...	Berk. and Curt., Linn. Journ. X. 288 (1869)	...	Barbed pleurote	...
189	160	IX. 199	P. sulciceps	...	Cooke and Mass., Grev. XVIII. 3 (1889)	...	Sulcate pleurote	...
190	182	V. 1510	P. tasmanicus	...	Berk., Fl. Tasm. II. 245 (1860) ..	...	Tasmanian pleurote	...
191	146	„ 1343	P. tephrophanus	...	Berk., Fl. Tasm. II. 244 (1860)	...	...	Ash-growing pleurote	...
192	171	„ 1445	P. Thozetii	...	Berk., Linn. Journ. XVIII. 383 (1881)	...	Thozet's pleurote	...

13. HYGROPHORUS.—Fries,

193	379	IX. 216	H. candidus	...	Cooke and Mass., Grev. XVIII. 4 (1889)	...	White hygrophore	...	...
194	381	V. 1634	H. ceraceus	...	Fries, Epicr. 330 (1838)	...	Waxy hygrophore	...	...
195	382	„ 1637	H. coccineus	...	Fries, Epicr. 330 (1838)	...	...	Scarlet hygrophore	...
196	384	„ 1638	H. conicus	...	Fries, Epicr. 331 (1838)	...	Conical hygrophore	...
197	376	„ 1599	H. flammans	...	Berk., Linn. Journ. XIII. 160 (1873)	...	Flaming hygrophore	...
198	378	IX. 224	H. gigasporus	...	Cooke and Mass., Grev. XVI. 31 (1887)...	...	Gigantic-spored hygrophore	...
199	375	„ 228	H. gilvus...	Kalch., Linn. Soc. N.S.W. 105 (1882)	...	Yellowish-tan hygrophore	...
200	873	V. 1570	H. hypothejus	...	Fries, Epicr. 324 (1838)	...	Sulphur-yellow hygrophore	...
201	387	IX. 227	H. Lewellinae	...	Kalch., Linn. Soc. N.S.W. 105 (1882)	...	Lewellin's hygrophore ...
202	383	V. 1659	H. miniatus	...	Fries, Epicr. 330 (1838)	...	Vermilion hygrophore ...
203	377	„ 1600	H. nigricans	...	Berk., Linn. Journ. XIII. 160 (1873)	...	Blackening hygrophore
204	372	„ 1555	H. porphyrius	...	Berk. and Br., Linn. Trans. II. 55 (1883)	...	Purple hygrophore	...
205	385	„ 1677	H. scarlatinus	...	Kalch., Grev. VIII. 159 (1880) ...	...	Scarlet hygrophore	...
206	380	„ 1628	H. sciophanus	...	Fries, Epicr. 329 (1838)	...	...	Shadowy hygrophore ...
207	386	IX. 229	H. subremotus	...	Cooke and Mass., Grev. XVI. 113 (1888)	...	Sequestered hygrophore ...
208	374	V. 1590	H. virgineus	...	Fries, Epicr. 327 (1838)	...	...	Virgin hygrophore	...

14. LACTARIUS.—D.C.,

209	392	V. 1737	L. pallidus	...	...	Fries, Epicr. 343 (1838)	...	Pale lactar	...
210	390	„ 1727	L. piperatus	...	Fries, Epicr. 340 (1838)	...	Peppery lactar
211	389	„ 1720	L. plumbeus	...	Fries, Epicr. 339 (1838)	...	Leaden lactar
212	388	„ 1694	L. stenophyllus	...	Berk., Fl. Tasm. II. 248 (1860) ...	...	Narrow-gill lactar	...
213	391	„ 1759	L. subtomentosus ...	Berk. and Rav., Ann. Nat. Hist. IV. 293 (1859) ...	Subtomentose lactar	...

13

OF AUSTRALIAN FUNGI—*continued.*

Number	Habitat						B.	Occurrence.	General Characters.
	W.A.	S.A.	T.	V.	N.S.W.	Q.			

S.M. I. 178 (1821).—Agaricus—*continued.*

169	W.A.	...	T.	V.	...	...	...	Stems, languid, but not dead, of *Grevillea* (Silky oak)	Phosphorescent. Fleshy, tawny, turning black. Stem solid, sometimes splitting.
170	...	...	...	...	...	Q.	...	Trunks ...	Small. Olive brown, or powdered with white.
171	...	...	...	...	N.S.W.	...	B	Trunks ...	Rather fleshy, white, tapering behind into rudimentary stem.
172	...	...	...	...	N.S.W.	...	...	Dead branches ...	Becoming livid purple, clad with a powdery down, kidney or fan-shaped.
173	...	...	...	...	N.S.W.	...	...	Wood ...	Rather fleshy, orange yellow. Stem hollow, thin, short, curved upwards.
174	...	...	...	...	...	Q.	B	Dead wood—Pines, Firs, and Larches	Rather fleshy, growing pale, kidney shaped. Stem lateral, compressed, dilated upwards with white scales. Common.
175	W.A.	...	...	...	...	...	...	Ground ...	Very large. Fleshy, reddish brown, cup shaped. Stem central.
176	...	...	...	V.	N.S W.	...	B	Trees ...	Soft, fleshy, shell-like, many overlapping, satiny, growing pale. Common. *Edible.*
177	W.A.	...	...	...	...	...	...	Trunks and branches	White, very delicate, tough, smooth.
178	...	...	...	V.	...	...	B	Ground ...	Fleshy, spoon shaped or fringed, disc woolly, brown to ashy buff. *Edible.*
179	...	...	T.	...	...	...	...	Roots of trees ...	Pale yellowish brown, funnel shaped. So phosphorescent that one was able to read books by its light, and even six days afterwards the light still served for reading.
180	S.A.	...	...	V.	...	...	...	Rotten wood	Fleshy, ochrey white, at length sulphur coloured, spotted with purple or sooty spots.
181	...	...	...	V.	N.S.W.	...	B	Trunks ...	Fleshy, greyish to tan colour, rather convex. Stem lateral, straight. *Edible.*
182	...	S.A.	...	V.	N.S.W.	...	B	Trunks, willow	Compact or spongy, shell shaped, white or grey. *Edible.*
183	...	...	...	V.	...	...	...	Rotten wood	White, sessile. Cap narrow or fan shaped, rough behind.
184	...	...	...	...	...	Q.	...	Wood ...	White. Cap half adherent, pale yellow. Stem lateral, short.
185	...	...	...	V.	...	...	Q.	Dead branches and leaves	Cap at first peziza-like, at length semisupine or half flattened out.
186	...	...	...	...	...	Q.	...	Wood ...	Dirty white, at first rather hairy, then becoming smooth. Stem obsolete.
187	...	...	...	...	...	Q.	B	Firwood twigs	Very delicate. Pale-ash colour, streaked, smooth.
188	...	...	...	V.	...	...	...	Rotten wood	Dark brown. Fan shaped, margin wavy. Stemless.
189	...	...	...	...	...	Q.	...	Rotten wood	Fleshy, sooty brown, disc darker, rather velvety. Stem thin, hollow.
190	...	...	T.	...	...	...	...	Rotten wood	Cap invested with gelatinous pellicle. Stem short, pure white, downy.
191	...	...	T.	...	...	...	...	Burnt wood	Excentric, funnel shaped, brown, powdery. Stem brown, hispid.
192	...	...	...	...	...	Q.	...	Dead leaves	Fan shaped and lobate, arising from rooting stem, whitish ochre.

Epicr. 320 (1838).—Agaricus.

193	...	...	...	V.	...	...	...	Ground ...	White. Fleshy, tinged with brown. Stem rather flexuous, stuffed.
194	...	S.A.	...	V.	...	...	B	Pastures, lawns, &c.	Small, wax coloured, viscid, brittle, lustrous, translucent.
195	...	...	...	V.	...	...	B	Pastures ...	Fragile. Scarlet, shaded with orange and yellow, turning pale. *Edible.*
196	...	...	...	V.	...	Q.	B	Pastures and sandy land	Fragile, rarely red, commonly yellow, usually turning black where bruised. Sub-membranaceous, conical. Very common.
197	...	...	...	V.	...	...	...	Moist rocks ...	Small, dark red, funnel-shaped. Stem dilated upwards.
198	...	...	...	V.	...	...	...	Horse dung, and around it	Fleshy, sooty brown, shining, viscid. Stem straight, elongated.
199	...	...	...	V.	...	...	...	Ground ...	Yellowish orange, funnel shaped. Stem paler, thickened upwards.
200	...	...	...	V.	...	...	B	Pine woods, on sandy soil	Fleshy, covered with olive evanescent gluten and yellow beneath. Very common.
201	...	...	...	V	...	...	...	Ground ...	Lilac, very elegant. Cap convex. Stem hollow, naked.
202	...	...	...	V.	...	Q.	B	Moist places	Fragile. Vermilion, changing colour, opaque. Stem scarlet.
203	...	...	...	V.	...	...	...	Ground ...	Small. Orange red, turning black. Stem thread-like.
204	...	...	...	V.	...	Q.	...	Among grass	Fleshy, rather viscid, purple. Stem swollen in middle.
205	...	...	...	V.	...	Q.	...	Ground ...	Small. Rather fleshy, margin bright scarlet. Stem hollow, rosy white.
206	...	...	...	V.	...	Q.	B	Mossy places, wood	Somewhat brick red, rather fleshy, viscid, margin streaked.
207	...	...	...	V.	...	...	...	Among grass	Yellow, disc becoming reddish, viscid. Stem elongated, hollow.
208	...	S A.	...	V.	...	...	B	Downs and grassy places	Small. Satiny white, becoming tinted, fleshy. Stem stuffed, firm, short. Common. *Edible.*

Fl. Fr. II. 141 (1805).—Agaricus.

209	...	...	...	V.	...	...	B	Woods	Fleshy, pallid, zoneless. Stem stuffed, then hollow. Milk mild, white. *Edible.*
210	...	...	...	...	...	Q.	B	Woods ...	White, turning black where bruised. Milk white, peppery. Common. *Edible.*
211	...	...	...	V.	...	...	B	Woods ...	Fleshy, yellowish to whitish, zoned. Gills very narrow, rather flesh coloured.
212	...	...	T.	...	...	...	...	Ground ...	Somewhat tomentose, umber. Milk white, turning yellow, acrid.
213	...	...	...	V.	N.S.W.	...	...	Ground, in swamps	Dingy to blackish brown. Stem hollow, white at base. Milk acrid, white, unchangeable.

Number	Cooke's Number	Saccardo's Number	Scientific Name	Authority for Name	English Name
					15. RUSSULA.—Pers. Obs.,
214	402	V. 1874	R. alutacea	Fries, Epicr. 362 (1838)	Buff-gilled russule
215	394	IX. 249	R. australiensis	Cooke and Mass., Grev. XVI. 32 (1887)	Australian russule
216	399A	V. 1842	R. Clusii ...	Fries, Hym. Eur. 449 (1874)	Clusius' russule
217	399	„ 1841	R. emetica	Fries, Epicr. 357 (1838)	Emetic russule
218	398	„ 1840	R. expallens	Gillet., Tab. 49 (1878) ...	Bleaching russule
219	400	„ 1852	R. fragilis	Fries, Epicr. 359 (1838)	Fragile russule
220	397	„ 1818	R. Linnaei	Fries, Epicr. 356 (1838)	Linnaeus russule
221	395	„ 1805	R. purpurea	Gillet., Tab. 47 (1878) ...	Purple russule
222	396	„ 1817	R. rubra ...	Fries, Epicr. 354 (1838)	Red russule
223	393	„ 1800	R. sanguinea	Fries, Epicr. 351 (1838)	Blood-red russule
224	401	IX. 259	R. subalbida	Bres., Fug., Myc. Austr. 4 (1890)	Lurid white russule
					16. CANTHARELLUS.
225	404	V. 1886	C. aurantiacus	Fries, S.M. I. 318 (1821)	Orange-coloured chantarelle
226	405	IX. 266	C. aureolus	Cooke and Mass., Grev. XVIII. 4 (1889) ...	Golden chantarelle
227	403	V. 1882	C. cibarius	Fries, S.M. I. 318 (1821)	Edible chantarelle
228	410	„ 1919	C. cinereus	Fries, S.M. I. 320 (1821)	Ash-grey chantarelle
229	412	„ 1934	C. concinnus	Berk., Linn. Journ. XVI. 38 (1878)	Elegant chantarelle
230	414	„ 1956	C. foliolum	Kalch., Grev. IX. 134 (1881)	Leaf-like chantarelle
231	411	„ 1929	C. leucophaeus	Nouel., Mem. Lille (1831)	White-looking chantarelle
232	413	„ 1952	C. lobatus	Fries, S.M. I. 323 (1821)	Lobed chantarelle
233	407	IX. 267	C. politus	Cooke and Mass., Grev. XVI. 32 (1887) ...	Polished chantarelle
234	409	V. 1914	C. pusio ...	Berk., Hook., Journ. VIII. 134 (1856)	Puny chantarelle
235	408	„ 1899	C. strigipes	Berk., Fl. Tasm. II. 248 (1860) ...	Hairy-stalked chantarelle
236	406	„ 1893	C. viscosus	Berk., Hook., Lond. Journ. IV. 49 (1845)	Viscid chantarelle
					17. MARASMIUS.—Fries,
237	444	V. 2225	M. acicutaeformis ...	Berk. and Curt., Linn. Journ. X. 297 (1869)	Needle-stalked marasmius
238	451	„ 2292	M. affixus ...	Berk., Fl. Tasm. II. 248 (1860) ...	Attached marasmius
239	442	„ 2218	M. bicolor ...	Sacc. and Cub., Syll. V. 555 (1887)	Two-coloured marasmius
240	438	„ 2187	M. calobates ...	Kalch., Grev. IV. 71 (1876)	Stilted marasmius
241	425	„ 2071	M. calopus ...	Fries, Epicr. 379 (1838)	Beautiful-stemmed marasmius ...
242	431	„ 2122	M. cauticinalis ...	Fries, Epicr. 383 (1838)	Craggy marasmius
243	418	„ 2046	M. confertus ...	Berk. and Br., Linn. Journ. XIV. 34 (1875)	Crowded marasmius
244	447	„ 2259	M. de Tonianus ...	Sacc. and Cub., Syll. V. 563 (1887)	De Toni's marasmius ...
245	450	„ 2286	M. emergens ...	Berk. in Cooke's Handb., Aust. Fungi, 88 (1892) ...	Emerging marasmius ...
246	445	„ 2239	M. epiphyllus ...	Fries, Epicr. 386 (1838)	Leaf marasmius
247	441	„ 2203	M. equicrinis ...	F. v. M., Grev. VIII. 153 (1880)	Horse-hair marasmius ...
248	420	„ 2051	M. erythropus	Fries, Epicr. 378 (1838)	Red-stalked marasmius
249	440	„ 2200	M. Eucalypti ...	Berk., Fl. Tasm. II. 249 (1860) ...	Eucalyptus marasmius
250	452	„ 2291	M. Exocarpi ...	Berk., Linn. Journ. XVIII. 384 (1881) ...	Native cherry marasmius
251	435	„ 2144	M. ferrugineus ...	Berk., Hook. Lond. Journ. II. 630 (1843)	Rust-red marasmius ...
252	429	„ 2063	M. floriceps ...	Berk. and Curt., Linn. Journ. X. 298 (1869)	Flower-capped marasmius
253	427	„ 2095	M. foetidus ...	Fries, Epicr. 380 (1838)	Foetid marasmius
254	434	„ 2143	M. haematocephalus ...	Mont., Syll. 109 (1856) ...	Blood-red capped marasmius
255	416	„ 2013	M. hepaticus ...	Berk., Hook., Lond. Journ. V. 1 (1846) ...	Liver-coloured marasmius
256	421	„ 2057	M. impudicus ...	Fries, Epicr. 377 (1838)	Impure marasmius ...
257	422	IX. 278	M. inauripes ...	Cooke and Mass., Grev. XVIII. 4 (1889) ...	Woolly-stalked marasmius
258	439	V. 2199	M. liguyodes ...	Berk., Linn. Journ. XVIII. 384 (1881)	Smoky marasmius ...
259	436	„ 2147	M. meloniformis ...	Berk., Fl. Tasm. II. 249 (1860) ...	Melon-shaped marasmius

OF AUSTRALIAN FUNGI—*continued.*

Number.	Habitat.						B.	Occurrence.	General Characters.
	W.A.	S.A.	T.	V.	N.S.W.	Q.			

I. 100 (1796).—Agaricus.

214	...	...	T.	...	...	...	B	Woods	Mild. Fleshy, dark lake, red, or purplish, margin flesh white. Gills bright buff. *Edible.*
215	...	...	...	V.	...	Q.	...	Ground	Acrid. Fleshy, red. Stem stuffed, then hollow, straw coloured.
216	...	...	...	V.	...	...	B	Woods	Blood red. flesh white to yellowish.
217	...	...	T.	V.	N.S.W.	...	B	Woods	Acrid. Fleshy, polished, shining, margin flesh white. Rose, varied with lilac or yellow.
218	...	...	...	V.	...	...	B	Under trees	Fleshy, viscid, bright, purple, becoming pale. Stem cylindrical, firm.
219	...	S.A.	...	V.	...	Q.	B	Woods ...	Very acrid, small. Fleshy, rose red, becoming pale, polished, slightly viscid.
220	...	...	...	V.	...	Q.	B	Woods ...	Mild. Fleshy, polished, dry, white. Stem spongy, stout, red.
221	...	...	...	V.	...	...	B	Under trees	Rather mild. Fleshy, viscid, dark purple. Stem white at top, rosy middle.
222	...	...	...	V.	N.S.W.	Q.	B	Grassy places	Acrid. Fleshy, polished, dry, deep dark vermilion. Stem white or red.
223	...	...	...	...	...	Q.	B	Woods ...	Acrid. Fleshy, generally blood red, glistening. Stem white or red.
224	...	...	...	...	...	Q.	...	Ground ...	Rather fleshy, margin lurid white. Stem stuffed, then hollow.

Pers. Tent. Disp. 26 (1797).—Agaricus.

225	...	...	...	...	...	Q.	B	Fir woods and heaths	Nearly orange colour or orange yellow. Fleshy, rather tomentose.
226	...	...	...	...	...	Q.	...	Ground ...	Golden. Thin, delicately downy. Stem slender, faintly streaked.
227	...	...	...	V.	N.S.W.	Q.	B	Woods ...	Apricot yellow and apricot scented. Fleshy. *Edible.*
228	...	...	T.	...	...	...	B	Woods ...	Sub-membranaceous, dingy black, hairy to scaly. Stem hollow.
229	...	...	...	...	N.S.W.	...	...	Ground ...	Small. Gills very narrow and forked. Stem thickened upwards.
230	...	...	...	...	...	Q.	...	Twigs, leaves, &c....	Membranaceous, whitish, pale ochre when dry or reddish brown. Colour and veins slightly prominent as in a dry leaf.
231	...	...	...	V.	...	...	B	Ground ...	Sub-membranaceous, umber. Stem stuffed, thin, of same colour.
232	...	...	...	V.	...	...	B	Mosses, in swamps	Membranaceous, gelatinous, sessile, dirty reddish brown.
233	...	S.A.	...	V.	...	...	...	Fern gully	Rather fleshy, viscid very shining, chestnut colour. Stem stuffed.
234	...	...	T.	V.	...	...	...	Ground ...	Becoming whitish. Funnel shaped, powdery, woolly. Stem brown when dry.
235	...	...	T.	...	...	...	...	Among ferns	Liver colour. Stem arising from tawny strigose hairs. Tapering upwards.
236	W.A.	...	...	...	...	...	...	Ground, among twigs	Beautifully yellow. Viscid, funnel shaped, somewhat wavy.

Epicr. 372 (1838).—Agaricus.

237	...	...	...	...	...	Q.	...	Rotten wood	Gregarious. Tawny. Stem hair-like, rigid, shining, brownish.
238	...	...	T.	...	...	Q.	...	Rotten wood	Whitish, mealy, tomentose, cup shaped, reflexed and attached by side.
239	...	...	...	...	...	Q.	...	Trunks ...	Small. White. Stem short, becoming red below, thread-like.
240	...	...	...	V.	...	Q.	...	Putrid leaves of *Banguinvillea*	Membranaceous, rust coloured, becoming brown. Stem horny, turning black.
241	...	...	...	V	...	...	Q.	B	Twigs, grass, roots, &c.	Inodorous. Rather fleshy. Stem shining, bay to red, hollow, not rooting.
242	...	...	...	V.	...	Q.	...	Ground, among leaves	Membranaceous, rust coloured, yellow, then ochre. Stem hollow, bay.
243	...	...	...	...	N.S.W.	Q.	...	Among dead vegetables	Sub-membranaceous, tawny. Stem of same colour, hollow.
244	...	...	...	...	N.S.W.	...	...	Branches...	Cap scarcely the size of a mustard seed, brownish. Stem hair-like.
245	...	...	T.	...	...	...	...	Wood ...	Very minute, white, bursting through. Stem shortened or elongated.
246	...	...	...	V.	...	...	B	Fallen leaves, twigs, &c.	Minute. Membranaceous, creamy. Stem rather horny, finely velvety.
247	...	...	...	V.	N.S.W.	Q.	...	Branches...	Whitish to tawny, small, membranaceous. Stem hair-like, rigid, black, shining, arising from black horsehair-like mycelium.
248	...	...	...	V.	...	Q.	B	Among leaves, near stumps	Inodorous. Rather fleshy. Stem dark red, hollow, streaked.
249	...	...	T.	...	...	...	...	Fruit and branches of *Eucalyptus*	Conical, brownish, silky. Stem hair-like, compressed, shining.
250	...	...	...	V.	...	Q.	...	Trunks of *Exocarpos latifolia* (Native Cherry)	White, wholly resupinate.
251	...	...	...	V.	...	...	...	Dead leaves, branches, &c.	Membranaceous, ferruginous yellow. Stem slender, twisted.
252	...	...	...	V.	...	...	...	Rotten wood	Conical, bright red brown. Stem twisted, hollow, shining.
253	...	...	...	...	...	Q.	B	Decayed twigs	Fœtid. Sub-membranaceous, pellucid, tawny chestnut or somewhat red.
254	...	...	...	V.	...	Q.	...	Dead leaves	Membranaceous, blood red. Stem horny, hair-like, umber.
255	...	...	T.	...	...	...	...	Among ferns	Rather fleshy, liver coloured. Stem of stringy fibres, thickened below.
256	...	...	...	V.	...	...	B	Dead trunks	Fœtid. Rather fleshy, chestnut red. Stem hollow, turning purple.
257	...	...	...	...	...	Q.	...	Rotten wood	Fleshy, lead colour or dirty dark blue. Stem red, densely velvety.
258	...	...	...	V.	...	...	...	Leaves of *Eucalyptus*	Furrowed. Stem black, shining, grooved.
259	...	...	T.	...	...	...	...	Leaves and branches of *Eucalyptus*	Minute. Bay brown, mealy. Stem thread-like, shining.

Number.	Cooke's Number.	Saccardo's Number.	Scientific Name.	Authority for Name.	English Name.
					17. MARASMIUS.—Fries,
260	419	V. 2049	M. Muelleri	Berk., Linn. Journ. XVIII. 383 (1881) ...	Mueller's marasmius ...
261	429	„ 2106	M. opacus	Berk. and Curt., Hook. Journ. I. 99 (1849)	Opaque marasmius
262	417	„ 2014	M. pilopus	Kalch., Grev. VIII. 153 (1880) ...	Downy-stalked marasmius
263	446	„ 2257	M. primulinus	Berk., Linn. Journ. XVI. 38 (1878)	Pale-yellow marasmius ...
264	433	„ 2133	M. putredinis	Berk. and Curt., Linn., Journ. X. 295 (1869)	Putrid marasmius
265	428	„ 2103	M. ramealis	Fries, Epicr. 381 (1838)	Twig marasmius
266	449	„ 2261	M. rhyticeps	Kalch. Grev. IV. 71 (1876)	Wrinkle-capped marasmius
267	437	„ 2150	M. rotula ..	Fries, Epicr. 385 (1838)	Collared marasmius ...
268	443	„ 2219	M. rufo-pallidus ...	Kalch., Grev. IV. 71 (1876)	Pale-red marasmius ...
269	424	„ 2070	M. scorodonius	Fries, Epicr. 379 (1838)	Shallot marasmius
270	426	„ 2084	M. stylobates	Berk. and Curt., Linn. Journ. X. 296 (1869)	Pillar-shaped marasmius
271	...		M. subroseus	Cooke and Mass., Grev. XXI. 37 (1892) ...	Somewhat rosy marasmius ...
272	448	„ 2260	M. subsupinus	Berk., Fl. Tasm. II. 249 (1860) ...	Subsupine marasmius ...
273	415	„ 1976	M. urens ...	Fries, Epicr. 373 (1838)	Acrid marasmius ...
					18. LENTINUS.—Fries,
274	461	V. 2332	L. blepharodes	Berk. and Curt., Linn. Journ. X. 301 (1869)	Eyelashed lentine ...
275	463	„ 2348	L. calvescens	Berk., Hook., Journ. VIII. 141 (1856) ...	Bald lentine ...
276A	487	„ 2483	L. castoreus, var. hirneoloides	Berk. and Br., Linn. Journ. X. 302 (1869)	Hirneola-like lentine ...
277	470	„ 2456	L. catervarius	Berk. and Br., Linn. Trans, II. 55 (1883)	Crowded lentine ...
278	476	„ 2415	L. cochleatus	Fries, Hym. Eur. 484 (1874)	Cochleate lentine
279	470	„ 2376	L. cretaceus	Berk. and Br., Linn. Journ. XIV. 42 (1875)	Chalky lentine
280	475	„ 2398	L. cyathus	Berk. and Br., Linn. Trans. I. 399 (1879)	Goblet lentine
281	459	„ 2325	L. dealbatus	Fries, Pl. Preiss. II. 133 (1846)	White-washed lentine ...
282	465	„ 2350	L. Dunalii	Fries, Epicr. 390 (1838)	Dunal's lentine
283	467	„ 2358	L. exasperatus	Berk. and Br., Linn. Trans. II. 55 (1883)	Rough lentine
284	484	„ 2472	L. exilis ..	Fries, Epicr. 393 (1838)	Thin lentine
285	458	„ 2317	L. fasciatus	Berk., Hook., Journ. II. 146 (1840)	Clustered lentine
286	483	„ 2471	L. fulvaster	Berk. and Cooke, Linn. Journ. XV. 373 (1877)	Yellowish lentine
287	455	„ 2312	L. fulvus...	Berk., Ann., Nat., Hist. X. 369 (1843)	Tawny lentine
288	456	„ 2315	L. fusco-purpureus	Kalch., Grev. VIII. 153 (1880) ...	Purple-brown lentine ...
289	486	IX. 322	L. fusipes	Cooke and Mass., Grev. XVI. 1 (1887) ..	Fusiform lentine
290	482	„ 301	L. gracilentus	Cooke and Mass., Grev. XVI. 73 (1888) ...	Slender lentine
291	480	V. 2458	L. Guilfoylei	Berk., Linn. Journ. XVIII. 384 (1881) ...	Guilfoyle's lentine
292	490	„ 2490	L. hepatotrichus	Berk., Fl. Tasm. II. 249 (1860) ...	Liver-coloured lentine ...
293	473	„ 2464	L. holopogonius	Berk., Grev. X. 63 (1881)	Bearded lentine
294	485	„ 2483	L. hyracinus	Kalch., Grev. VIII. 153 (1880) ...	Hyrax-coloured lentine
295	474	„ 2394	L. Kurzianus	Curr., Linn. Trans. I. 120 (1876)	Kurz's lentine
296	478	„ 2449	L. lœviceps	Kalch., Grev. VIII. 153 (1880) ...	Even-headed lentine ...
297	489	IX. 317	L. lasiophyllus	Cooke and Mass., Grev. XVI. 1 (1887) ...	Hairy-gilled lentine ...
298	481	V. 2459	L. lateritius	Berk., Linn. Journ. XVIII. 384 (1881) ...	Brick-red lentine ...
299	453	„ 2308	L. Lecomtei	Fries, Epicr. 368 (1838) ...	Lecomte's lentine
300	466	„ 2351	L. lepideus	Fries, Epicr. 390 (1838)	Scaly lentine ...
301	469	„ 2371	L. manipularis	Berk. and Br., Linn. Journ. XIV. 43 (1875)	Tufted lentine
302	493	„ 2499	L. pelliculosus	Fries, Epicr. 393 (1838)	Thin-skinned lentine ...
303	477	„ 2439	L. perganus	Lev., Champ., Mus. 117 (1846) ...	Parchment lentine
304	494	„ 2500	L. pulvinulus	Berk., Fl. Tasm. II. 250 (1860) ...	Pulvinate lentine
305	491	„ 2495	L. punctaticeps	Berk. and Br., Linn. Trans. II. 55 (1883)	Punctate-headed lentine ...
306	462	„ 2395	L. radicatus	Cooke and Mass. Grev. XIV., 118 (1886)	Rooting lentine ...
307	472	„ 2512	L. Schomburgkii ...	Berk., Linn. Trans. XX. 111 (1851)	Schomburgk's lentine ...
308	462	„ 2533	L. siparius ...	Berk. and Curt., Linn. Journ. X. 301 (1869)	Curtain lentine ...
309	454	„ 2311	L. strigosus	Fries, Epicr. 388 (1838)	Strigose lentine
310	492	„ 2497	L. subdulcis	Berk., Hook., Journ. II. 46 (1851) ...	Sweet-scented lentine ...
311	468	„ 2361	L. subnudus	Berk., Hook. Lond. Journ. VI. 492 (1847)	Somewhat-naked lentine ...
312	460	„ 2330	L. tener ...	Klotseh in Linn. Epicr. 389 (1838)	Slender lentine
313	464	„ 2349	L. tigrinus	Fries, Epicr. 389 (1838)	Tiger-tuft lentine
314	457	„ 2316	L. villosus	Klotseh in Linn. 479 (1843)	Villous lentine
315	488	„ 2486	L. vulpinus	Fries, Mon. Hym. II. 238 (1857)	Fox-coloured lentine ...

OF AUSTRALIAN FUNGI—continued.

Number	W.A.	S.A.	T.	V.	N.S.W.	Q.	B.	Occurrence.	General Characters.

Epicr. 372 (1838).—Agaricus—continued.

Number	W.A.	S.A.	T.	V.	N.S.W.	Q.	B.	Occurrence.	General Characters.
260	...	...	...	V.	...	Q.	...	Ground ...	Tawny, delicately tomentose. Stem thin, dilated at base.
261	...	...	...	...	N.S.W.	Q.	...	Leaves and twigs ...	Slender. Opaque, powdery, whitish. Stem mealy towards base.
262	...	...	...	...	N.S.W.	Q.	...	Wood ...	Leathery to membranaceous, yellowish tan. Stem with powdery ochrey down.
263	...	...	...	...	N.S.W.	...	...	...	Pale yellow, powdery. Stem short, slender, mealy.
264	...	...	...	V.	...	Q.	...	Eucalyptus	Thin, reddish yellow or grey. Stem of same colour, solid.
265	...	...	...	...	...	Q.	B	Dry dead branches	Inodorous. Somewhat fleshy, opaque, white, disc somewhat red. Very common.
266	...	...	...	...	...	Q.	...	Passion-flower twigs	Membranaceous, tawny to reddish brown. Stem thread-like, velvety.
267	...	...	...	V.	...	Q.	B	Fallen twigs, &c. ...	Minute. Membranaceous, whitish. Stem horny, shining, blackish, with collar from separating gills.
268	...	...	...	...	...	Q.	...	Ground, about trunks	Membranaceous, pale red. Stem thread-like, fixed at base by white mycelium.
269	...	...	...	V.	...	...	B	Dry ground ...	Small. Strong scented, oniony odour. Somewhat fleshy, red becoming white, or red buff. Edible.
270	...	...	...	...	...	Q.	...	Wood ...	White. Thin, smooth. Stem arising from circular base, smooth, hollow.
271	...	...	...	V.	...	...	...	Rotten wood ...	Membranaceous, pale-tan colour, tinged with pink. Stem horny, hollow.
272	...	...	T.	...	...	Q.	...	Rotten wood ...	Small. Mealy, adhering behind. Stem short, mealy.
273	...	...	...	...	...	Q.	B	Wood ...	Acrid, odourless. Fleshy to leathery. Usually ochrey tan. Stem fibrous.

Pl. Homon. 77 (1825).—Agaricus.

Number	W.A.	S.A.	T.	V.	N.S.W.	Q.	B.	Occurrence.	General Characters.
274	...	...	...	...	...	Q.	...	Dead branches ...	Brown, hispid, margin ciliate. Stem velvety.
275	...	...	...	...	...	Q.	...	Rotten trunks ...	Pale, at first woolly then bald, margin lobed. Stem short, nearly naked.
276A	...	...	...	...	N.S.W.	...	...	Rotten logs, in woods	Pale, tawny. Gills paler. cap thin.
277	...	...	...	...	...	Q.	...	Trunks ...	Golden yellow. Cap convex, then flattened. Stem cylindrical.
278	W.A.	...	...	V.	...	Q.	B.	Trunks and ground	Tough, flaccid. Fleshy, reddish brown, somewhat lobed or contorted.
279	...	...	...	...	...	Q.	...	Ground ...	White, orbicular. Stem thin, at length furrowed.
280	...	...	...	...	...	Q.	...	Dense scrubs ...	Ochrey, with velvety brown lines radiating from centre, funnel shaped.
281	W.A.	...	...	...	...	Q.	...	Rotten trunks ...	Becoming white, leathery, woolly to hairy, zoneless. Stem short.
282	...	...	...	...	...	Q.	B.	Trunks -- Willows and Poplars	Small, tufted. Fleshy to leathery, yellow white with brown scales.
283	...	...	...	...	...	Q.	...	Trunks ...	Rough, with rigid warts, ferruginous, powdery. Stem thickened downwards.
284	...	S.A.	...	V.	...	Q.	...	Rotten wood ...	Papery, pale tawny.
285	W.A.	...	T.	...	N.S.W.	Q.	...	Trunks ...	Thin, leathery, wine-glass shape, pale ochre. Stem velvety, tawny.
286	...	...	...	...	...	Q.	...	Deadwood ...	Orbicular, white, becoming tawny when dry. Stem slender, smooth.
287	...	...	...	V.	N.S.W.	Q.	...	Rotten wood ...	Deeply funnel shaped, bay brown, somewhat zoned. Stem rough or downy.
288	...	...	...	...	N.S.W.	Q.	...	Wood ...	Leathery, funnel shaped, purple brown, hairy. Stem tall, bristly.
289	...	...	...	...	...	Q.	...	Rotten wood ...	Fleshy, white, downy. Stem lateral, spindle shaped, rooting.
290	...	...	...	V.	...	Q.	...	Rotten wood ...	Rather membranaceous, ochrey, funnel shaped. Stem slender, brown.
291	...	...	...	...	N.S.W.	Q.	...	Rotten wood ...	Umber, brown. Stem curved, of same colour.
292	...	...	T.	...	...	Q.	...	Bark of Eucalyptus	Hoof-like. liver coloured, becoming smooth in front, hispid behind.
293	...	...	...	...	...	Q.	...	Stumps ...	Dirty white, funnel shaped, densely hispid. Stem hispid.
294	...	...	...	...	N.S.W.	Q.	...	Wood ...	Orbicular or semi-orbicular, full red brown, downy, wrinkled behind.
295	...	...	...	...	...	Q.	...	Ground ...	Funnel shaped, mealy, tawny. Stem short, rusty brown.
296	...	...	...	...	...	...	...	...	Fleshy to leathery, yellowish white. Stem solid, scaly at base.
297	...	...	...	V.	N.S.W.	...	...	Stumps ...	Thin, ochrey, shining, lobed at margin. Stems discoid, downy.
298	...	...	...	...	...	...	...	Wood ...	Brick red, and quite smooth. Stem of same colour, rigid.
299	...	...	...	...	N.S.W.	Q.	...	Rotten wood ...	Leathery, funnel shaped, fawn colour. Stem hairy, of same colour.
300	...	...	...	...	...	Q.	B.	Stumps of firs, &c.	Fleshy, compact, pale ochrey, broken up into darker spot-like scales.
301	...	...	...	...	...	Q.	...	Dead wood ...	Tufted, orbicular, white, clad with mealy scales. Stem curved.
302	...	...	...	...	N.S.W.	...	...	Rotten trunks ...	Sessile, tough, very thin, tawny fawn colour, kidney shaped.
303	...	...	...	...	...	Q.	...	Ground ...	White, leathery to membranaceous, funnel shaped. Stem solid.
304	...	...	T.	...	...	...	...	Rotten wood ...	Cushion shaped. pale, margin furrowed, ochrey.
305	...	...	...	...	...	Q.	...	Trunks ...	Punctately hispid, presenting sponge-like appearance. Stem yellow.
306	...	...	...	...	...	Q.	...	Ground ...	Fleshy, funnel shaped, pale ochrey, shortly velvety. Stem velvety.
307	...	...	...	...	...	Q.	...	Wood ...	Leathery when dry, fawn coloured, broadly funnel shaped.
308	...	...	...	...	...	Q.	...	Rotten wood ...	Orange, brown when dry. Woolly with erect rigid hairs intermixed.
309	...	...	...	...	N.S.W.	Q.	...	Trunks ...	Reddish-fawn colour. Roughly hairy. Stem excentric and hairy.
310	...	...	...	...	...	Q.	...	Deadwood ...	White, sweet scented. Cap fleshy smooth. Stem obsolete or spurious.
311	...	S.A.	...	V.	...	Q.	...	Fallen trees ...	Rather funnel shaped, at first clad with mealy scales. Stem slender.
312	...	...	...	...	...	Q.	...	Rotten wood ...	Membranaceous, bay brown. Stem very slender, hollow.
313	...	...	...	...	...	Q.	B.	Trunks ...	Fleshy to leathery, whitish to yellow white, with tawny scales. Edible.
314	...	...	...	...	...	Q.	...	Rotten wood ...	Brown, leathery, with fibrous bristles. Stem solid, tawny.
315	...	...	...	...	N.S.W.	Q.	B.	Stumps ...	Sessile, imbricated. Fleshy, conchate, tan coloured, corrugated or woolly.

Number.	Cooke's Number.	Saccardo's Number.	Scientific Name.	Authority for Name.	English Name.
					19. PANUS.—Fries,
316	511	...	P. angustatus	Berk. in Cooke's Handb. Aust. Fungi 9s (1892)	Narrow panus
317	303	V. 2567	P. arenicola	Berk., Linn. Journ. XVIII. 384 (1881)	Sand-growing panus
318	501	„ 2552	P. carbonarius	Cooke and Mass., Grev. XV. 94 (1887)	Charcoal panus
319	505	„ 2569	P. cinnabarinus	Fries, Pl. Preiss. II. 133 (1846)	Vermilion panus
320	495	„ 2519	P. conchatus	Fries, Epicr. 398 (1838)	Conch panus
321	500	„ 2547	P. coriaceus	Berk., Linn. Journ. XIII. 160 (1873)	Coriaceous panus
322	507	„ 2573	P. eugrammus	Fries, Nov. Symb. 40 (1851)	Well-lined panus
323	498	„ 2541	P. incandescens	Berk. and Br., Linn. Trans. II. 55 (1883)	Incandescent panus
324	506	IX. 329	P. lateritius	Sacc. Hedw. 125 (1889)	Brick-red panus
325	508	„ 328	P. olivaceo-flavidus	Cooke and Mass., Grev. XVI. 1 (1887)	Yellowish-olive panus
326	510	V. 2588	P. patellaris	Fries. Epicr. 400 (1838)	Cup-shaped panus
327	497	„ 2540	P. rivulosus	Berk., Linn. Journ. XVIII. 384 (1881)	Cracked panus
328	509	„ 2578	P. saccharinus	Berk., Fl. Tasm. II. 260 (1860)	Saccharine panus
329	502	„ 2557	P. stypticus	Fries, Epicr. 399 (1838)	Styptic panus
330	499	„ 2542	P. suborbicularis	Berk. and Br., Linn. Trans. II. 56 (1883)	Sub-orbicular panus
331	496	„ 2521	P. tortulosus	Fries. Epicr. 397 (1838)	Twisted panus
332	504	„ 2568	P. viscidulus	Berk. and Br., Linn. Trans. II. 56 (1883)	Gelatinous panus
					20. XEROTUS.—Fries,
333	516	V. 2609	X. albidus	Berk. and Br., Linn. Trans. II. 56 (1883)	Whitish xerote
334	520	„ 2617	X. Archeri	Berk., Fl. Tasm. II. 250 (1860)	Archer's xerote
335	517	„ 2611	X. Berterii	Mont. Chil. VII. 353 (1852)	Bertero's xerote
336	522	...	X. Drummondi	Berk. in Cooke's Handb. Aust. Fungi 100 (1892)	Drummond's xerote
337	521	...	X. fulvus	Berk. and Br. in Cooke's Handb. Aust. Fungi 100 (1892)	Tawny xerote
338	523	V. 2596	X. griseus	Berk., Hook., Lond. Journ. VI. 497 (1847)	Grey xerote
339	518	„ 2613	X. lateritius	Berk. and Curt. Linn. Journ. X. 303 (1869)	Brick-red xerote
340	514	„ 2606	X. papposus	Kalch., Grev. VIII. 154 (1880)	Papuan xerote
341	519	„ 2616	X. papyraceus	Berk., Fl. Tasm. II. 250 (1860)	Papery xerote
342	513	„ 2601	X. proximus	Berk. and Br., Linn. Trans. II. 56 (1883)	Approximate xerote
343	515	„ 2607	X. rawakensis	Pers. in Fries., Epicr. 401 (1838)	Rawak xerote
344	512	„ 2599	X. tener	Berk. and Br., Linn. Journ. XIV. 45 (1875)	Tender xerote
					21. TROGIA.—Fries,
345	524	V. 2627	T. crispa	Fries, Mon. Hym. II. 244 (1857)	Crisped trogia
					22. LENZITES.—Fries,
346	529	V. 2638	L. abietina	Fries, Epicr. 407 (1838)	Larch lenzites
347	532	„ 2634	L. acuta	Berk., Hook. Lond. Journ. I. 146 (1842)	Acutely-margined lenzites
348	534	„ 2657	L. applanata	Fries, Epicr. 404 (1838)	Depressed lenzites
349	535	„ 2638	L. aspera	Fries, Epicr. 405 (1838)	Rough lenzites
350	536	„ 2664	L. Beckleri	Berk., Linn. Journ. XIII. 161 (1873)	Beckler's lenzites
351	530	„ 2651	L. Berkeleyi	Lev., Ann. Sci. Nat. V. 122 (1846)	Berkeley's lenzites
352	526	„ 2630	L. betulina	Fries, Epicr. 405 (1838)	Birch lenzites
352A	„	„ 2651	L. betulina, var. velutina	Berk. Ann. Nat. Hist. III. 381 (1839)	Velvety lenzites
353	...	...	L. bifasciatus	Cooke and Mass., Grev. XXI. 37 (1892)	Bifasciate lenzites
354	533	V. 2656	L. deplanata	Fries, Epicr. I. 404 (1838)	Levelled lenzites
355	540	„ 2685	L. faventina	Cald., Erb. Crit. Ital. No. 89 (1878, &c.)	Honeycombed lenzites
356	527	„ 2631	L. flaccida	Fries, Epicr. 406 (1838)	Flaccid lenzites
357	539	„ 2682	L. Guilfoylei	Berk. Grev. X. 61 (1881)	Guilfoyle's lenzites
358	538	„ 2670	L. nivea	Cooke. Grev. XV. 94 (1887)	Snow-white lenzites
359	541	„ 2687	L. Palisotii	Fries, Epicr. 405 (1838)	Palisot's lenzites
360	542	„ 2689	L. repanda	Fries, Epicr. 406 (1838)	Repand lenzites
361	528	„ 2636	L. sepiaria	Fries, Epicr. 407 (1838)	Chocolate lenzites
362	531	„ 2653	L. striata	Fries, Epicr. 406 (1838)	Striated lenzites
363	537	„ 2665	L. torrida	Kalch., Grev. VIII. 154 (1880)	Torrid lenzites

OF AUSTRALIAN FUNGI—continued.

Number	W.A.	S.A.	T.	V.	N.S.W.	Q.	H.	Occurrence	General Characters
								Epicr. 396 (1838).—Agaricus, Lentinus.	
316	...	...	...	...	...	Q.	...	Logs ...	Spoon shaped, tawny, nearly sessile, with a few scattered hairs.
317	...	...	...	V.	...	...	...	Sandy soil ...	Brown, spoon shaped. Stem and cap covered with particles of sand.
318	...	S.A.	...	...	...	...	...	Among ferns where burnt	Fleshy, umber, fan or funnel shaped. Stem short, pale.
319	W.A.	...	...	...	...	Q.	...	Base of trunks ...	Leathery, vermilion, sessile, kidney shaped.
320	...	...	...	...	N.S.W.	...	B.	Trunks — Poplar, Beech, Birch	Largish. Fleshy, cinnamon, becoming pale, conchate. *Edible.*
321	...	...	...	V.	...	...	...	Bark ...	Very beautiful. Leathery, brown behind or black when young, sessile.
322	...	...	...	...	...	Q.	...	Bark ...	Sessile, imbricate. Leathery to membranaceous, pale, kidney shaped.
323	...	S.A.	...	V.	N.S.W.	Q.	...	Buried wood, but apparently on soil	Sometimes funnel shaped, smooth. Very luminous at night.
324	...	S.A.	...	...	...	...	...	Rotten wood of *Eucalyptus*	Membranaceous, tan coloured, sprinkled with brick-red point-like threads.
325	...	...	...	V.	N.S.W.	...	...		Sooty brown, densely velvety, with yellowish-olive down, sessile.
326	...	...	...	...	...	Q.	B.	Branches — Beech. Cherry, &c.	Leathery, mealy to downy, flat cup shaped.
327	...	...	...	V.	...	...	...	Trunks ...	Ochrey, striately cracked. Stem excentric, similarly cracked.
328	...	...	T.	...	...	...	...	Rotten wood	Rather fleshy. Stem short, mealy or obsolete. Edge of gills as if dusted with sugar.
329	...	...	...	V.	...	...	B.	Stumps ...	Small. Leathery, kidney shaped, cinnamon turning pale. Astringent taste.
330	...	...	...	...	...	Q.	...	Old trunks	Sub-orbicular, white, delicately downy. Stem obsolete.
331	...	...	...	...	...	Q.	B.	Trunks ...	Fleshy to leathery, flesh coloured to ochrey pink. *Edible.*
332	...	...	...	V.	N.S.W.	Q.	...	Rotten trunks, decaying bark	Upper layer gelatinous, rather viscid, dull slate coloured. Stem short, lateral, arising from spongy base.
								Elench. I. 48 (1828).	
333	...	...	...	...	...	Q.	...	Wood ...	Whitish, kidney shaped. Stem lateral, smooth or slightly velvety.
334	...	...	T.	V.	...	...	...	Sticks ...	Kidney shaped, reddish brown. Stem very short, lateral, powdery.
335	...	...	...	...	...	Q.	...	Fallen branches ...	Gregarious. Leathery to membranaceous, rust coloured, kidney shaped.
336	...	...	...	V.	...	...	...	Twigs ...	Gregarious. Kidney shaped, lobed or crispate, rust coloured.
337	...	...	...	...	...	Q.	...	Wood ...	Tawny ochrey. Membranaceous, kidney shaped. Stem lateral, thin.
338	...	...	...	V.	...	...	...	Old wood...	Funnel shaped, splitting, grey. Stem compressed, wedge shaped.
339	...	...	...	...	...	Q.	...	Dead bark	Sub-orbicular, brick red, furrowed.
340	...	...	...	...	N.S.W.	...	...	Bark ...	Ochrey tan colour. Membranaceous to leathery, radiately furrowed.
341	...	...	T.	V.	...	...	...	Rotten wood ...	Papery, pale. Stem very short or obsolete.
342	...	...	...	V.	...	Q.	...	Branches...	White, then yellowish brown. Sub-orbicular, delicately powdery.
343	...	...	...	...	...	Q.	...	Wood ...	Smooth, cinnamon. Leathery to membranaceous. Stem solid, short.
344	...	...	...	...	...	Q.	...	Dead wood	Kidney shaped, membranaceous, umber. Stem very short.
								Epicr. 402 (1838).—Merulius, Cantharellus.	
345	...	...	...	V.	...	Q.	B.	Twigs—Birch, Beech, &c.	Tough, cup shaped, reddish yellow. Gills crisp, plaited, forked.
								Epicr. 403 (1838).—Agaricus, Dædalea.	
346	...	S.A.	...	...	...	Q.	B.	Wood ...	Leathery, clothed with umber down, at length smooth and whitish.
347	...	...	...	...	...	Q.	...	Wood ...	Kidney shaped, leathery, greyish umber. Stem distinct and disc shaped.
348	...	...	...	...	...	Q.	...	Wood ...	Kidney shaped, corky, whitish, zoneless, downy.
349	...	...	...	...	N.S.W.	Q.	...	Dead wood	Thick, spongy to corky, rough, pale, concentrically furrowed.
350	...	...	...	...	N.S.W.	Q.	...	Trunks ...	Woody, whitish, rather thick, margin ochrey.
351	...	...	...	...	...	Q.	...	Trunks ...	Leathery, flexible, hairy, brownish, sessile, somewhat kidney shaped.
352	...	...	...	V.	...	Q.	B.	Trunks ...	Corky to leathery, pale, concentrically grooved, downy.
352a	...	...	...	...	...	Q.	...	Trunks ...	Hard, sessile, lobed, deeply zoned, tawny, velvety.
353	...	...	...	V.	...	...	...	Bark ...	Kidney or shell shaped, leathery, greyish-fawn colour, silky.
354	...	...	...	...	N.S.W.	...	...	Trunks ...	Corky, kidney shaped, tan colour to whitish, downy, zoneless.
355	...	...	...	...	...	Q.	...	Trunks, chiefly Poplar	Corky, white, at length turning ash coloured, sessile, zoneless.
356	...	...	...	V.	N.S.W.	...	B.	Stumps of beech, &c.	Leathery, flaccid, hairy, pale, zoned.
357	...	...	...	...	N.S.W.	Q.	...	Trunks ...	Shell shaped, smoky behind, pale in front, dotted with tubercles.
358	...	...	...	V.	...	...	...	Trunks ...	Snow white, corky to leathery, pitted, rather discoid behind.
359	...	...	...	...	N.S.W.	Q.	...	Trunks ...	Corky, ochrey to white, hemispherical, zoned, margin lobed.
360	...	...	...	...	N.S.W.	Q.	...	Trunks ...	Corky, white to pale, margin slightly waved.
361	...	...	...	V.	...	...	B.	Wood, pine	Leathery, bright brown, zoned, margin yellowish. Common.
362	...	S.A.	...	V.	...	Q.	...	Trunk ...	Leathery, soft, downy, rust coloured, obsoletely zoned.
363	...	...	...	V.	N.S.W.	...	...	Wood ...	Entirely white. Compact, woody, sessile, concentrically furrowed.

SYSTEMATIC ARRANGEMENT

Number	Cooke's Number	Saccardo's Number	Scientific Name	Authority for Name	English Name
			23. SCHIZOPHYLLUM.—Fries,		
364	525	V. 2705	S. commune	... Fries, S.M. I. 333 (1821)	... Common schizophyllum
365	„	„ 2706	S. multifidum	... Fries, S.M. I. 333 (1821)	... Multifid schizophyllum
			24. METRARIA.—Cooke and Mass.,		
366	190	IX. 348	M. insignis	.. Cooke and Mass., Grev. XIX. 105 (1891)	... Remarkable metraria
			25. VOLVARIA.—Fries,		
367	191	V. 2712	V. bombycina	... Fries, S.M. I. 277 (1821)	... Silky volvar
368	195	„ 2740	V. parvula	... Weinm., Ross 238 (1836)	... Little volvar
369	194	„ 2735	V. speciosa	... Fries, S.M. I. 278 (1821)	... Beautiful volvar
370	192	„ 2717	V. Taylori	... Berk., Outl. 140 (1860)	... Taylor's volvar
371	193	„ 2733	V. xanthocephala	... Berk., Hook., Lond. Journ. IV. 45 (1845)	... Yellow-headed volvar
			26. ANNULARIA.—Schulz. Verh. Zool. Bot.		
372	196	IX. 350	A. insignis	... Cooke and Mass., Grev. XVIII. 3 (1889)	... Remarkable annularia
			27. PLUTEUS.—Fries,		
373	197	V. 2747	P. cervinus	... Fries, Epicr. 140 (1838)	... Fawn pluteus
374	198	„ 2806	P. Wehlianus	... F. v. M., Grev. XV. 93 (1887)	... Wehl's pluteus
			28. ENTOLOMA.—Fries,		
375	202	V. 2828	E. Bloxami	... Berk. and Br., Outl. 143 (1860)	... Bloxam's entolome
376	200	IX. 354	E. galbineum	... Cooke and Mass., Grev. XVII. 7 (1888)	... Yellowish entolome
377	201	„ 359	E. lacticolor	... Cooke and Mass., Grev. XVI. 31 (1887)	... Bright-coloured entolome
378	199	„ 360	E. melaniceps	... Cooke and Mass., Grev. XVI. 31 (1887)	... Black-headed entolome
379	203	V. 2863	E. panniculus	... Berk., Fl. Tasm. II. 245 (1860)	... Ragged entolome
			29. CLITOPILUS.—Fries,		
380	204	V. 2900	C. cancrinus	... Fries, Epicr. 150 (1838)	... Crab-like clitopile
381	...	...	C. cyathoideus	... Cooke and Mass., Grev. XXI. 36 (1892)	... Goblet-like clitopile
			30. LEPTONIA.—Fries,		
382	207	V. 2945	L. aquila	... Fries, Epicr. 151 (1838)	... Eagle leptonia
383	205	„ 2923	L. lampropoda	... Fries, S.M. I. 203 (1821)	... Brilliant-stalked leptonia
384	208	...	L. melanura	... Cooke and Mass., Grev. XIX. 89 (1891)	... Black-tailed leptonia
385	206	IX. 372	L. quinquecolor	... Cooke and Mass., Grev. XVII. 7 (1888)	... Five-coloured leptonia
			31. NOLANEA.—Fries,		
386	210	V. 2967	N. mammosa	... Fries, Mon. Hym. I. 293 (1857)	... Papillate nolanea
387	209	„ 2960	N. pascua	... Fries, S.M. I. 205 (1821)	... Pasture nolanea
388	211	„ 2980	N. rufo-carnea	... Berk. Outl. 148 (1860)	... Reddish-brown nolanea
			32. ECCILIA.—Fries,		
389	212	V. 3030	E. rhodocylix	... Lasch. in Fries, Hym. Eur. 213 (1874)	... Rose-cup eccilia
			33. CLAUDOPUS.—Smith,		
390	213	V. 3037	C. variabilis	Fries, Hym. Eur. 213 (1874)	... Variable claudopus

OF AUSTRALIAN FUNGI—*continued.*

Number	W.A.	S.A.	T.	V.	N.S.W.	Q.	B.	Occurrence	General Characters.

Obs. I. 103 (1815).—Agaricus.

364	W.A.	S.A.	T.	V.	N.S.W.	Q.	B.	Dead wood	... Dry, white or greyish, scarcely exceeding inch in diameter. Cosmopolitan.
365	...	...	...	V.	...	...	...	Wood ...	... Deeply cut into numerous lobes, becoming pale yellow.

Grev. XIX. 104 (1891).

366	...	...	...	V.		...	...	Woods ...	... Margin cream colour, disc darker and reddish brown, viscid, shining when dry.

S.M. I. 277 (1821).—Agaricus, Amanita.

367	...	...	...	V.	...	...	B.	Decayed wood	... Large. Fleshy, silky, fawn to brown, globose and viscid at first. *Edible.*
368	...	...	...	...	...	Q.	B.	Pastures, after stormy weather	White. Rather fleshy, downy, conical at first. Stem stuffed, silky.
369	...	...	...	V.	...	...	B.	Dung-hills, road-sides, &c.	Large. Fleshy, smooth, viscid or polished, grey. Stem rather bulbous.
370	...	...	T.	V.	...	...	B.	Ground ...	... Thin, livid. Stem pale, solid, smooth.
371	W.A.	...	...	...	...	...	...	Ground ...	... Golden yellow, spotted with white from remains of volva. Stem bulbous.

Gesell. 49 (1868).—Agaricus, Chamæota.

372	...	...	...	V.		...	...	Ground ...	... Fleshy, pale, cuticle broken up into broad darker scales. Stem short, thick.

Epicr. 110 (1838).—Agaricus.

373	...	...	T.	V.	...	...	B.	Trunks of trees	... Large. Fleshy, dull tawny, smooth, then clad with evanescent scales.
374	...	S.A.	...	V.	...	...	...	Rotten wood on ground	Fleshy, shining, ochrey to white, disc darker; stem 6 to 8 inches long, thick.

Epicr. 143 (1838).—Agaricus.

375	...	S.A.	...	V.	...	...	B.	Open exposed pastures	Fleshy, compact, blackish blue. Stem slightly tapering upwards.
376	...	...	...	V.	...	...	...	Ground ...	... Sulphur colour. Rather fleshy, almost saffron colour. Stem hollow.
377	...	S.A.	...	V.	...	...	...	Sandy soil	... Rather fleshy, shining, amethyst colour. Stem thin, nearly solid.
378	...	S.A.	...	V.	...	...	...	Ground ...	... Fleshy, compact, dark sooty brown. Stem solid, short, pale.
379	...	...	T.	...	...	...	...	Among ferns	... Thin, bell shaped, dark violet. Stem thickened at base, and downy.

Epicr. 148 (1838).—Agaricus.

380	...	S.A.	...	V.		Q.	B.	Grass fields	... Small and beautiful. Fleshy to membranaceous, flesh colour to white.
381	...	...	...	V.				Under burnt logs ...	Rather thin, pale, finally funnel shaped. Stem hollow, white and woolly at base.

S.M. I. 201 (1821).—Agaricus.

382	...	S.A.	...	...	...	...	...	Ground ...	... Rather membranaceous, bay brown. Stem short, stuffed.
383	...	...	...	V.	...	...	B.	Pastures ...	... Rather fleshy, mouse coloured, or steel grey, or sooty. Common.
384	...	...	...	V.	...	...	...	Ground ...	... Bell shaped, shining black, silky. Stem cylindrical.
385	...	...	...	V.	...	...	...	Black loam	... Membranaceous, disc brownish brick red, margin yellowish.

S.M. I. 204 (1821).—Agaricus.

386	...	...	...	V.	...	...	B.	Meadows	... Sub-membranaceous, papillate, tawny. Stem hollow, polished.
387	...	...	T.	...	...	...	B.	Pastures ...	... Membranaceous, shining like silk when dry. Stem silky fibrous.
388	...	...	...	V.	...	...	B.	Heaths ...	... Small. Sub-membranaceous, red brown, indistinctly scaly.

S.M. I. 207 (1821).—Agaricus.

389	...	...	...	V.	...	...	B.	Rotten wood	... Membranaceous, tawny, when dry flocculose, grey.

Seem. Journ.—Agaricus.

390	...	...	...	V.	...	Q.	B.	Sticks, &c.	... Very small. Sub-membranaceous, silky with white down. Common.

Number	Cooke's Number.	Saccardo's Number.	Scientific Name.	Authority for Name.	English Name.

34. PHOLIOTA.—Fries,

Number	Cooke's	Saccardo's	Scientific Name	Authority	English Name
391	226	V. 3103	P. allantopoda	Berk., Hook., Lond. Journ. IV. 45 (1845)	Sausage-stalked pholiota
392	218	„ 3053	P. blattaria	Fries, S.M. I. 246 (1821)	Cockroach-like pholiota
393	224	„ 3094	P. congesta	Kalch., Grev. IX. 147 (1881)	Congested pholiota
394	220	...	P. disrupta	Cooke and Mass., Grev. XIX. 89 (1891)	Disrupted pholiota
395	225	V. 3104	P. effusa	Kalch., Grev. IX. 147 (1881)	Expanded pholiota
396	215	„ 3050	P. erebia	Fries, S.M. I. 246 (1821)	Lurid pholiota
397	230	„ 3128	P. eriogena	Fries, Pl. Preiss. II. 132 (1846)	Woolly pholiota
398	227	„ 3109	P. flammans	Fries, S.M. I. 244 (1821)	Flame-coloured pholiota
399	229	„ 3130	P. marginata	Fries, Hym. Eur. 225 (1874)	Margined pholiota
400	228	„ 3129	P. mutabilis	Fries, S.M. I. 245 (1821)	Changeable pholiota
401	232	„ 3137	P. mycenoides	Fries, S.M. I. 246 (1821)	Mycena-like pholiota
402	222	„ 3071	P. phylicigena	Berk., Linn. Journ. XV. 52 (1877)	Phylica-growing pholiota
403	219	„ 3055	P. praecox	Fries, Hym. Eur. 217 (1874)	Precocious pholiota
404	221	„ 3065	P. pudica	Fries, Hym. Eur. 218 (1874)	Modest pholiota
405	231	„ 3135	P. pumila	Fries, El. 29 (1828)	Dwarfish pholiota
406	217	IX. 394	P. recedens	Cooke and Mass., Grev. XVIII. 25 (1889)	Receding pholiota
407	223	V. 3102	P. spectabilis	Fries, El. 28 (1828)	Notable pholiota
408	216	„ 3052	P. togularis	Fries, Hym. Eur. 216 (1874)	Gowned pholiota

35. LOCELLINA.—Gill. Champ.

409	214	V. 3141	L. cyenopotamia	Sacc. Syll., V. 762 (1887)	Swan river locellinia

36. INOCYBE.—Fries,

410	234	V. 3149	I. cincinnata	Fries, S.M. I. 256 (1821)	Curly inocybe
411	236	„ 3165	I. flocculosa	Fries, Hym. Eur. 229 (1874)	Flocculous inocybe
412	238	...	I. gigaspora	Cooke, Handb. Aust. Fungi 47 (1892)	Large-spored inocybe
413	237	V. 3235	I. gomphodes	Kalch., Grev. VIII. 152 (1880)	Pap-like inocybe
414	235	„ 3155	I. lanuginosa	Fries, S.M. I. 257 (1821)	Woolly inocybe
415	233	„ 3148	I. plumosa	Fries, Mon. Hym. (1857)	Downy inocybe
416	239	IX. 421	I. Victoriæ	Cooke and Mass., Grev. XVI. 72 (1888)	Victorian inocybe

37. HEBELOMA.—Fries,

417	245	IX. 426	H. arenicolor	Cooke and Mass., Grev. XVII. 7 (1888)	Sand-coloured hebeloma
418	240	V. 3259	H. fastibile	Fries, Epicr. 178 (1838)	Disagreeable hebeloma
419	241	„ 3260	H. glutinosum	Lindgr., Bot. Not. 199 (1845)	Glutinous hebeloma
420	...		H. griseum	Cooke and Mass., Grev. XXI. 36 (1892)	Grey hebeloma
421	242	V. 3268	H. mesophæum	Fries, Epicr. 179 (1838)	Dusky-centred hebeloma
421A	„	„ „	H. mesophæum, var. holophæum	Fries, Hym. Eur. 240 (1874)	Dusky hebeloma
422	244	„ 3291	H. nudipes	Fries, Epicr. 181 (1838)	Naked-stalked hebeloma
423	243	„ 3275	H. olidum	Cooke and Mass., Grev. XV. 93 (1887)	Strong-smelling hebeloma
424	246	„ 3319	H. petiginosum	Fries, S.M. I. 259 (1821)	Scabby hebeloma

38. FLAMMULA. —Fries,

425	252	IX. 437	F. avellanea	Cooke and Mass., Grev. XVIII. 3 (1889)	Nut-brown flammula
426	253	V. 3344	F. Baileyi	Berk. and Br., Linn. Trans. II. 54 (1883)	Bailey's flammula
427	258	„ 3369	F. flavida	Fries, S.M. I. 250 (1821)	Yellowish flammula
428	257	„ 3363	F. fusa	Fries, Hym. Eur. 247 (1874)	Fusiform flammula
429	261	„ 3382	F. hybrida	Fries, Mon. Hym. I. 360 (1857)	Hybrid flammula
430	251	IX. 438	F. hyperiou	Cooke and Mass., Grev. XVI. 72 (1888)	Hyperion flammula
431	260	V. 3373	F. inopoda	Fries, S.M. I. 251 (1821)	Fibril-stalked flammula

OF AUSTRALIAN FUNGI—continued.

Number	W.A.	S.A.	T.	V.	N.S.W.	Q.	B.	Occurrence	General Characters

S.M. I. 240 (1821).—Agaricus.

Number	W.A.	S.A.	T.	V.	N.S.W.	Q.	B.	Occurrence	General Characters
391	W.A.	...	...	V.	...	...	...	Ground	Fleshy, golden yellow. Stem elongatedly bulbous at base.
392	...	...	...	V.	...	...	B.	Ground	Elegant, small. Rather fleshy, rust coloured, margin grooved.
393	...	...	...	V.	...	...	...	Trunks	Fleshy, the size of a pea, rather mealy, brownish.
394	...	...	...	V.	...	...	...	Ground	Fleshy, creamy white, at first smooth, then cracked deeply.
395	...	...	...	V.	...	...	...	Wood	Fleshy, white, breaking into polygonal wart-like spaces.
396	...	...	...	V.	...	...	B.	Grassy places ...	Fleshy, rather viscid, lurid. Stem hollow, fibrillose to scaly.
397	W.A.	...	...	...	...	...	...	Trunks	Fleshy, rust coloured. Stem with dense woolly mycelium at base.
398	...	...	...	...	...	Q.	B.	Pinewoods ...	Fleshy, tawny, with sulphureous scales. Stem stuffed, thin, hollow.
399	...	...	...	V.	...	...	B.	Ground, among firs	Rather fleshy, cinnamon, margin striate. Stem hollow, not scaly.
400	...	...	T.	V.	...	...	B.	Trunks or ground ...	Fleshy, cinnamon, becoming pale. Stem rough with scales. *Edible.*
401	...	...	...	V.	...	...	B.	Ground, in damp places	Membranaceous, rust coloured, tawny or pale when dry.
402	...	...	...	...	...	Q.	...	Trunks of *Phylica*...	Fleshy, tawny. Stem thick below, tapering upwards.
403	W.A.	...	...	...	...	...	B.	Gardens and pastures	Fleshy, soft, white to yellowish. Stem downy or mealy. *Edible.*
404	...	...	...	V.	...	...	B.	Trunks, &c. ...	Largish. Fleshy, dry, smooth, whitish buff, modestly coloured. *Edible.*
405	...	...	...	V.	N.S.W.	...	B.	Woods	Somewhat fleshy, hemispherical. Stem hollow, slender.
406	...	...	...	V.	...	...	...	Ground	Rather fleshy, golden tawny. Stem elongated, cylindrical
407	...	...	...	V.	N.S.W.	...	B.	Dead stumps—Oak, &c.	Large, compact. Golden orange or tawny buff, scales silky and broad. Odour bad.
408	...	...	...	V.	...	...	B.	Grassy places, &c....	Fleshy, pale ochre. Stem hollow, cracking. Ring hanging like toga.

Fr. 428 (1874).—Agaricus, Acetabularia.

Number	W.A.	S.A.	T.	V.	N.S.W.	Q.	B.	Occurrence	General Characters
409	W.A.	...	...	...	...	...	...	Ground	Gills pale fawn colour, leaving a free space round stem. Very rare.

S.M. I. 254 (1821).—Agaricus, Hebeloma.

Number	W.A.	S.A.	T.	V.	N.S.W.	Q.	B.	Occurrence	General Characters
410	...	...	...	V.	...	...	B.	Shady woods ...	Rather fleshy, scaly. Stem solid, thin, scaly.
411	...	...	...	V.	...	...	B.	Bare soil and among grass	Somewhat fleshy, tawny brown. Stem pale reddish.
412	...	...	...	V.	...	...	...	Ground	Rather fleshy, yellow brown. Stem abruptly rooting.
413	...	...	...	...	N.S.W.	...	...	Ground	Rather fleshy, with globose pap-like apex, tawny.
414	W.A.	...	...	...	...	...	B.	Ground	Rather fleshy, umber, becoming yellowish. Stem solid, thin, scaly.
415	...	...	...	V.	...	...	B.	Moist pine woods ...	Rather fleshy. Odour weak, not unpleasant. Stem slender.
416	...	...	...	V.	...	...	...	Grassy ground ...	Rather fleshy, viscid, whitish, shining. Stem stuffed, white.

S.M. I. 249 (1821).—Agaricus.

Number	W.A.	S.A.	T.	V.	N.S.W.	Q.	B.	Occurrence	General Characters
417	...	...	...	V.	...	...	...	Ground	Fleshy, rather viscid, dingy ochre or sand colour. Stem cylindrical.
418	...	...	...	V.	...	...	B.	Woods	Compact, viscid, yellowish tan or tan colour. Stem solid, white.
419	...	...	...	V.	...	...	B.	Among dead leaves	Fleshy, viscous with a tenacious gluten, yellowish white. Stem stuffed.
420	...	...	...	V.	...	...	...	Ground in woods ...	Fleshy, mouse grey or pale silver grey, glutinous, smooth, shining when dry. Odour unpleasant.
421	...	...	...	V.	...	...	B.	Woods, &c. ...	Rather fleshy, viscid, yellowish tan, disc bay. Stem slender, white. Common.
421A	...	...	...	V.	...	...	...	Ground	Dark brown, veil resembling a ring. Stem becoming brown.
422	...	...	...	...	...	Q.	B.	Ground	Fleshy, slightly viscid, tan coloured, thin, pale. Stem solid, white.
423	...	S.A.	...	...	...	...	...	Stony ground ...	Fleshy, viscid, full red brown. Stem hollow, smooth. Odour fœtid.
424	...	...	...	V.	...	...	B.	Ground in shady places	Rather fleshy, dry, brown, circumference silky grey. Stem slender, powdery, brick red.

S.M. I. 250 (1821).—Agaricus, Paxillus.

Number	W.A.	S.A.	T.	V.	N.S.W.	Q.	B.	Occurrence	General Characters
425	...	...	...	V.	...	Q.	...	Sandy ground ...	Fleshy, nut brown. Stem tapering upwards, grooved.
426	...	...	...	V.	...	Q.	...	Rotten wood ...	Orange. Bell shaped to hemispherical, woolly, sprinkled with reddish-yellow mealy particles.
427	...	...	...	V.	N.S.W.	...	B.	Trunks—Pine, &c.	Fleshy, yellow, smooth, moist. Stem yellow, then rusty.
428	...	S.A.	...	V.	...	...	B.	Ground and fallen logs	Compact, rather viscid, flesh becoming yellow. Odour not unpleasant.
429	...	...	...	V.	...	...	B.	Fir stumps ...	Fleshy, moist, at first cinnamon brown then golden tawny. Veil forming ring.
430	...	...	...	V.	...	...	...	Stumps (?) ...	Fleshy, golden tawny, then darker. Stem tapering downwards, furrowed.
431	...	...	...	V.	...	...	B.	Pine trunks ...	Fleshy, moist, honey tan colour, becoming pale. Stem fibrillose.

24

SYSTEMATIC ARRANGEMENT

Number.	Cooke's Number.	Saccardo's Number.	Scientific Name.	Authority for Name.	English Name.

38. FLAMMULA.—Fries,

132	259	V. 3379	F. limonia	Cooke and Mass., Grev. XV. 94 (1887) ...	Lemon-coloured flammula
133	369	„ 3326	F. paradoxa	Kalch., Fung. Hung. t. 16, f. 1 (1873) ...	Paradoxical flammula ...
134	262	„ 3381	F. penetrans	Fries, Hym. Eur. 230 (1874)	Penetrating flammula ...
134A	„	„ „	F. penetrans, var. australis	F. v. M., Linn. Journ. XIII. 158 (1873) ...	Southern flammula ...
135	255	„ 3346	F. peregrina	Fries, Elen. l. 31 (1828) ...	Foreign flammula ...
136	265	„ 3389	F. picrea ...	Fries, Hym. Eur. 251 (1874) ...	Bitter flammula ...
137	254	IX. 446	F. prasina	Cooke and Mass., Grev. XVIII. 3 (1889)...	Leek-green flammula ...
138	260	V. 3393	F. purpureo-nitens	Cooke and Mass., Grev. XV. 94 (1887) ..	Shining-purple flammula ...
139	249	IX. 445	F. rubra ...	Cooke and Mass., Grev. XIX. 46 (1890) ...	Red flammula
140	263	V. 3385	F. sapinea	Fries, Epicr. 189 (1838) ...	Pine-wood flammula ...
141	256	„ 3358	F. spumosa	Fries, S.M. l. 252 (1821)	Frothy flammula ...
142	248	...	F. veluticeps	Cooke and Mass., Grev. XIX 89 (1891) ...	Velvet-capped flammula ...
143	247	„ 3323	F. vinosa...	Fries, Hym. Eur. 244 (1874) ...	Wine-coloured flammula
144	250	...	F. xanthophylla ...	Cooke and Mass., Handb. Aust. Fungi 50 (1892) ...	Yellow-gilled flammula

39. NAUCORIA.—Fries,

145	267	V. 3412	N. anguinea	Fries, Epicr. 193 (1838) ...	Snake-like naucoria ...
146	281	„ 3506	N. Bowmani	Berk., Linn. Journ. XIII. 158 (1873)	Bowman's naucoria ...
147	269	„ 3435	N. cerodes	Fries, Epicr. 195 (1838) ...	Wax-like naucoria ...
148	278	„ 3484	N. Drummondi	Berk., Hook., Lond. Journ. IV. 46 (1845)	Drummond's naucoria ...
149	282	„ 3514	N. escharoides	Fries, S.M. l. 260 (1821) ...	Scabby naucoria ...
150	275	IX. 458	N. fraterna	Cooke and Mass., Grev. XVI 31 (1887) ...	Fraternal naucoria ...
151	283	V. 3495	N. frosticola	Berk., Linn. Journ. XIII. 158 (1873) ...	Tufted naucoria ...
152	270	„ 3437	N. melinoides	Fries, S.M. l. 266 (1821) ...	Honey-like naucoria ...
153	272	„ 3426	N. nasuta	Kalch., Grev. VIII. 152 (1880) ...	Long-nosed naucoria ...
154	276	„ 3469	N. pediades	Fries, S.M. I. 290 (1821) ...	Field naucoria ...
155	271	„ 3440	N. pusiola	Fries, S.M. i. 264 (1821) ...	Little naucoria ...
156	268	„ 3427	N. russa ...	Cooke and Mass., Grev. XV. 94 (1887) ...	Red naucoria...
157	273	„ 3450	N. scolecina	Fries, Epicr. 194 (1838) ...	Worm-eaten naucoria ...
158	277	„ 3470	N. semiorbicularis...	Fries, Mon. Hym. I. 376 (1857) ...	Hemispherical naucoria ...
159	280	„ 3507	N. sivaria	Fries, S.M. I. 261 (1821) ...	Veiled naucoria ...
160	279	„ 3486	N. tenuulenta	Fries, S.M. I. 268 (1821) ...	Dripping naucoria ...
161	274	„ 3457	N. triscopoda	Fries, Mon. Hym. I. 375 (1857)...	Hair-stalked naucoria ...

40. GALERA.—Fries,

462	286	V. 3568	G. hypnorum	Fries, S.M. l. 267 (1821)	Moss galera ...
463	287	„ 3574	G. minuta	Quel., Champ. Jura III. 10 (1873) ...	Minute galera ...
464	285	„ 3549	G. peroxydata	Berk., Hook., Lond. Journ. II. 411 (1843) ...	Peroxide galera
465	284	„ 3537	G. tenera	Fries, Hym. Eur. 267 (1874) ...	Delicate galera ...

41. TUBARIA.—Smith,

466	288	V. 3584	T. furfuracea	Fries, Hym. Eur. 272 (1874) ...	Mealy tubaria ...
467	289	„ 3597	T. inquilina	Fries, Hym. Eur. 274 (1874)	Little tubaria
467A	„	„ „	T. inquilina, var. cebola	Fries, Hym. Eur. 275 (1874) ...	Clay-coloured tubaria ...
468	...		T. strigipes	Cooke and Mass., Grev. XXI. 36 (1892)	Rough-stalked tubaria

42. CREPIDOTUS.—Fries,

469	292	V. 3599	C. alveolus	Lasch., in Fries Epicr. 210 (1838)	Alveolate crepidotus ...
470	304	„ 3665	C. auricula	Berk., Fl. Tasm. II. 246 (1860) ...	Eared crepidotus ...
471	300	„ 3627	C. cassia-color	Berk., Fl. Tasm. II 246 (1860) ...	Cinnamon crepidotus ...
472	299	„ 3601	C. epigaeus	Berk. and Br., Ann. Nat. Hist. IX. 179 (1882)	Earth-borne crepidotus ...
473	291	„ 3610	C. globigera	Berk., Linn. Journ. XIII. 158 (1873)	Globose-spored crepidotus
474	298	„ 3620	C. haustellaris	Fries, S.M. l. 274 (1821)	Damp-loving crepidotus ...

OF AUSTRALIAN FUNGI—continued.

Number.	W.A.	S.A.	T.	V.	N.S.W.	Q.	B.	Occurrence.	General Characters.

S.M. I. 250 (1821).—Agaricus Paxillus—continued.

Number.	W.A.	S.A.	T.	V.	N.S.W.	Q.	B.	Occurrence.	General Characters.
432	...	S.A.	...	V.	N.S.W.	...	...	Rich soil	Fleshy, moist, sulphur coloured. Stem stuffed, yellowish white.
433	...	...	...	V.	...	Q.	B.	Ground	Fleshy, dry, downy, red umber. Stem solid, yellow or reddish.
434	...	S.A.	...	V.	N.S.W.	...	...	Wood	Fleshy, dry, yellow tawny or golden. Stem silky, with fleeting veil.
434A	...	...	...	V.	N.S.W.	...	...	Soil (probably covering pine chips)	Orange tawny, with stem and gills paler.
435	W.A.	...	...	V.	...	...	...	Trunks	Fleshy, rust coloured, corrugated. Stem solid, smooth.
436	...	...	...	V.	...	Q.	B.	Dead trunks of Encephalartos Denisonii	Rather fleshy, moist, red to bay cinnamon, becoming pale. Stem thin, almost umber, tapering upwards, without veil.
437	...	...	...	V.	...	...	...	Ground	Fleshy, dry, silky, leek green. Stem straight, stuffed, lemon yellow.
438	W.A.	...	...	V.	...	Q.	...	Wood	Fleshy, shining, purple brown. Stem ascending, solid, paler.
439	...	...	...	V.	...	...	...	Ground	Fleshy, shining, red, with tinge of purple. Stem hollow, paler.
440	...	S.A.	...	V.	N.S.W.	Q.	B.	Fallen branches and chips	Compact, golden tawny, fluffy to scaly, then cracked. Strong odour.
441	...	S.A.	...	V.	...	...	B.	Woods	Small, stem tall. Fleshy, viscid, yellow. Very common.
442	...	...	...	V.	...	...	...	Among grass on hill-sides	Densely and shortly velvety, bay brown. Stem expanded upwards into cap, rather short, and of same colour.
443	...	S.A.	...	V.	...	...	B.	Ground	Fleshy, dry, rusty-fawn colour. Stem solid, delicately fluffy.
444	...	S.A.	...	V.	...	...	...	Wood	Ochrey yellow. Fleshy, compact, hard when dry. Stem lateral, short.

S.M. I. 260 (1821).—Agaricus.

Number.	W.A.	S.A.	T.	V.	N.S.W.	Q.	B.	Occurrence.	General Characters.
445	...	...	...	...	...	Q.	B.	Ground ...	Slightly fleshy, yellowish to tan colour. Stem with white fibrils.
446	...	...	...	...	...	Q.	...	Ground ...	Small. Rough with woolly tufts. Stem slender, fluffy.
447	...	...	...	V.	...	...	B.	Burnt soil ...	Rather fleshy, orbicular, ochrey. Stem naked, yellow, rusty at base.
448	W.A.	...	...	V.	...	...	...	Rotten wood ...	Viscid, when young very white. Stem mealy above, downy at base.
449	...	...	...	...	...	Q.	B.	Bare ground ...	Gregarious, fragile. Rather fleshy, whitish-tan colour, scabby scales.
450	...	...	...	V.	...	...	...	Logs ...	Tawny ferruginous. Stem elongated, thin, hollow, of same colour.
451	...	S.A.	...	...	N.S.W.	...	...	Roots of grass. &c.	Densely tufted. Tawny. Stem slender, mealy, thickened downwards.
452	W.A.	...	...	V.	...	Q.	B.	Among grass ...	Somewhat fleshy, tawny, ochrey when dry. Stem hollow, yellow.
453	...	...	...	...	N.S.W.	...	...	Ground ...	Rather fleshy, ochrey, with elongated tent-like mubo
454	...	S.A.	...	V.	...	...	B.	Pastures ...	Somewhat fleshy, yellow ochrey to tan colour. Very common.
455	...	...	...	V.	...	...	B.	Ground ...	Slightly fleshy, rather viscid, tawny yellow. Stem thread-like, shining.
456	...	...	...	V.	...	...	...	Ground ...	Thin, brick red. Stem nearly of same colour, whitish downy below.
457	...	S.A.	...	...	...	...	B.	Moist ground ...	Rather fleshy, ferruginous bay. Stem rusty, sprinkled with white meal.
458	...	...	...	V.	...	Q.	B.	Lawns and pastures	Rather fleshy, hemispherical, somewhat viscid, tawny or ochrey.
459	...	...	...	V.	...	...	B.	Soil, fern stems, &c.	Rather fleshy, with downy scales, red to rust colour.
460	...	...	...	V.	...	...	B.	Moist woods ...	Sub-membranaceous, rust colour, tan colour when dry. Stem polished.
461	...	...	...	V.	...	...	B.	Old wood ...	Rather fleshy, bay brown, ochrey when dry. Stem hair-like, rusty.

S.M. I. 264 (1821).—Agaricus.

Number.	W.A.	S.A.	T.	V.	N.S.W.	Q.	B.	Occurrence.	General Characters.
462	...	S.A.	...	V.	...	...	B.	Among moss ...	Minute. Membranaceous, bell shaped, sub-papillate, tawny. Common.
463	...	...	...	V.	...	...	B.	Decayed wood ...	Membranaceous, tawny, streaked. Stem almost hair-like.
464	...	...	...	...	...	Q.	...	Ground ...	Membranaceous, reddish brown, bell shaped. Stem very thin.
465	...	S.A.	T.	V.	...	...	B.	Grassy places, manure, &c.	Small, delicate. Sub-membranaceous, nearly conical, buff. Common.

Secm. Journ. (1870).—Agaricus.

Number.	W.A.	S.A.	T.	V.	N.S.W.	Q.	B.	Occurrence.	General Characters.
466	...	S.A.	T.	V.	...	Q.	B.	Chips. &c. ...	Small. Somewhat fleshy, at first clothed with silky evanescent scales, rich umber.
467	...	...	...	...	N.S.W.	...	B.	Chips ...	Minute. Sub-membranaceous, brown. Stem hollow, dark brown. Common.
467A	...	...	...	...	N.S.W.	...	B.	Grass roots ...	Clay coloured. Stem rooting; gills crowded, rusty.
468	...	...	...	V.	...	...	...	In tufts among grass	Hemispherical, tawny yellow, with conical spreading scales. Stem slender.

S.M. I. 272 (1821).—Agaricus.

Number.	W.A.	S.A.	T.	V.	N.S.W.	Q.	B.	Occurrence.	General Characters.
469	...	...	...	V.	...	...	B.	Old stumps ...	Fleshy, soft, ochrey brown, contracted, downy to shaggy behind.
470	...	...	T.	...	...	...	...	Dead wood ...	Sessile, shell shaped, cream colour. Flesh thick, brittle when dry.
471	...	S.A.	T.	...	...	...	...	Rotten bark ...	Mealy, cinnamon. Stem very short, slender, white, downy.
472	...	S.A.	...	V.	...	...	B.	Ground ...	Fragile, reddish grey, kidney shaped; base shaggy, whitish.
473	...	...	...	V.	...	...	...	Wood ...	Kidney shaped, tapering at base. About an inch long and wide.
474	...	S.A.	...	...	...	...	B.	Rotten trunks of Eucalyptus viminalis	Rather fleshy, flaccid, tan coloured. Stem tapering upwards, hairy.

SYSTEMATIC ARRANGEMENT

Number.	Cooke's Number.	Saccardo's Number.	Scientific Name.	Authority for Name.	English Name.
					42. CREPIDOTUS.—Fries.
475	295	V. 3602	C. hepatochrous	Berk., Hook., Lond. Journ. VII. 574 (1848)	Liver-coloured crepidotus ...
476	301	.. 3628	C. insidiosus	Berk., Hook., Lond. Journ. VII. 574 (1848) ...	Insidious crepidotus ... ...
477	296	., 3603	C. interceptus	Berk., Fl. Tasm. II. 246 (1860)...	Interposed crepidotus ... ...
478	303	,. 3663	C. leptomorphus ...	Berk., Fl. Tasm. II. 246 (1860)... ...	Delicate crepidotus ... ...
479	302	,. 3641	C. lepton ...	Berk.. Hook., Lond. Journ. IV. 46 (1845) ...	Thin crepidotus ... ...
480	293	,, 3600	C. mollis ...	Fries, S.M. I. 274 (1821)	Soft crepidotus ... ...
481	294	.. 3598	C. palmatus ...	Fries, Mon. Hym. I. 395 (1857)	Palmate crepidotus ... ...
482	290	IX. 481	C. phaeton ...	Cooke and Mass., Grev. XV. 99 (1887) ...	Brilliant crepidotus ... ...
483	297	V. 3655	C. stromaticus ...	Cooke and Mass., Grev. XV. 94 (1887) ...	Stromate crepidotus ... ...
484	305	,, 3664	C. turbidulus ...	Berk.. in Cooke's Handb. Aust. Fungi 60 (1892)...	Turbid crepidotus ... ...
					43. CORTINARIUS.—Pers.
485	361	V. 3763	C. Archeri	Berk., Fl. Tasm. II. 247 (1860)... ...	Archer's cortinar ... ...
486	366	., 3903	C. bovinus	Fries. Epicr. 297 (1838)	Ox cortinar ... ...
487	365	.. 3848	C. cinnabarinus	Fries. Epicr. 298 (1838)	Vermilion cortinar ...
488	360	., 3740	C. decoloratus	Fries, Epicr. 270 (1838)	Discoloured cortinar ...
489	362	...	C. erythraeus	Berk., Hook., Lond. Journ. IV. 48 (1845)	Blood-red cortinar ... ...
490	364	V. 3849	C. sanguineus	Fries, Epicr. 288 (1838)	Dark-red cortinar ... ...
491	363	., 3788	C. violaceus	Fries, Epicr. 279 (1838)	Violet-coloured cortinar ...
					44. PAXILLUS.—Fries.
492	370	V. 4020	P. crassus ...	Fries. Hym. Eur. 404 (1874)	Thick paxil ... ...
493	368	,, 4010	P. Eucalyptorum ...	Berk., Hook.. Lond. Journ. IV. 49 (1845)	Eucalypt paxil ... ..
494	367	., 4008	P. Muelleri	Berk., Linn. Journ. XIII. 159 (1873)	Mueller's paxil ... ...
495	371	., 4021	P. panuoides	Fries, Hym. Eur. 404 (1874) ..	Panus-like paxil ... ...
					45. AGARICUS.—Linn. Sp. Pl.
496	306	V. 4039	A. arvensis	Schaeff, Icon. t. 310. 311 (1762)...	Field agaric ... ...
497	307	,, 4053	A. campestris	Linn., Sp. Pl. 1173 (1753)	Pasture agaric ... ...
498	,,	,, 4054	A. silvicola	Vitt. Mang. (1835)	Wood agaric ... ...
499	310	IX. 559	A. elatior	Cooke and Mass., Grev. XVIII. 3 (1889)	Tall agaric ... ...
500	308	V. 4061	A. silvaticus	Schaeff, Icon. t. 242 (1762) ...	Sylvan agaric ...
501	309	., 4081	A. versipes	Berk. and Br., Linn. Trans. II. 54 (1883)	Twisted-stalked agaric
					46. STROPHARIA.—Fries, Summ. Veg.
502	311	V. 4130	S. coronilla	Fries, Hym. Eur. 285 (1874) ...	Crowned stropharia ... ...
503	314	,, 4144	S. merdaria	Fries, Hym. Eur. 286 (1874) ...	Dung-borne stropharia ...
504	313	,, 4151	S. semiglobata	Fries, Hym. Eur. 287 (1874) ...	Hemispherical stropharia ...
505	312	,, 4124	S. squamosa	Fries. Hym. Eur. 285 (1874) ...	Scaly stropharia ... ...
					47. HYPHOLOMA.—Fries,
506	317	IX. 565	H. adustum ...	Cooke and Mass.. Grev. XVIII. 3 (1889) ...	Scorched hypholoma ... ...
507	318	V. 4212	H. Candolleanum ...	Fries, S.M. I. 296 (1821)	De Candolle's hypholoma ...
508	...	..	H. discretum ...	Cooke and Mass., Grev. XXI. 37 (1892) ...	Separated hypholoma ...
509	316	V. 4182	H. dispersum ...	Fries, Epicr. 222 (1838)	Scattered hypholoma ... ...
510	315	,, 4179	H. fasciculare ...	Fries, S.M. I. 288 (1821)	Tufted hypholoma ... ...

27

OF AUSTRALIAN FUNGI—*continued.*

Number.	Habitat.						B.	Occurrence.	General Characters.
	W.A.	S.A.	T.	V.	N.S.W.	Q.			

S.M. I. 272 (1821).—Agaricus—*continued.*

475	...	...	T.	...	...	...	...	Bark	...	Rather fleshy, liver coloured. Cap globose at first, with short central stem.
476	...	...	T.	...	...	...	...	Bark ...	Membranaceous, margin downy, yellowish brown when dry. Stem short and slender.
477	...	...	T.	V.	...	...	...	Bark ...	Kidney shaped, ochrey white. Cap of three layers, the middle one, white, interposed between two darker ones.
478	...	...	T.	...	...	...	...	Dead wood	Sessile, whitish, downy, fixed at apex by a few white threads.
479	W.A.	...	...	...	...	...	...	Bark ...	Tawny ochre. Stem obsolete, extremely short if present.
480	W.A.	S.A.	...	V.	...	Q.	B.	Old stumps	Gelatinous to fleshy, flaccid, pale. Stem obsolete.
481	...	...	T.	...	...	...	...	Trunks ...	Fleshy, compact, rust coloured. Stem excentric or lateral.
482	...	...	...	V.	...	...	...	Ground (?)	Sub-membranaceous, brick red. Stem lateral, elongated.
483	W.A.	...	...	...	...	...	...	Bark ...	Sessile, flaccid, tan colour, arising from white woolly stroma.
484	...	...	T.	...	...	...	...	Wood ...	Sessile, kidney shaped, ochrey, smooth.

Syn. 16 (1801).—Agaricus.

485	...	...	T.	...	...	...	...	Ground ...	Fleshy, violet brown. Stem stout, viscid, violet.
486	...	...	...	V.	...	...	B.	Woods ...	Fleshy, watery cinnamon. Stem stout, spongy, bulbous, grey.
487	...	...	...	V.	...	Q.	B.	Under trees	Fleshy, silky, vermilion, shining. Stem stuffed, short, vermilion.
488	...	...	...	V.	...	...	B.	Woods ...	Fleshy, viscid, soon dry, floccose and discoloured. Stem tapering from base.
489	W.A.	...	...	...	...	...	...	Ground ...	Small, blood red. Cap clothed with thick gelatinous coat. Stem short, viscid.
490	...	...	...	V.	...	...	B.	Woods ...	Entirely dark red. Fleshy, silky, or scaly. Stem stuffed, thin, hollow.
491	...	...	...	V.	...	...	B.	Woods ...	Dark violet. Fleshy, woolly to scaly. Stem bulbous, spongy, shaggy. *Edible.*

Gen. Hym. 8 (1836).—Agaricus, Merulius.

492	...	...	...	...	...	Q.	B.	Ground ...	Fleshy, rust coloured. Stem stuffed, excentric, very short.
493	W.A.	...	...	...	...	...	...	Under Eucalyptus trees	Thick and fleshy, compact, tawny yellow. Stem transversely scaly.
494	...	...	...	V.	N.S.W.	Q.	...	Meadows ...	Dark brown, convex. Stem tawny, frosted.
495	...	...	...	V.	...	...	B.	Cellars, on sawdust, &c.	Fleshy, shell-shaped, dirty yellow or whitish ochre, sessile.

(1753).—Psalliota, Pratella.

496	W.A.	...	T.	V.	N.S.W.	...	B.	Meadows, &c.	Very large, expanding late. Fleshy, flesh turning slightly yellow where bruised. *Edible.*
497	W.A.	S.A.	T.	V.	N.S.W.	Q.	B.	Rich pastures	Fleshy, silky floccose, or scaly. Stem stuffed, ring median. *Edible.*
498	...	...	...	V.	...	...	B.	Woods ...	Smooth, shining white. Stem stuffed, elongated, somewhat bulbous.
499	...	...	...	V.	...	...	...	Ground ...	Thinly fleshy, brown, scaly. Stem cylindrical, silky, whitish.
500	...	...	...	V.	...	...	B.	Woods ...	Fleshy, thin, bell shaped, fibrous or scaly. Stem hollow, whitish. *Edible.*
501	...	...	...	...	...	Q.	...	Roots of bamboos ...	White, smooth, like chamois leather. Stem loosely stuffed, tapering at base.

Scan. II. 295 (1849).—Agaricus.

502	...	S.A.	...	V.	...	...	B.	By waysides	Fleshy, viscid, ochrey; margin whitish, fluffy. Stem white, stuffed.
503	...	...	...	V.	...	...	B.	Among grass	Moist, somewhat cinnamon colour, dry ochrey. Stem hollow, short.
504	W.A.	S.A.	T.	V.	N.S.W.	Q.	B.	Dung ...	Small. Somewhat fleshy, hemispherical, mottled yellowish. Very common.
505	...	S.A.	...	V.	N.S.W.	...	B.	Woods	Fleshy, thin, somewhat viscid, yellowish tawny, sprinkled with superficial concentric scales.

S.M. I. 287 (1821).—Agaricus.

506	...	...	...	...	...	Q.	...	Ground ...	Fleshy, dark brown, variegated with darker scales, yellowish within.
507	...	...	...	V.	N.S.W.	...	B.	Dead stumps	Somewhat fleshy, ochrey and whitish. Stem hollow, fragile, white.
508	...	...	...	...	...	...	...	Ground ...	Bell shaped, tawny yellow. Stem slender, faintly streaked.
509	W.A.	S.A.	T.	V.	...	...	...	Stumps and ground	Somewhat fleshy, tawny honey colour, margin silky. Stem thin.
510	...	S.A.	T.	V.	...	...	B.	Old stumps, &c. ...	Fleshy, yellowish, with greenish tinge. Stem hollow, flesh yellow.

Number.	Cooke's Number.	Saccardo's Number.	Scientific Name.	Authority for Name.	English Name.
					48. PSILOCYBE.—Fries,
511	324	IX. 568	P. ceres ...	... Cooke and Mass., Grev. XVI. 72 (1888) ...	... Ceres psilocybe
512	322	V. 4269	P. cernua	... Fries, S.M. I. 298 (1821)	... Nodding psilocybe
513	320	., 4259	P. compta	... Fries, Hym. Eur. 301 (1874)	... Ornamented psilocybe
514	319	„ 4235	P. ericæa	... Fries, S.M. I. 291 (1821)	... Heath-growing psilocybe
515	323	„ 4275	P. fœnisecii	... Fries, S.M. I. 295 (1821)	... Lawn psilocybe
516	321	„ 4267	P. spadicea	... Fries, Epicr. 225 (1838)	... Date-brown psilocybe
					49.—DECONICA.—Smith,
517	325	V. 4293	D. atro-rufa	... Sacc. Syll V. 1059 (1887)	... Dark-red deconica
518	326	„ 4294	D. nuciseda	... Sacc. Syll. V. 1059 (1887)	... Nutty deconica
					50. PSATHYRA.—Fries.
519	327	V. 4297	P. conopilea	... Fries, S.M. I. 504 (1821)	... Cone-capped psathyra
520	330	„ 4341	P. fatua ...	Fries, S.M. I. 296 (1821)	... Tasteless psathyra
521	331	„ 4349	P. gossypina	Fries, S.M. I. 310 (1821)	... Cottony psathyra
522	329	„ 4329	P. obtusata	Fries, S.M. I. 293 (1821)	... Obtuse psathyra
523	328	„ 4311	P. Sonderiana	Berk., Linn. Journ. XIII. 159 (1873)	... Sonder's psathyra
					51. BOLBITIUS.—Fries,
524	...	...	B. candidus	... Cooke and Mass., Grev. XXI. 37 (1892) ...	... White bolbitius
525	358	V. 4357	B. conocephalus	... Fries, Epicr. 205 (1838)	... Cone-headed bolbitius
526	356	„ 4355	B. fragilis	... Fries, Epicr. 254 (1838)	... Fragile bolbitius
527	357	„ 4358	B. titubans	Fries, Epicr. 254 (1838)	... Tottering bolbitius
					52. COPRINUS.—Pers.
528	345	V. 4374	C. comatus	Fries, Epicr. 242 (1838)	... Maned coprin
529	351	„ 4129	C. deliquescens ...	... Fries, Epicr. 249 (1838)	... Deliquescent coprin
530	354	„ 4480	C. ephemerus	... Fries, Epicr. 252 (1838)	... Ephemeral coprin
531	347	„ 4404	C. fimetarius ...	... Fries, Epicr. 245 (1838)	... Dung coprin ...
531A	347	„	C. fimetarius, var. macrorhizus	(Fries), Hym. Eur. 324 (1874) ...	... Large-rooting coprin
532	349	„ 4416	C. micaceus ...	... Fries, Epicr. 247 (1838)	... Glistening coprin
533	353	„ 4477	C. murinus ...	... Kalch., Grev. VIII. 152 (1880) ...	... Mouse-coloured coprin
534	348bis	„ 4407	C. niveus ..	... Fries, Epicr. 246 (1838)	... Snowy coprin
535	346	„ 4394	C. picaceus ...	... Fries, Epicr. 244 (1838)	... Variegated coprin
536	355	„ 4490	C. plicatilis	... Fries, Epicr. 252 (1838)	... Plaited coprin
537	352	„ 4465	C. stercorarius	... Fries, Epicr. 251 (1838)	... Dung-borne coprin
538	348	„ 4406	C. tomentosus	... Fries, Epicr. 246 (1838)	... Tomentose coprin
539	350	„ 4420	C. truncorum	... Fries, Epicr. 248 (1838)	... Trunk coprin
					53. PANÆOLUS.—Fries,
540	338	V. 4544	P. campanulatus ...	... Fries, Hym. Eur. 311 (1874)	... Bell panæolus
541	333	IX. 598	P. eburneus	... Cooke and Mass., Grev. XVIII. 4 (1889) ...	... Ivory-white panæolus
542	340	V. 4555	P. fimicola	... Fries, Hym. Eur. 312 (1874)	... Dung-borne panæolus
543	332	IX. 596	P. ovatus	... Cooke and Mass., Grev. XVIII. 4 (1889)	... Ovate panæolus
544	339	V. 4547	P. papilionaceus	... Fries, Epicr. 235 (1838)	... Butterfly panæolus
545	335	„ 4556	P. phalænarum	... Fries, Epicr. 235 (1838)	... Moth panæolus
546	336	„ 4539	P. retirugis	... Fries, Epicr. 235 (1838)	... Wrinkled panæolus
547	337	IX. 597	P. veluticeps	... Cooke and Mass., Grev. XVIII. 4 (1889) ...	... Velvet-capped panæolus

OF AUSTRALIAN FUNGI—*continued.*

Number	W.A.	S.A.	T.	V.	N.S.W.	Q.	B.	Occurrence.	General Characters.

S.M. I. 289 (1821).—Agaricus.

Number	W.A.	S.A.	T.	V.	N.S.W.	Q.	B.	Occurrence.	General Characters.
511	...	...	...	V.	...	...	...	Ground	Thin, brick red. Stem elongated, ochrey, downy downwards.
512	...	...	...	V.	...	...	B.	Chips, decayed wood, &c.	Somewhat fleshy, wrinkled when dry, white. Stem hollow, white.
513	...	...	...	V.	...	...	B.	Woods	Pale ochrey, grooved, with scattered shining spots. Stem shining, silky.
514	W.A.	...	...	V.	...	...	B.	Exposed pastures after rain	Fleshy, rather viscid when moist, shining when dry, ferruginous tawny.
515	...	S.A.	...	...	...	...	B.	Among grass, lawn	Somewhat fleshy, dark brown, hemispherical or bell shaped. Stem pale red.
516	...	...	T.	V.	...	...	B.	Dead stumps, ground, &c., in woods	Fragile, rigid. Fleshy, bay to umber, moist. Stem hollow, tough, pale.

Seem. Journ. (1876).—Agaricus, Psilocybe.

Number	W.A.	S.A.	T.	V.	N.S.W.	Q.	B.	Occurrence.	General Characters.
517	W.A.	...	...	...	...	...	...	Ground in woods ...	Dark red or purple brown. Rather fleshy, discoloured when dry.
518	...	...	...	...	...	Q.	...	Chips ...	Rather fleshy, yellowish, silky when dry.

S.M. I. 11 (1821).—Agaricus.

Number	W.A.	S.A.	T.	V.	N.S.W.	Q.	B.	Occurrence.	General Characters.
519	...	...	...	V.	...	...	B.	Ground ...	Large, graceful. Sub-membranaceous, growing pale. Stem tall. Common.
520	...	...	...	...	...	...	B.	Gardens, &c.	Very fragile. Sub-membranaceous, clay coloured, rugged. Stem smooth.
521	...	...	...	...	...	Q.	B.	Ground ...	Ochrey to clayey. Sub-membranaceous, downy, becoming smooth.
522	W.A.	...	...	V.	...	...	B.	Oak trunks and ground	Sub-membranaceous, wrinkled, rather shining, obtuse. Stem rigid.
523	...	S.A.	...	...	...	...	...	Ground ...	Pale and dirty yellowish, acutely convex. Stem white, silky.

Epicr. 253 (1838).—Agaricus.

Number	W.A.	S.A.	T.	V.	N.S.W.	Q.	B.	Occurrence.	General Characters.
524	...	...	...	V.	...	...	...	Stable refuse ...	Membranaceous, white, bell shaped. Stem long, hollow.
525	...	...	...	V.	N.S.W.	...	...	Moist ground	Fragile, graceful, from livid to clay white. Membranaceous, conical, rather viscid.
526	W.A.	...	...	V.	...	Q.	B.	Dung	Small, but rather tall. Sub-membranaceous, viscid, pellucid, yellow, becoming pale.
527	...	...	...	V.	...	...	B.	Among grass	Small, tall, very fragile, trembling and tottering. Membranaceous, yellow.

Tent. disp. 62 (1797).—Agaricus.

Number	W.A.	S.A.	T.	V.	N.S.W.	Q.	B.	Occurrence.	General Characters.
528	...	...	...	V.	...	...	B.	Sides of roads, pastures	Large and tall, white. Rather fleshy, cylindrical. Cuticle broken and feathery. Edible.
529	...	...	...	...	...	Q.	B.	Old stumps	Large. Sub-membranaceous, livid, top papillate. Stem hollow.
530	...	...	...	...	...	Q.	B.	Dung-hills ...	Small. Very thin, splitting, somewhat mealy. Stem slender.
531	...	...	...	...	...	Q.	B.	Dung-heaps ...	Sub-membranaceous, soon torn, disc livid. Stem scaly.
531A	...	S.A.	...	...	...	...	B.	Dung-heaps ...	Leathery to scaly. Stem rooting, shaggy.
532	...	S.A.	...	V.	...	...	B.	About old stumps ...	Small. Sub-membranaceous, brown, covered with glittering particles.
533	...	...	...	V.	N.S.W.	...	...	Ground ...	Small. Sub-membranaceous, with prominent papilla at apex, grey.
534	...	...	...	V.	...	...	...	Horse-dung	Small. Sub-membranaceous, clad with dense white down.
535	...	...	...	...	...	Q.	B.	Road-sides, &c.	Sub-membranaceous, deep black, variegated, with broad white superficial scales.
536	...	S.A.	...	V.	N.S.W.	Q.	B.	Pastures	Small, delicate. Very thin, splitting, furrowed and folded, grey.
537	...	...	T.	V.	N.S.W.	...	B.	Rich soil and dung	Very thin, ovate, covered with a dense white micaceous meal.
538	...	...	...	V.	...	Q.	B.	Dung and rich pastures	Sub-membranaceous, cylindrical to conical, woolly to downy, whitish grey.
539	...	...	...	...	...	Q.	...	Wood ...	Membranaceous, deliquescent, ferruginous ochrey, at first densely micaceous.

Epicr. 234 (1838).—Agaricus.

Number	W.A.	S.A.	T.	V.	N.S.W.	Q.	B.	Occurrence.	General Characters.
540	W.A.	...	...	V.	N.S.W.	...	B.	Rich soil	Fragile. Somewhat fleshy, bell shaped, shining, dry, red brown.
541	...	...	...	V.	...	Q.	B.	Dung ...	Rather fleshy, ivory white, shining. Stem fragile, elongated.
542	...	...	...	V.	...	Q.	B.	Dung, rich pastures &c.	Somewhat fleshy, marked near margin with a narrow brown zone. Stem fragile, elongated.
543	...	...	...	V.	...	...	...	Manure	Rather fleshy, ovate, at length cracked, white. Stem erect, silky.
544	...	S.A.	...	V.	...	...	...	Dung, rich pastures, &c.	Somewhat fleshy, pale tan, conico-convex, when dry cracked and scaly.
545	...	...	...	V.	...	...	B.	Dung ...	Rather fleshy, viscid, dirty clay colour. Veil fleeting.
546	...	...	...	V.	...	...	B.	Dung ...	Somewhat fleshy, reticulated with raised ribs, flesh to tan colour.
547	...	S.A.	...	...	...	Q.	...	In garden among grass	Velvety, grey, convex or bell shaped. Stem elongated, hollow, silvery grey.

Consec. Number.	Saccardo's Number.	Scientific Name.	Authority for Name.	English Name.
				54. ANELLARIA.—Karst.
334	V. 4561	A. fimiputris	... | *Karst.* Hattsv. I. 518 (1879, &c.)	... | Putrid dung anellaria ... ... |
				55. PSATHYRELLA.—Fries,
343	V. 4595	P. crenata	... Lasch., in Fries Hym. Eur. 315 (1874)	... Crenate psathyrella ... ...
344	„ 4597	P. disseminata	... *Fries,* Hym. Eur. 316 (1836) ...	... Scattered psathyrella ... ...
341	„ 4572	P. biascens	... Fries, Hym. Eur. 314 (1874)	... Gaping psathyrella ... ...
342	„ 4575	P. trepida	... Fries, Epicr. 235 (1838)	... Trembling psathyrella ...
				ADDITIONS TO
...	...	Amanita Forrestiæ	... Kalch., Linn. Soc. N.S.W, VII. 638 (1882)	... Forrest's amanita ...
...	V. 92	Lepiota acute-squamosa	... Weium., Syll. I. 70 (1836)	... Acute-scaly lepiota ...
44		L. megalotheles	... Kalch., Linn. Soc. N.S.W. VII. 563 (1882)	... Large-nippled lepiota ...
		Tricholoma carneo-flavidum	Kalch., Linn. Soc. N.S.W. VII. 639 (1882)	... Fleshy-yellow tricholome ...
	V. 474	T. panæolum	... Fries. Epicr. 49 (1838)	... Variegated tricholome ...
...	...	T. plagiotum	... | Kalch., Linn. Soc. N.S.W. VII. 639 (1882)	Oblique tricholome ... ...
...	V. 496	T. sordidum	... Fries, S.M. I. 51 (1821)	... Sordid tricholome ... ...
...	...	T. turbinipes	... Kalch.. Linn. Soc. N.S.W. VII. 639 (1882)	... Turbinate-stalked tricholome ...
...	V. 630	Clitocybe catiua ...	... Fries, Epicr. 72 (1838)	Bowl-shaped clitocybe ...
...	„ 673	C. ditopoda	... Fries. S.M. I. 171 (1821)	Variable-stalked clitocybe ...
...	„ 529	C. ochro-purpurea ...	... Berk., Hook. Lond. Journ. Bot. IV. 299 (1845)	Ochrey-purple clitocybe ...
...	...	Collybia muscipula	... Cooke and Mass., Grev. XXII. 26 (1893)...	Mousey collybia ...
...	V. 1109	Mycena epipterygia	... *Fries,* S.M. I. 155 (1821) ...	Winged mycena ... ...
...	„ 962	M. luteo-alba ...	... *Fries,* S.M. (1821) ...	Yellowish-white mycena ...
...	„ 1005	M. polygramma ...	... *Fries,* S.M. I. 116 (1821)	Many-lined mycena ...
...	„ 1137	M. pterigena ...	... Fries, S.M. I. 160 (1821)	Pteris-borne mycena ...
...	„ 992	M. ræborrhiza ...	... Lasch., Linn. 539 (1829)	Crooked-root mycena ...
	„ 1467	Pleurotus acerinus	... Fries, Epicr. 134 (1838) ...	... Maple pleurote ...
	„ 1502	P. cyphellæformis	... Berk., Mag. Zool. and Bot. 511 (1837) ...	... Cyphella-like pleurote...
	„ 1339	P. lignatilis ..	... *Fries,* Epicr. 132 (1838)	... Wood-growing pleurote ...
	„ 1561	Hygrophorus discoideus	... *Fries,* Epicr. 323 (1838) ...	... Discoid hygrophore ... ...
	„ 1651	H. puniceus	... Fries, Mon. Hym. II. 21 (1857) ...	... Purple hygrophore ...
	„ 1738	Lactarius quietus ...	... *Fries,* Epicr. 343 (1838)	... Mild lactar ... ...
	„ 1821	Russula xerampelina	... *Fries,* Epicr. 356 (1838)	... Dark-red russule ... ...
	„ 1989	Marasmius badius...	... Berk. and Curt., Linn. Journ. X. 294 (1869)	... Bay-brown marasmius ...
	„ 1996	M. pellucidus ...	... Berk. and Br., Linn. Journ. XIV. 35 (1875) ...	Pellucid marasmius ... ...
	„ 1991	M. rhyssophyllus ...	... Mont. in Berk and Curt., Linn. Journ. X. 294 (1869)	Wrinkle-gilled marasmius ...

OF AUSTRALIAN FUNGI—*continued.*

Number	Habitat						B.	Occurrence.	General Characters.
	W.A.	S.A.	T.	V.	N.S.W.	Q.			

Hattsv. I. 25 (1879).—Agaricus, Panæolus.

| 548 | ... | ... | ... | V. | ... | ... | B. | Dung, &c. | ... | Sub-membranaceous, viscid, dark grey. Stem slender, equal. |

Epicr. 237 (1838) —Agaricus.

549	...	S.A.	...	...	...	B.	Grassy ground ...	Fragile. Membranaceous, ochrey to pale red, margin notched.
550	W.A.	...	T.	V.	Q.	B.	About trunks of trees and on ground	Densely clustered. Membranaceous, pearly white, scurfy. Stem lax, fragile.
551	...	...	...	...	Q.	B.	Ground ...	Membranaceous, fissured and furrowed, becoming yellow. Stem brittle
552	...	S.A.	...	V.	...	B.	Ground ...	Very fragile. Membranaceous, sooty, bell shaped, densely streaked. Stem nearly straight, transparent.

AGARICACEÆ.

553	W.A.	...	...	...	...	...	Fleshy, convex to plane, smooth, naked, white to ashy grey or brown. Stem stout, stuffed, white, fibrous, not bulbous.
554	...	...	...	V.	...	B.	Grassy places	Fleshy, at first woolly hairy, then scaly, acute, dark tan. Stem somewhat stuffed, bulbous.
555	...	...	...	...	Q.	...	River-side	Fleshy, bell shaped, smooth, with brown adpressed scales. Stem somewhat hollow, dilated at base, naked, from white to brown.
556	W.A.	...	...	...	...	...	Fleshy, hemispherical, woolly scaly, fleshy yellow as if peach coloured. Stem solid, thick, of same colour
557	W.A.	...	...	...	...	H.	Grassy places	Small, spongy to compact, convex to plane, variegated with grey frosted spots. Stem solid, fibrous to striate.
558	W.A.	...	...	...	...	...	Fleshy, plane, depressed, smooth, pale reddish brown. Stem excentric, stuffed, cylindrical, base slightly thickened.
559	...	...	...	V.	...	B.	Meadows, dung-heaps, &c.	Somewhat fleshy, from bell shaped and convex to plane and depressed, finally squalid. Stem stuffed, base thickened.
560	W.A.	...	...	...	...	...	Large, pale reddish. Fleshy, irregularly convex. Stem top shaped, thick.
561	...	...	...	V.	...	B.	Among dead leaves	White, discoloured, odour pleasant. Fleshy, plane, then funnel shaped, dry. Stem stuffed, thick, elastic, white.
562	...	...	...	V.	...	B.	Woods ...	Strong smelling, like new meal. Rather fleshy, dingy, brownish grey, smooth. Stem hollow, almost smooth, of same colour.
563	...	...	...	...	...	...	Clayey soil in woods	Somewhat hemispherical, at length depressed, fleshy, compact, pale tan, becoming slightly purple. Stem thick, swollen in middle.
564	...	...	...	...	Q.	Ground ...	Fleshy, smooth, mouse grey or brown, wrinkled. Stem stuffed, tapering downwards and rooting, striate lengthwise.
565	...	...	...	...	Q.	B.	Among moss and leaves	Membranous, bell shaped, striate, very viscid and easily separable, usually grey. Stem elongated, tough, rooting, yellowish.
566	...	...	...	V.	...	B.	Among moss, &c. ...	Membranous, bell shaped, slightly grooved, pale yellow. Stem thread-like, shining, smooth, becoming yellow.
567	...	...	...	V.	...	B.	Trunks ...	Rather membranous, conical to bell shaped, dry, grooved. Stem rigid, longitudinally furrowed and grooved, shining, rooting.
568	...	...	...	V.	...	B.	Dead fern stems, veins of leaves, &c.	Very elegant, delicate, rosy. Bell shaped, obtuse, and stem wavy, very thin, with disc at base.
569	...	...	...	V.	...	...	About trunks	Somewhat membranous, acutely bell shaped, dry, rather tawny or pale. Stem firm, thick, rooting.
570	...	...	...	...	Q.	B.	Trunks ...	White, firm. Fleshy, thin, unequal, silky hairy. Stem almost lateral, slender or nearly obsolete, downy.
571	...	...	...	...	Q.	B.	Moss and dead stems of herbaceous plants	Gregarious, small. Rather fleshy, sessile, cup shaped, grey, margin paler, delicately downy.
572	...	...	...	V.	...	B.	Trunks, rotten wood, &c.	Odour mealy. Fleshy, tough, convex to plane, dingy white. Stem stuffed, then hollow, slender base, rooting and downy.
573	...	...	...	V.	...	B.	Grassy places	Gregarious. Fleshy, smooth, very glutinous, yellowish tan, disc somewhat rusty. Stem stuffed, soft, viscid, pale white.
574	...	...	...	V.	...	B.	Mossy meadows, &c.	Very large, very showy, fragile. Bell shaped, viscid, scarlet to blood red. Stem hollow, thick, bulging, base white.
575	...	...	...	...	Q.	B.	Woods, &c.	Fleshy, viscid at first, slightly cinnamon, then dry and slightly silky, somewhat zoned. Stem spongy to stuffed, finally rusty red.
576	...	...	...	V.	...	B.	Woods ...	Mild. Fleshy, compact, dry, opaque, rose purple. Stem strong, firm, finally spongy to soft.
577	...	...	...	...	Q.	Wood ...	Reddish when fresh. Convex, striate, smooth, margin incurved. Stem frosted, becoming smooth.
578	...	...	...	...	Q.	Dead branches, &c.	Convex, pellucid. Stem equal, brown, delicately powdery.
579	...	...	...	...	Q.	Wood ...	Fibrous, smooth, pale yellow. Stem same colour, smooth, with rough-haired base.

Saw anto.s Number.	Scientific Name	Authority for Name.	English Name.
			ADDITIONS TO
V. 2379	Lentinus descendens	Fries, Epicr. 290 (1838)	Descending lentinus
„ 2924	Leptonia æthiops ...	Fries, Epicr. 152 (1838)	Æthiopian leptonia
„ 3406	Nolanea sub-globosa	Cooke, Grev. XVII. 38 (1888) ...	Sub-globose nolanea ...
„ 3107	Pholiota adiposa	Fries, S.M. I. 242 (1821)	Glutinous pholiota ...
...	P. bicincta	Kalch., Linn. Soc. N.S.W. VII. 639 (1882)	Twice-girt pholiota
V. 3064	P. radicosa	Fries. S.M. I. 242 (1821)	Rooting pholiota
...	Hebeloma Kirtoni ...	Kalch., Linn. Soc. N.S.W. VII. 564 (1882)	Kirton's hebeloma ...
V. 3359	Flammula carbonaria	Fries, S.M. I. 252 (1821)	Charcoal-loving flammula
„ 3322	F. gymnopodia	Fries, Hym. Eur. 218 (1874) ...	Naked-stalked flammula
„ 3365	F. sapinea, var. terrestris	Fries, S.M. I. 239 (1821)	Terrestrial flammula ...
„ 3433	Naucoria abstrusa ...	Fries, Epicr. 194 (1838)	Concealed naucoria ...
„ 3509	N. conspersa	Fries, S.M. I. 260 (1821)	Besprinkled naucoria ...
„ 3499	N. sobria ...	Fries, Epicr. 200 (1838)	Sober naucoria
„ 3481	N. tenax ...	Fries, Epicr. 198 (1838)	Firm naucoria
„ 3605	Crepidotus applanatus	Fries, Mon. Hym. I. 399 (1857) ...	Depressed crepidotus ...
	Cortinarius Walkeri	Cooke and Mass., Grev. XXII. 36 (1893)	Walker's cortinarius
IX. 553	Paxillus hirtulus ...	F.v.M., Linn. Soc. N.S.W. VIII. 175 (1883)	Hairy paxillus
	Hypholoma peltastes	Kalch., Linn. Soc. N.S.W. VII. 564 (1882)	Shield-like hypholoma
V. 4261	Psilocybe semilanceata	Fries, Obs. II. 178 (1818)	Semi-pointed psilocybe
„ 4560	Anellaria separata	Karst., Hatt-v. I. 517 (1879)	Separate anellaria
„ 4566	Psathyrella impatiens	Fries, S.M. I. 302 (1821)	Impatient psathyrella
V. 1960	A. cupularis	Fries, S.V. S. 312 (1849)	**56 ARRHENIA.—Fries**, Cupular arrhenia

Number.	Habitat.						B.	Occurrence.	General Characters
	W.A	S.A.	T.	V.	N.S.W.	Q.			

AGARICACEÆ—*continued*.

580	...	...	...	...	...	Q.	...	Ground	Somewhat woody, funnel shaped, tan coloured, spotted with minute scales. Stem solid, very hard rooting.
581	...	...	...	V.		...	B.	Among grass, &c. ...	Fleshy, depressed, smooth, shining, sooty black. Stem slender, smooth, blackish brown, with black points towards top.
582	...	...	...	V.	...	...	B.	Ground ...	Fleshy, sub-globose, viscid, yellow. Stem thin, becoming hollow, longitudinally striate.
583	...	...	...	...		Q.	B.	Trunks	Tufted, very showy and large. Fleshy, glutinous, yellow, scaly. Stem stuffed, glutinous, base somewhat bulbous.
584	W.A.	...	...	...		...	...	...	Convex, obtuse, light umber, scaly. Stem solid, bulbous, fibrillose, about middle and above base doubly girt.
585	...	...	...	V.	...	...	B.	Woods, near to stumps	Large, beautiful, sweet odour. Fleshy, dry, smooth, spotted, almost clay colour. Stem solid, passing into tapering root.
586	...	...	...	...	N.S.W.	...	...	...	Compact, fleshy, convex to plane, smooth. Stem solid, fleshy, fibrillose, white to silky.
587	...	...	...	V.		...	B.	Burnt ground, charcoal, &c.	Gregarious. Fleshy, firm, viscid, tawny yellow, often depressed in centre. Stem rigid, scaly, pale, base usually darker.
588	...	...	...	...		Q.	B.	Ground ...	Often tufted, rusty brown. Fleshy, bell shaped to convex, scaly. Stem solid, almost smooth.
440A	...	...	...	...	...	Q.	B.	Ground ...	Tufted, stem elongated, rooting in a spindle-shaped manner.
589	W A.	...	...	...	...	...	B.	Damp earth, leaf-soil, &c.	Rather fleshy, smooth, viscid, rusty tan. Stem cartilaginous, tough, hollow, polished, rusty.
590	...	...	...	V.		...	B.	Ground, among leaves, &c.	Gregarious, fragile. Fleshy, with scurfy scales, bay cinnamon when moist, ochrey when dry. Stem fibrillose, brownish cinnamon.
591	...	S.A.	...	...		...	B.	In moist woods or scrubs	Fleshy, slightly viscid, somewhat silky, honey colour when moist, not absorbent of moisture, hence the name. Stem thick, hollow, rusty brown below.
592	...	...	...	V.		...	B.	Woods, among grass	Somewhat fleshy, bell shaped then expanded, smooth, slightly viscid, cinnamon when moist, ochrey when dry. Stem stuffed then hollow, dusky yellow.
593	...	...	...	V.	...	...	B.	Rotten wood	Fleshy, soft, fragile, kidney to wedge shaped, whitish, at length depressed behind. Stem very short, whitish downy.
594	...	...	...	...	N.S.W.	...	...	Ground ...	Convex, then expanded, minutely silky, pale green then bluish green. Stem slightly thickened at base, stuffed, reddish.
595	...	...	...	...		Q.	...	Ground ...	Convex to depressed, becoming darkly lurid. Stem thickened downwards, base abruptly rooting, hairy.
596	...	...	...	...	N.S.W.	...	...	...	Fleshy, viscid, scutiform, becoming brown. Stem solid, thickened downwards, naked, white.
597	...	...	...	V.	...	...	B.	Among grass	Gregarious. Somewhat membranous, acutely conical, almost pointed, slightly viscid, pale yellow when dry. Stem tough, wavy, silky fibrous, shining.
598	...	...	...	...		...	B.	Dung	Fleshy, bell shaped, viscid, ochrey, then whitish and wrinkled when old, shining. Stem long, straight, shining, whitish, tapering upwards with persistent ring.
599	...	...	...	V.		...	...	Moist woods	Membranous, bell shaped, convex, smooth. Stem weak, smooth, white.

Summ. Veg. Scand. 312 (1849).—Cantharellus, Mernlius.

600	...	...	...	...		Q.	...	Young pinnate leaf	Small, resupinate, soft, circular, shaggy, grey.

ORDER II.—POLYPORACEÆ, FRIES.

57. Boletus, *Linn.*
58. Strobilomyces, Berk.
59. Fistulina, Bull.
60. Polyporus, *Linn.*

61. Fomes, Fries.
62. Polystictus, Fries.
63. Poria, Pers.
64. Trametes, Fries.

ORDER II.—POLYPORACEÆ
57. BOLETUS.—*Linn.*, Sp. Pl. 1176 (1753).

Number.	Cooke's Number.	Saccardo's Number.	Scientific Name.	Authority for Name.	English Name.
601	561	VI. 4749	B. æreus	Bull., Champ. 321 (1798)	Bronze bolet...
602	562	„ 4755	B. æstivalis	Fries., Epicr. 422 (1838)	Summer bolet
603	551	„ 4673	B. alliciens	Berk.. Hook., Lond. Journ. IV. 50 (1845)	Attractive bolet
604	550	„ 4671	B. arenarius	Fries, Pl. Preiss. II. 134 (1846)...	Sand-loving bolet
605	552	„ 4674	B. australis	Cooke and Mass., Grev. XVI. 32 (1887)	Southern bolet
606	547	„ 4653	B. badius	Fries, Elench. 126 (1828)	Bay-brown bolet
607	557		B. brunneus	Cooke and Mass., Grev. XIX. 90 (1891)...	Brown bolet ...
608	565	VI. 4761	B. cæsareus	Fries, Pl. Preiss. II. 134 (1846)...	Imperial bolet
609	558	„ 4726	B. calopus	Fries, S.M. I. 390 (1821)	Red-stalked bolet
610	553	„ 4680	B. chrysenteron ...	Fries, Epicr. 415 (1838)	Red-crack bolet
611	560	„ 4748	B. edulis...	Bull., Champ. 60 (1798)	Edible bolet ...
612	541	„ 4642	B. elegans	Schum., Saell. II. 374 (1801)	Elegant bolet
613	569	„ 4801	B. fellens	Bull., Champ. 379 (1798)	Bitter bolet ...
614	545	„ 4645	B. flavus...	Wither., Fries, Epicr. 410 (1838)	Yellow bolet...
615	555	„ 4833 & IX.633	B. fruticicola	Berk., Hook., Lond. Journ. VII. 574 (1848)	Shrub-growing bolet ...
616	546	VI. 4648	B. granulatus	Linn., Sp. Pl. 1177 (1753)	Granulated bolet
617	572	„ 4847	B. bædinus	Berk. and Br.. Linn., Trans. II. 57 (1883)	Kid-like bolet
618	564	„ 4760	B. infractus	Fries, Pl. Preiss. II. 134 (1846)	Fractured bolet
619	571	IX. 641	B. lacunosus	Cooke and Mass., Grev. XVIII. 5 (1889)	Pitted bolet ...
620	566	VI. 4768	B. luridus	Schaeff., Fung. 107 (1762)	Lurid bolet ...
621	543	„ 4641	B. luteus	Linn., Sp. Pl. 1177 (1753)	Yellow bolet...
622	567	„ 4794	B. marginatus	Drum., Berk., Hook., Lond. Journ., Bot. IV, 50 (1845)	Margined bolet
623	570	„ 4803	B. megalosporus ...	Berk., Fl. Tasm. II. 251 (1860)...	Large-spored bolet
624	573	„ 4830	B. napipes	F. v. M., Linn. Journ. XIII. 161 (1873) ...	Turnip-stalked bolet ...
625	559	„ 4728	B. pachypus	Fries, S.M. I. 390 (1821)	Thick-stalked bolet ...
626	563	„ 4756	B. portentosus	Berk. and Br.. Linn. Journ. XIV, 46 (1875)	Monstrous bolet
627	568	„ 4800	B. prunicolor	Cooke and Mass., Grev. XVI. 32 (1887)...	Plum-coloured bolet ...
628	548	„ 4656	B. sanguineus	With., Arr. IV. 319 (1796)	Blood-red bolet
629		„ 4792	B. scaber	Fries, S.M. I. 293 (1821)	Rough bolet ...

65. Sclerodepsis, Cooke.
66. Hexagonia, Fries.
67. Dædalea, Pers.
68. Ceriomyces, Corda.

69. Favolus, Fries.
70. Laschia, Fries.
71. Campbellia, Cooke.
72. Merulius, Hall.

Number	Habitat.						B.	Occurrence.	General Characters.
	W.A.	S.A.	T.	V.	N.S.W.	Q.			

FRIES, PL. HOM. 79 (1825).

Tubiporus, Agaricus.

Number	W.A.	S.A.	T.	V.	N.S.W.	Q.	B.	Occurrence.	General Characters.
601	...	...	...	...		Q.	B.	Woods ...	Cap smooth, olive brown, turning blackish. Stem stout and yellow. Pores sulphur yellow. Rare. *Edible*.
602	...	...	...	...		Q.	B.	Woods, and in pastures under trees	Largest of this genus. Cap smooth, whitish. Stem very thick and yellowish. Pores greenish yellow. *Edible*.
603	W.A.	...	...	...		...	...	Ground ...	Cap smooth, yellow, viscid. Stem downy. Pores yellow. *Edible*.
604	W.A.	...	...	...		...	...	Sandy soil	Cap flattened, viscid, excentric. Stem elongated, pale above, black below. Pores cinnamon.
605	...	S.A.	...	V.		...	...	Ground ...	Cap viscid, umber. Stem flesh colour. Pores hexagonal, sulphur colour.
606	...	...	...	...		Q.	B.	Woods ...	Cap soft, viscid, bay brown. Stem solid, with brownish bloom. Pores from yellowish white to greenish. *Edible*.
607	...	...	...	V.		...	...	Ground ...	Cap somewhat downy, reddish brown. Stem short, stout. Pores rather large, greenish grey.
608	W.A.	...	...	...		...	...	Sandy soil	Cap fleshy, blood red to purple. Stem stout, sulphur colour. Pores rounded, yellow, with oblique openings.
609	...	...	...	...	N.S.W.	...	B.	Woods	Cap globose, somewhat downy, olive. Stem firm and thick, scarlet. Pores yellow.
610	...	...	...	V.		Q.	B.	Woods, &c. ...	Cap, dull brown with red cracks. Stem rigid, crimson or yellow. Pores greenish yellow.
611	...	...	...	...		Q.	D.	Woods ...	Cap smooth, moist, brownish. Stem stout, pale brown. Pores lemon to yellowish-green. *Edible*.
612	...	...	...	...		Q.	B.	Woods ...	Golden yellow entirely. Cap viscid. Stem firm. Pores sulphur colour. *Edible*.
613	...	...	...	V.		Q.	B.	Woods	Cap soft, smooth, brown or reddish grey. Stem solid, stout. Pores angular, flesh pink.
614	...	...	...	V.		...	B.	Woods	Large, entirely yellow. Cap compact, viscid. Stem spotted with brown and with fugacious rings. Pores angular, yellow
615	...	...	T.	...		...	...	Ground at roots of *Pleurandra riparia*	Cap fleshy, smooth, red. Stem nearly smooth. Pores pale orange yellow.
616	...	...	...	V.		Q.	B.	Grassy places ...	Cap slimy. Stem covered with milky drops drying into brown granules. Pores granulated. *Edible*.
617	...	...	...	...		Q.	...	Ground ...	Cap thick, tan coloured. Stem similarly coloured. Pores pale.
618	W.A.	...	...	...		...	...	Ground ...	Cap smooth, purple, with margin much broken. Stem very short and tuberous. Pores sulphur colour.
619	...	...	...	...		Q.	...	Sandy ground ...	Cap soft, somewhat viscid, tawny. Stem deeply pitted. Pores angular whitish to flesh colour.
620	...	...	...	V.		Q.	B.	Ground ...	Large. Cap viscid, soft, olive brown or tawny. Stem stout, tall, vermilion. Pores orange, red, crimson. Common.
621	W.A.	...	...	V.		Q.	B.	Ground in Pine woods	Large. Cap viscid, soft, dingy yellow. Stem tall, firm, with broad dingy ring. Pores yellow. Common. *Edible*.
622	W.A.	...	...	V.		...	...	Ground ...	Cap compact, delicately velvety, margin thin and distinct from hymenium. Stem short, black. Pores internally pallid.
623	...	...	T.	...		...	...	Ground in woods ...	Cap somewhat tan coloured. Stem warty. Pores flesh colour.
624	...	...	...	V.		...	...	Meadows	Cap reddish brown, at length blackish. Stem obconical. Pores lemon yellow.
625	...	...	...	...	N.S.W.	Q.	B.	Woods ...	Very large. Cap brownish, then pale tan colour. Stem thick, firm, yellow variegated with red, very bulbous. Pores round, yellow. Common.
626	...	...	...	V.		...	...	Ground ...	Very large. Cap depressed in centre. Stem thick and dilated at base. Pores lemon yellow.
627	...	...	...	V.		...	...	Ground ...	Cap soft, viscid, plum coloured. Stem club shaped, pale. Pores rounded, pale.
628	...	...	...	...		Q.	B.	Woods ...	Very small. Cap smooth, viscid, blood red. Stem yellow and red. Pores large, orange yellow. Rare.
629	...	...	...	V.		...	B.	Woods ...	Large. Dull brown, very rough. Cap cushion shaped, viscid. Stem solid, tall, scurfy. Very common. *Edible*.

D 2

Number.	Cooke's Number.	Sac. and's Number.	Scientific Name.	Authority for Name.	English Name.

57. BOLETUS.—Linn., Sp. Pl. 1176 (1753).

630	519	VI. 4670	B. subsimilis ...	Fries, Pl. Preiss. II. 134 (1846)	Simulating bolet ... ...
631	554	„ 4682	B. subtomentosus...	... Linn., Sp. Pl. 1178 (1753) ...	... Downy bolet... ... ...
632	556	„, 4703	B. Thozetii	..., Berk., Linn. Journ. XVIII. 384 (1881) ...	Thozet's bolet ...

58.—STROBILOMYCES.—Berk. in Hook.,

633	581	VI. 4838	S. ananæceps	... Sacc., Syll. VI. 50 (1888)	Pine-apple-headed strobilomyces
634	577	...	S. fasciculatus	... Cooke, Grev. XX. 4 (1891)	Fasciculate strobilomyces ...
635	579	VI. 4835	S. floccopus	... Vahl., Fl. Dan. t. 1252 (1764) ...	Woolly-stalked strobilomyces ...
636	578		S. ligulatus	Cooke, Grev. XX. 4 (1891)	Ligulate strobilomyces
637	574	VI. 4837	S. nigricans	... Berk., Hook. Journ. 139 (1852)	Blackening strobilomyces ...
638	575	IX. 645	S. pallescens	Cooke and Mass., Grev. XVIII. 5 (1889)	... Bleaching strobilomyces ...
639	576	„, 646	S. rufescens	Cooke and Mass., Grev. XVIII. 5 (1889)	Reddish-brown strobilomyces ...
640	580	„ 644	S. velutipes	... Cooke and Mass., Grev. XVIII. 5 (1889)	Velvety-stalked strobilomyces ...

59. FISTULINA.—Bull.

| 641 | 582 | VI. 4849 | F. hepatica | Fries, S.M. I. 396 (1821) | Liver-coloured Fistulina, Beef-steak fungus |

60. POLYPORUS.—Adans, Fam. II. 10 (1763).—

642	646	VI. 5116	P. adustus	Fries, S.M. I. 363 (1821)	... Scorched polypore ...
643	591	„ 4913	P. alveolarius	Bose, Berl. Mag. IV. (1811)	... Depressed polypore ...
644	666	„ 5253	P. anebus	Berk., Hook., Lond. Journ. VI. 504 (1847)	... Beardless polypore
645	632	„ 5080	P. angustus ...	Berk., Fl. Tasm. II. 253 (1860) ...	Narrow-capped polypore ...
646	622	„ 5045	P. anthracophilus	Cooke, Grev. XII. 16 (1884) ...	Burnt-ground polypore ...
647	591	„ 4903	P. arcularius	Fries, S.M. I. 342 (1821)	Convex polypore ...
648	636	„ 5107	P. argentatus	Cooke, Grev. XV. 20 (1886) ...	Silvery polypore ..
649	667	„ 5255	P. ascobuloides ...	... Berk., Linn. Journ. XIII. 162 (1873)	Ascobolus-like polypore ...
650	651	„ 5207	P. betulinus ...	... Fries, S.M. I. 358 (1821)	Birch polypore ...
651	599	„ 1944	P. biennis	... Fries, Epicr. 433 (1838)	Biennial polypore ...
652	...	„ 5166	P. bircinum	... Kalch., Hedw. XV. 114 (1876)	Tawny polypore ...
653	652	„ 5187	P. borealis	Fries, S.M. I. 366 (1821)	Northern polypore
654	589	„ 4885	P. brumalis	Fries, S.M. I. 348 (1821)	Wintry polypore ...
655	637	„ 5106	P. campylus	Berk., Fl. Tasm. II. 252 (1860)	... Curved polypore ...
656	659	„ 5232	P. cartilagineus ...	Berk. and Br., Linn. Journ. XIV. 49 (1875)	... Cartilaginous polypore ...
657	635	„ 5093	P. chioneus ...	Fries, S.M. I.359 (1821)	... Snowy polypore ... ...

Number.	Habitat.						B.	Occurrence.	General Characters.
	W.A.	S.A.	T.	V.	N.S.W.	Q.			

Tubiporus, Agaricus—*continued*.

630	W.A.	...	...	...	...	Q.	...	Ground ...	...	Cap fleshy, viscid, shining brown. Stem solid, sulphur yellow. Pores very short, sulphur yellow. Simulating *B. lividus*.
631	W.A.	S.A.	...	V.	N.S.W.	Q.	B.	Woods. &c.	...	Cap soft, dry, finely tomentose, bronze, with yellow cracks. Stem tall, yellow variegated with red. Pores yellow.
632	...	...	...	...		Q.	...	Barren soil	...	Yellow. Cap with granular warts. Stem slender and flexuous. Pores free.

Kew Misc. III. 78 (1851).—Boletus.

633	...	...	...	V.	N.S.W.	Q	...	Ground ...		Cap broken up into flat, thick, broad, scaly warts.
634	...	...	...	V.	...	...	...	Ground ...	...	Cap reddish brown, scaly, with fascicles of strap-like scales. Stem paler. Pores angular, yellowish.
635	...	...	...	...		Q.	...	Ground ...	...	Ash coloured, becoming black. Cap soft, scaly, and veil silky. Stem stout and downy. Pores large, greyish white. Rare.
636	...	...	...	V.			...	Ground ...	...	Cap hemispherical, brown, with darker ligulate scales. Stem paler. Pores angular, yellowish, or tinted with red.
637	...	...	...	...	...	Q.	...	Woods ...	...	Small. Cap rough in centre, with hexagonal warts. Stem solid, with woolly scales, like margin of cap.
638	...	...	...	...		Q.	...	Base of trees	...	Cap rosy purple, turning pale, with thick obtuse conical warts overlapping. Stem thick, reddish. Pores large, angular, yellowish.
639	...	...	...	...		Q.	...	Base of trees	...	Entirely reddish brown. Cap hemispherical, thickly covered with overlapping conical warts. Stem solid, pale above, reddish brown below. Pores large, angular, tawny.
640	...	...	...	...		Q.	...	Ground ...	...	Blackening. Cap with thick irregular warts overlapping. Stem velvety. Pores large, angular.

Champ. I. 314 (1798).—Boletus.

| 641 | W.A. | ... | ... | V. | | | B. | Trunks of living trees | | Fleshy and red juiced. Cap flesh colour to blood red and liver colour, roundish, attached by broad base, internally streaked. *Edible*. |

Polystictus, Fomes, Favolus, Trametes, Poria, Boletus, Dædalea.

642	...	...	...	V	N.S.W.	Q.	B.	Trunks, stumps, &c.	Cap fleshy, tough, pale ash colour, with margin blackening. Pores minute, white to grey, becoming black.
643	...	...	...	...		Q.	...	Trunks ...	...	Cap fleshy, leathery, depressed, brown. Stem firm and thickened at base. Pores hexagonal, white.
644	...	...	...	...		Q.	...	Wood	Pale fawn colour. Cap thin, leathery, delicately velvety. Pores small, round, short.
645	...	...	T.	...			...	Rotten wood	...	Cap narrow, downy, brown. Pores small, angularly punctiform.
646	W.A.	...	...	V.	...	Q.	...	Burnt ground	...	Imbricate and much divided, very leathery and hardening. Caps growing together, overlapping, bay brown. Pores angular, white.
647	...	...	...	V.	N.S.W.	Q.	...	Trunks ...	...	Cap tough and leathery, without zones, brown to yellowish. Stem short greyish brown. Pores oblong rhomboid, large, whitish.
648	...	...	...	V.		Q.		Trunks ...	...	Cap fleshy to leathery, shell shaped, slightly silky, white in front, ashy brown behind. Pores white, rounded.
649	...	...					...	Trunks ...	...	Cap circular, thickish, white, downy. Pores hexagonal, small.
650	...	...				Q.	B.	Birch, &c.	...	Cap fleshy, then corky, hoof shaped, smooth. Pores minute, short, white, or brownish.
651	...	...	...	...		Q.	...	Ground, near trunks	Cap spongy then corky to leathery, white to rusty colour. Stem short, thick, rust coloured, woolly. Pores torn and toothed, dull white to brownish.
652	...	...	...	...	N.S.W.		...	Trunks ...	...	Hemispherical, sessile, spongy to powdery, tawny, cinnamon or bay brown.
653	...	...	...	V.			B.	Trunks ...	White to yellowish. Cap spongy to corky, hairy. Pores unequal, torn, white.
654	...	...	...	...		Q.	B.	Trunks ...	Cap tough, fleshy to leathery, sooty brown. Stem thin, hairy, scaly. Pores angular, toothed, white.
655	...	...	T.	V.			...	Rotten wood	Cap palmate, lobed, white, smooth. Hymenium concave. Pores small, irregular.
656	...	...	...	...		Q.	...	Deal wood	...	Cap red brown to sooty brown, cuticle cartilaginous. Pores minute.
657	...	...	...	V.	N.S.W.	Q.	B.	Trunks and stumps	White. Cap fleshy, soft, smooth. Pores short, minute, rounded. Smell rather acid.

38

Number.	Cooke's Number.	Saccardo's Number.	Scientific Name.	Authority for Name.	English Name.

60. POLYPORUS.—Adans. Fam. II. 10 (1763).—

Number.	Cooke's Number.	Saccardo's Number.	Scientific Name.	Authority for Name.	English Name.
658	620	VI. 5019	P. confluens	*Fries*, S.M. I. 355 (1821) ...	Confluent polypore ... ...
659	628	„ 5072	P. corrivalis ...	Berk., Linn. Journ. XIII. 162 (1873) ...	Overgrowing polypore ...
660	663	„ 5241	P. cubensis ...	Mont., Cuba 404 (1838)	Cuban polypore ... ...
661	590	„ 4902	P. cupuliformis	Berk. and Curt., Grev. I. 38 (1872)	Cup-shaped polypore ...
662	641	„ 5140	P. demissus	Berk., Hook., Lond. Journ. IV. 52 (1845) ...	Hood-shaped polypore ...
663	647	„ 5152	P. dichrous ...	Fries, S.M. I. 364 (1821)	Two-coloured polypore ...
664	613	„ 4982	P. dictyopus ...	Mont., Fl. Fern. 14 (1835) ...	Net-stalked polypore ...
665	616	„ 5008	P. dorcadideus ...	Berk. and Br., Linn. Trans. II. 57 (1883) ...	Fawn-coloured polypore ...
666	609	„ 4971	P. elegans	*Fries*, Epicr. 440 (1838) ...	Elegant polypore ... ...
666A	„ „	„ „	P. elegans, var. nummularius	*Fries*, S.M. (1821) ...	Coin-like polypore ...
667	627	„ 5067	P. epileucus	Fries, Epicr. 452 (1838) ...	Whitish polypore ...
668	656	„ 5216	P. Eucalyptorum	Fries, Pl. Preiss. II. 135 (1846) ...	Eucalyptus polypore ...
669	639	„ 5123	P. foedatus	Berk., Linn. Journ. XVI. 41 (1878) ...	Dirty polypore ... ...
670	633	„ 5082	P. fragilis	Fries, Elench. 86 (1828) ...	Fragile polypore ...
671	618	„ 5015	P. frondosus ...	*Fries*, S.M. I. 355 (1821) ...	Frondose polypore ... ...
672	643	„ 5169	P. fruticum	Berk. and Curt., Linn. Journ X. (1869)	Shrub-growing polypore ...
673	617	„ 5009	P. fusco-lineatus	Berk. and Br., Linn. Trans, I. 401 (1879)	Tawny-lined polypore ...
674	641	„ 5129	P. gilvus ...	Schw., Carol. 897 (1822) ...	Yellowish-tan polypore ...
675	610	„ 4974	P. glabratus ...	Kalch., in Hedw. XV. 114 (1876)	Smooth polypore ...
676	614	„ 5005	P. grammocephalus ...	Berk., Hook., Lond. Journ. I. 148 (1842) ...	Line-headed polypore ...
676A	„ „	P. grammocephalus, var. Emerici	Berk., Grev. X. 96 (1882) ...	Emericus polypore ...
676B	„ „	„ „	P. grammocephalus, var. Muelleri	Kalch., Grev. X. 97 (1882)	Mueller's polypore ...
677	611	„ 4976	P. Guilfoylei	Berk. and Br., Linn. Trans. II. 58 (1883)	Guilfoyle's polypore ...
678	631	„ 5079	P. Gunnii	Berk., Fl. Tasm. II. 253 (1860) ...	Gunn's polypore ...
679	585	„ 4866	P. Hartmanni	Cooke, Grev. XII. 14 (1884)	Hartman's polypore ...
680	648	„ 5165	P. hispidus	*Fries*, S.M. I. 362 (1821)	Bristly polypore ...
681	IX. 629 / VI. 5179	P. hispidans	Berk., in Fries, Nov. Sym. 37 (1851)	Bristling polypore ...
682	658	VI 5227	P. hypopolius	Kalch., Grev. X. 99 (1882)	Hoary polypore ...
683	602	„ 4952	P. hystriculus	Cooke, Grev. XV. 16 (1886)	Porcupine-like polypore ...
684	607	„ 4965	P. infernalis	Berk., Hook., Lond. Journ. II. 637 (1843)	Infernal polypore ...
685	619	„ 5017	P. intybaceus	Fries, Epicr. 446 (1838)	Endive polypore ...
686	623	„ 5047	P. laetus	Cooke, Grev. XII. 16 (1884)	Bright-coloured polypore ...

OF AUSTRALIAN FUNGI—continued.

Number.	Habitat.						B.	Occurrence.	General Characters.
	W.A.	S.A.	T.	V.	N.S.W.	Q.			

Polystictus, Fomes, Favolus, Trametes, Poria, Boletus, Daedalea—continued.

658	...	...		...	N.S.W.	Q	...	Wood	...	Branching, firmly fleshy, fragile. Caps thick, overlapping, confluent, smooth, flesh colour to yellowish or dark. Stems very short, confluent. Pores short, minute, white.
659	...	S.A.			N.S.W.	Q.	...	Trunks ...	...	Imbricate. Caps shell shaped, whitish, downy. Pores small, angular.
660	...	...			...	Q.		Trunks ...	...	Pale white. Cap sessile, fleshy to corky, smooth. Pores round, small, white, then red brown.
661	...	...		V.		...	...	Trunks ...	...	Cap cup shaped, at first reddish brown, downy. Stem very short. Pores small, red brown.
662	W.A.	...	...	...		...	...	Rotten wood	...	Caps overlapping, hood shaped, corky, dependent, spongy, downy, pale yellow. Pores roundish, minute.
663	...	...		V.			...	Trunks ...	...	Cap fleshy, tough, silky, white. Pores short, minute, round, cinnamon brown. Pretty species.
664	...	...		V	...	Q.	...	Trunks ...	...	Cap fleshy to leathery, rigid, smooth, bay brown. Stem lateral, thick, reticulately wrinkled, bay to black. Pores minute, rounded, pale.
665	...	...			...	Q.		Trunks ...	...	Cap fan shaped, lobed, rich umber or fawn colour, with velvety bloom. Stem short, downy. Pores hexagonal.
666	...		T.			Q.	B.	Trunks ...	...	Cap fleshy, soon hardening and becoming woody, flat. Stem excentric or lateral. Pores minute, roundish, yellowish white.
666A	...	...			N.S.W.		B.	Trunks ...	...	Smaller, thinner, rather regular. Stem equal, excentric.
667	...	...	...		...	Q.	B.	Rotten trunks	...	Cap soft and cheesy, then firm, roughly hairy, whitish, semicircular. Pores minute, round, white.
668	W.A.	S.A.		V.		...	...	Eucalyptus trunks		Cap semicircular, hoof shaped, thick, very soft, white, invested with thin evanescent dark-brown crust. Pores short, small, falling away.
669	...	...		...		Q.	...	Trunks ...	...	Cap rather thin, kidney shaped, umber to sooty brown. Pores punctiform, pale cinnamon.
670	...	...		V.		...	B.	Wood ...	...	White, spotted with brown when touched. Cap fleshy, fragile, kidney shaped, rough. Pores very thin.
671	...	...	T.	...		...	B.	On trunks and at base		Elaborately branched, fibrously fleshy, tough. Caps very numerous, semicircular, rough, lobed, greyish to sooty brown. Stems growing together. Pores small, white.
672	...	...		...		Q.	...	Branches	...	Cap thin, soft, semicircular, rhubarb colour. Pores small, angular, toothed.
673	...	...		...		Q.	...	Trunks ...	...	Cap, thin, tough, ochrey, streaked with radiating brown, hispid lines. Stem ochrey, thicker above. Pores irregular, tawny brown
674	W.A.					Q.	B.	Trunks ...	...	Cap fleshy, tough, yellowish tan. Pores minute, yellowish tan to rusty brown.
675	...	...		V.	...	...	...	Trunks ..	...	Cap excentric, fleshy, smooth, dark brown. Stem solid, tapering, becoming brown. Pores minute, round, white to yellowish.
676	...	...			N.S.W.	Q.	...	Trunks ...	...	Cap at first wedge shaped, then kidney shaped, flattened, pale umber. Stem lateral. Pores short, umber.
676A	...	...			...	Q.	...	Trunks ...	...	Whitish. Cap spoon shaped or kidney shaped. Pores angular.
676B	...	...			N.S.W.		...	Trunks ...	...	Whitish tan when dry. Cap thin, rigid. Pores short, angular, unequal.
677	...	...	...	...		Q	...	Trunks ...	...	Cap spoon shaped, lateral, fine, powdery. Stem black, cartilaginous. Pores punctiform.
678	...	...	T.	V.			...	Branches	...	Cap fan shaped, thin, whitish, downy, rough. Pores irregular, of medium size.
679	...	...				Q.	...	Ground	...	Cap fleshy, rather fragile, finely velvety, reddish brown. Stem swollen, thick, reddish. Pores small, round, pale. Elegant species.
680	...	...		...		Q.	B.	Trunks ...	...	Large, brown, juicy. Cap compact, spongy to fleshy, semicircular, thick set with bristly down. Pores minute, rounded, pale.
681	...	...	...			...	...	Trunks ...	...	Cap semicircular, fleshy to leathery, fawn or dusky, bristly. Pores angular.
682	...	...	...			Q.	...	Trunks ...	...	Cap leathery to woody, somewhat shell shaped, with rigid chestnut crust, rough with thick tubercles. Pores short, angular, white to hoary.
683	...	...		V.			...	About root	...	Cap tough, bristly, dark brown. Stem thick, shortened. Pores large, angular, torn or toothed.
684	...	...		V.		Q.	...	Trunks ...	...	Cap fan shaped, smooth, blackish, liver coloured. Stem short, lateral, black. Pores minute, round, very short.
685	...	...		V.		Q	B.	Trunks, and at foot of trees		Very much branched, fleshy, rather fragile. Caps very numerous, yellowish to tawny. Stems confluent into a very short trunk. Pores firm, white to tawny.
686	...	...		V.			...	Trunks ...	...	Imbricated and much divided, leathery, orange tawny. Caps grown together and converging behind into stem. Pores pale.

Number.	Cooke's Number.	Saccardo's Number.	Scientific Name.	Authority for Name.	English Name.
				60. POLYPORUS.—Adans, Fam. II. 10 (1763).—	
687	588	VI. 4884	P. lentus	Berk., Outl. 237 (1860)...	Tough polypore ... ...
688	662	„ 5240	P. lignosus	Klotsch., in Fries, Epicr. 471 (1838)	... Woody polypore ... ...
689	604	„ 4958	P. melanopus	Fries, S.M. I. 347 (1821) ...	... Black-stalked polypore ...
690	587	„ 4870	P. myelodes	Kalch., Grev. IV. 73 (1875)	... Marrow-like polypore...
691	1351	...	P. Mylittæ	(Cooke and Mass., Grev. XXI. 37 (1892)	... Mylitta polypore ... }
692	638	VI. 5116	P. nidulans	(Sacc., Hedw. 56 (1893) ... ... / Fries, S.M. I. 364 (1821)	... Native bread polypore } / Nest polypore ...
693	583	„ 4858	P. ovinus	Fries, S.M. I. 346 (1821)	Sheep polypore
694	630	„ 5190	P. pelliculosus ...	Berk., Hook., Lond. Journ. VII. 575 (1848)	... Cuticular polypore ...
695	...	IX. 675	P. Pentzkei	Kalch., Proc. Linn. Soc. N.S.W. VIII. 175 (1883)	Pentzke's polypore
696	584	VI. 4862	P. pes-capræ	Pers., Champ. Com. (1818)	... Goat's-foot polypore ... ...
697	613 bis	„ 4990	P. petaloides	... Fries, Epicr. 444 (1838)	... Petal-like polypore ...
698	...	„ 4999	P. phlebophorus	Berk., Fl. N.Z. 177 (1855) ...	... Vein-bearing polypore ...
699	605	„ 4966	P. picipes	Fries, S.M. I. 353 (1821)	... Pitch-stalked polypore
700	596	„ 4923	P. pisiformis	Kalch., Grev. X. 98 (1882)	Pea-shaped polypore ...
701	615	„ 5007	P. platotis	Berk. and Br., Linn. Trans I. 401 (1879)	Broad polypore
702	664	„ 5247	P. plebeius	Berk., Fl. N.Z. II. 179 (1855)	Plebeian polypore
703	655	„ 5212	P. portentosus	Berk., Hook., Journ. 188 (1844)...	... Monstrous polypore ...
704	601	„ 4951	P. proteiporus	Cooke, Grev. XII. 15 (1884)	Variable-pored polypore
705	625	„ 5054	P. retiporus	Cooke, Grev. XII. 15 (1884)	... Net-pored polypore
706	645	„ 5141	P. rhinocephalus ...	Berk., Fl. Tasm. II. 253 (1860) ...	Rough-headed polypore
707	640	„ 5121	P. rubidus	Berk., Hook.. Journ. 500 (1847),...	Reddish polypore
708	690	„ 1216	P. rufescens	Fries, S.M. I. 351 (1821)	Reddening polypore
709	621	„ 5026	P. scabriusculus	Berk., Linn. Journ. XVIII. 384 (1881) ...	Roughish polypore
710	597	„ 4938	P. Schweinitzii	Fries, S.M. I. 351 (1821)	Schweinitz's polypore ...
711	632	„ 5130	P. scrupusus ...	Fries, Epicr. 473 (1838)	Rugged polypore
711A	643	„ 5131	P. scrupusus, var. isidioides...	Cooke, Grev. XIII. 87 (1885)	Coral-like polypore
712	630	„ 5078	P. semidigitaliformis	Berk., Linn. Journ. XVI. 39 (1878)	Finger-like polypore ...
713	593	„ 4910	P. similis	Berk., Hook., Lond. Journ. II. 635 (1843)	Similar polypore
714	651	„ 5181	P. spiculifer	Cooke, Grev. XV. 20 (1886)	Spiculate polypore
715	...	„ 5186	P. spumeus	Fries, S.M. I. 358 (1821)	... Frothy polypore
716	605	„ 4953	P. squamosus	Fries, S.M. I. 343 (1821) ...	Scaly polypore
717	595	„ 4923	P. stipitarius	Berk and Curt., Linn. Journ. X. 304 (1869)	... Stalked polypore
718	606	IX. 667	P. Strangeri	F. v. M., Linn. Soc. N.S.W. 106 (1882) ...	... Stranger's polypore
719	657	VI. 5220	P. strumosus	Fries, Epicr. 462 (1838)	Swollen polypore

OF AUSTRALIAN FUNGI—*continued.*

Number	W.A.	S.A.	T.	V.	N.S.W.	Q.	B.	Occurrence	General Characters
			Habitat.				B.	Occurrence.	General Characters.

Polystictus, Fomes, Favolus, Trametes, Poria, Boletus, Daedalea—*continued.*

Number	W.A.	S.A.	T.	V.	N.S.W.	Q.	B.	Occurrence	General Characters
687	...	...	...	V.	...	...	B.	Rotten stems, &c.	Cap fleshy to tough and leathery, pale ochrey. Stem short, rough haired and mealy. Pores irregular, white.
688	...	...	...	V.		...	...		Trunks ...	Cap fleshy to corky or woody, pale to yellowish. Pores long, small, very thin.
689	...	...	...	V.		Q.	B.	Ground and branches	Cap fleshy, tough, white to yellowish brown. Stem excentric, velvety at first, black. Pores rounded, small, white to yellowish
690	...	...			...	Q.	...	Ground, at base of trunks	Cap fleshy, fragile, tan or pale brown. Stem solid, obconical. Pores short, minute, whitish. *Edible.*
691		S.A.	T.	V.	N.S.W.	Q.	...	Sclerotium, known as *Mylitta australis*	Cap fleshy, tough, elastic, minutely velvety, white. Stem short, solid. Pores white, somewhat angular.
692	...	...	...	...		Q.	B.	Trunks ..	Cap fleshy, very soft, yellowish tan. Pores elongated, angular, tawny, brick red.
693	...	...	...	V.		...	...	Woods ...	Cap fleshy, fragile, whitish. Stem short, white. Pores minute, rounded, white to lemon yellow.
694	...	S.A.	T.	V.		Q.	...	Wood	Dark red, juicy, densely gregarious, shell shaped, at first with dense rough hairs. Pores rather angular.
695	...	...	...			Q.		... ...	Cap slender, leathery, base wedge shaped, smooth, zoned, ochrey, becoming dark. Stem cylindrical, pale.
696	...	...	...	V.		...	...	Pine woods	In tufts. Cap fleshy, fragile, hay brown to dark brown. Stem deformed, yellowish white. Pores broad, yellowish white.
697	...	...	...	V.		...	B.	Trunks ...	Cap membranaceous, spoon shaped, chestnut brown. Stem lateral, compressed, whitish. Pores very short, small, white.
698						Q.	...	Stems of *Eucalyptus hemiphloia*	Small, white. Cap fan shaped. Stem short, smooth, with undulating furrows.
699	...			V.		Q.	B.	Trunks ...	Cap fleshy to leathery, smooth, depressed. Stem excentric and lateral, firm, black. Pores rounded, small, white to yellowish.
700	...			V.			...	Wood ...	Entirely white, globose, sessile, size of pea or less. Pores minute, punctiform. Probably young stage of known species.
701	...	...	...			Q.	...	Wood	Cap club shaped to funnel shaped, fragile, ochrey. Stem elongated and thickened upwards. Pores angular, ochrey.
702	...	...		V.		Q.	...	Rotten wood	Pale, imbricate. Cap semicircular, delicately downy, ochry. Pores minute, punctiform.
703	W.A.	S.A.	T.	V.	N.S,W.	Q.	...	Trunks ..	Sessile, very large. Cap fleshy, smooth, with thick tan cuticle. Pores small, externally brown, internally pale.
704	...	...		V.		Q.	...	Ground ...	Cap whitish, fleshy, tough, delicately downy and scurfy. Stem short, whitish. Pores angular, irregular, pale umber.
705	...	...		V.		Q.	...	Trunks ...	Tufted, forming clumps of juicy cheesy consistence, ochrey. Caps very broad. Pores angular, very short, with net-like partitions.
706	...	...	T.	...			...	Rotten wood	Cap semicircular, shell shaped, whitish, rough veined. Pores white to ashy.
707	...	...			N.S,W.	Q.	...	Wood ...	Reddish. Cap thin, leathery, kidney shaped, silky. Pores small, short, punctiform.
708	W.A.	...		V.	N.S,W.	Q.	B.	Grassy ground about old trunks	Flesh coloured. Cap spongy, hairy. Stem short, deformed. Pores large, sinuous, white to flesh colour.
709	...					Q.	...	Trunks ...	Caps spoon shaped, delicately downy, roughish. Stem cylindrical, repeatedly branched, central, sometimes nearly obsolete.
710	...		...			Q.	B.	Pine stumps	Very large. Cap thick, spongy to corky, bay brown. Stem thick, very short or obsolete, rusty brown. Pores large, sulphury to greenish.
711	W.A.		T	V.	N.S,W.	Q.	...	Dead wood	Cap corky, rough and rugged, umber. Pores minute, rounded, rusty brown.
711A	W.A.	...					...	Trunks and at foot of trees	Corky or woody, sessile, yellowish tan to rusty, rough, with thick tubercles.
712					N.S.W.		...	Trunks ...	Gregarious. Cap hoof shaped, whitish, rough, and downy. Pores large.
713						Q.	...	Trunks ...	Cap tough, leathery, smooth. Stem thickened downwards, velvety, becoming smooth. Pores small.
714				V.			...	Trunks ...	Cap fleshy, soft, and watery, sooty brown to black, beset with scattered obtuse spicules. Pores minute.
715	...	...		V.			B.	Old trunks of Eucalypts	White. Cap fleshy to spongy, compact, cushion shaped, rough haired, base stem-like.
716	...	...		V.		Q.	B.	Trunks ...	Cap fleshy to tough, fan shaped, ochrey, marked with brown scales. Stem stout, excentric, lateral at base. Very common.
717	...					Q.	...	Wood, &c.	White. Cap circular, thin, smooth. Stem slender.
718	...				N.S,W.	...	...	Trunks ...	Cap corky to leathery, kidney shaped, umber, turning blackish. Stem short, cylindrical, altogether black.
719	...			V.			...	Trunks ...	Fleshy, tough, afterwards very hard, sooty brown, margin acute, turning black.

Number	Cooke's Number	Saccardo's Number	Scientific Name	Authority for Name	English Name
					60. POLYPORUS.—Adans, Fam. II. 10 (1763).—
720	634	VI. 5092	P. stypticus	Fries, S.M. I. 359 (1821)	Astringent polypore
721	653	„ 5191	P. substuppeus	Berk. and Cooke, Linn. Journ. XV. 380 (1877)	Tow-like polypore
722	661	IX. 696	P. subzonalis	Cooke, Grev. XIX. 44 (1890)	Slightly-zoned polypore
723	624	VI. 5050	P. sulphureus	Fries, S.M. I. 357 (1821)	Sulphur-coloured polypore
724	598	„ 4939	P. tabulæformis	Berk., Hook., Lond Journ. IV. 302 (1845)	Flat polypore
725	626	„ 5064	P. tephronotus	Berk., Fl. Tasm. II. 252 (1860)	Ashy polypore
726	665	„ 5248	P. testudo	Berk. and Broome, Linn. Trans. II. 59 (1883)	Tortoise polypore
727	592	„ 4908	P. tricholoma	Mont., Syll. I. 53 (1856)	Hairy polypore
728	586	IX. 650	P. tumulosus	Cooke, Grev. XVII. 55 (1889)	Buried polypore
729	...	VI. 4907	P. umbilicatus	Berk., Hook., Journ. 79 (1851)	Umbilicate polypore
730	...	„ 5222	P. ungulatus	Cooke, Grev. XIII. 116 (1885)	Hooked polypore
731	608	„ 4968	P. varius...	Fries, S.M. I. 352 (1821)	Variable polypore
732	...	„ 5178	P. Weinmanni	Fries, Epicr. 459 (1838)	Weinmann's polypore
733	660	„ 5237	P. zonalis	Berk., Ann. Nat. Hist. X. 375 (1842)	Zoned polypore
					61. FOMES.—Fries,
734	672	VI. 5300	F. amboinensis	Cooke, Grev. XIII. 118 (1885)	Amboina fomes
734A	672	„ 5302	F. amboinensis, var. gibbosus	Cooke, Grev. XIII. 118 (1885)	Swollen fomes
735	708	„ 5487	F. annosus	Cooke, Grev. XIV. 20 (1885)	Aged fomes
736	683	„ 5397	F. applanatus	Cooke, Grev. XIV. 18 (1885)	Flattened fomes
737	681	„ 5394	F. australis	Cooke, Grev. XIV. 18 (1885)	Southern fomes
737A	681	IX. 723	F. australis, var. arculatus	Bres., Pug. Myc. Austr. (1890)	Bow-shaped fomes
738	723	VI. 5529	F. bistratosus	Cooke, Grev. XIV. 21 (1885)	Stratose fomes
739	717	„ 5509	F. carneus	Cooke, Grev. XIV. 21 (1885)	Flesh-coloured fomes
740	682	„ 5396	F. chilensis	Cooke, Grev. XIV. 18 (1885)	Chilian fomes
741	718	„ 5512	F. cinereo-fuscus	Cooke, Grev. XIV. 21 (1885)	Ashy-brown fomes
742	709	„ 5491	F. compressus	Cooke, Grev. XV. 51 (1886)	Compressed fomes
743	679	IX. 705	F. concavus	Cooke, Grev. XIX. 44 (1890)	Concave fomes
744	680	VI. 5385	F. conchatus	Cooke, Grev. XIV. 18 (1885)	Shell-shaped fomes
745	710	„ 5485	F. connatus	Cooke, Grev. XIV. 20 (1885)	Connate fomes
746	694	„ 5451	F. contrarius	Berk. and Curt., Grev. XV. 21 (1886)	Contrary fomes
747		„ 5524	F. cryptarum	Fries, S.M. I. 376 (1821)	Crypt fomes
748	705	„ 5481	F. Curreyi	Berk., Grev. XV. 21 (1886)	Currey's fomes
749	678	„ 5359	F. dochmius	Cooke, Grev. XIV. 17 (1885)	Oblique fomes
750	704	„ 5478	F. endapalus	Cooke, Grev. XIV. 20 (1885)	Soft fomes
751	693	„ 5450	F. exotephrus	Cooke, Grev. XIV. 19 (1885)	Ashy fomes
752	716	„ 5499	F. fasciatus	Cooke, Grev. XIV. 21 (1885)	Banded fomes

OF AUSTRALIAN FUNGI—continued.

Number.	Habitat.						B.	Occurrence.	General Characters.
	W.A.	S.A.	T.	V.	N.S.W.	Q.			

Polystietus, Fomes, Favolus, Trametes, Poria, Boletus, Dædalea—*continued.*

720	W.A.	...	...	...	...			Trunks ...	Fleshy to corky, cushion shaped, fragile, whitish ; margin obtuse, somewhat reddish.
721	...	...	...	...		Q.		Wood	Semicircular, decurrent behind, rough, short, invested with tow-like wool, pale ochrey when dry.
722	...	...	...	...		Q.		Wood	Corky, rather thin, sessile, entirely cream coloured, kidney shaped, faintly concentrically zoned.
723	...	...	T.	...		Q.	B.	Trunks ...	Large, tufted, and much divided ; of juicy, cheesy, or doughy consistence. Caps very broad, overlapping, reddish or lemon yellow.
724	...	...	...	...	...	Q.		Trunks ...	Cap circular, thick in centre, thin at margin, somewhat lobed, slightly zoned, velvety, rusty bay. Stem short, central or lateral.
725	...	...	T.	...	N.S.W	...		Rotten wood	Soft, downy, snow white, brown behind. Hymenium white, then turning slightly ashy.
726	...	...	...	...		Q.		Trunks ...	Caps overlapping, rigid, powdery, obscurely streaked and rough here and there.
727	...	...	...	...		Q.		Fallen branches ...	Cap leathery to membranous, rigid, convex, then funnel shaped, yellowish, with fringe of stiff brown hairs. Stem thin, yellowish brown.
728	...	...	...	...		Q.		Ground ...	Cap fleshy, firm, pale, with darker sunken scales. Stem short, thick, solid, ochrey.
729	...	...	T.	...		...		Wood ...	Cap fleshy to tough, membranous, then somewhat funnel shaped, snow white or cream coloured. Stem swelling above, and below minutely scaly.
730	...	S.A.	...	...		...		Trunks ...	Cap hard, whitish, shortly hooked behind, delicately downy, margin obtuse, wrinkled.
731	W.A.	...	T.	V.		Q.	B.	Trunks ...	Cap fleshy, tough, becoming woody, variable in form. Stem excentric and lateral, finally black.
732	...	...	...	V.		...		Trunks ...	White to red. Cap fleshy to spongy, firm, with reddish-brown hairs.
733	...	...	...	V.		Q.		Wood ...	Corky, thin, overlapping, sessile, semicircular, lobed, repeatedly zoned, covered with bloom, fawn colour.

Nov. Symb. 31 (1851).--Agaricus, Polyporus, Boletus, Trametes, Ganoderma.

734	...			...		Q.		Trunks ...	Cap corky to woody, somewhat ear-shaped, rough, pimpled. Stem lateral, very long, turning black.
734A	...	...	...	...		Q.		Trunks ...	Cap somewhat kidney shaped, and stem reddish brown.
735	...	...	...	...		Q.	B.	Trunks ...	Woody, rough, for the current year brown and silky; for previous season with blackened crust.
736	...	...	...	V.		Q.	B.	Trunks ...	Flattened, tuberculous, obsoletely zoned, powdery, cinnamon to hoary.
737	...	...	T.	V.	N.S.W.	Q.		Trunks ...	Very hard, semicircular, sessile, wavy, incrusted, somewhat bay brown.
737A	...	...	...	...	...	Q.		Trunks ..	Margin thicker, low shaped.
738	...	...	...	...		Q.		Wood	Spread out, umber, with very little substance. Pores stratose, punctiform.
739	...	S.A.	...	V.		Q.	B.	Trunks ...	Expanded and reflexed, woody, hard, thin, without zones, flesh coloured without and within.
740	...		...	...		Q.		Trunks ...	Corky, hoof shaped and dilated, with elevated ridges, brick red, turning pale.
741	...	...	...	...		Q.		Trunks ...	Semicircular, woody, very hard. Margin thin, ashy brown.
742	W.A	...	...	V.		Q.		Wood ...	Small, obliquely compressed, hoof shaped, zoned, light brown at first, dark brown afterwards.
743	...	...	...	...		Q.		Trunks	Very hard, semicircular, turning nearly black, comparatively thin and covered with hard crust.
744	...	...	...	V.		Q.	B.	Trunks ...	Corky to tindery, thin, spread out, somewhat shell shaped, bay brown.
745	...	...	...	...		Q.	B.	Trunks ...	Corky to woody, spread out, overlapping and growing into each other, downy white or ashy.
746	...	...	...	...		Q.		Trunks ...	Somewhat zoned, brown at first, ochrey, thin, rigid, downy, finally smooth.
747	...	...	...	...		Q.		Logs and rotten wood	Corky to tow like, zoneless, silky, reddish to rusty brown, but variable.
748	...	...	...	...		Q.		Trunks ...	Rigid, corky to leathery, semicircular, kidney shaped, brown, with concentric elevated zones.
749	...	...	...	...		Q.		Wood	Semicircular, oblique, hard, zoned, banded.
750	...	...	...	...	N.S.W.	Q.		...	Overlapping each other, leathery, bay brown, longitudinally rough in lines, delicately downy, substance soft.
751	...	...	...	...		Q.		Trunks ...	Hard, zoned, delicately downy at first, then smooth. Margin furrowed, lobed.
752	...	...	...	...		Q.		Trunks ...	Woody, thin, flattened, rusty brown with black bands.

Number	Cooke's Number	Saccardo's Number	Scientific Name	Authority for Name	English Name
					Gl. FOMES.—Fries, Nov. Symb. 31 (1851).—
753	713	VI. 5501	F. ferreus	Cooke, Grev. XIV. 21 (1885)	Hard fomes ...
754	686	„ 5409	F. fomentarius	Cooke, Grev. XIV. 18 (1885)	Tender fomes
755	689	„ 5417	F. fulvus	Cooke, Grev. XIV. 18 (1885)	Brown fomes
756	703	„ 5477	F. Gourlici	Cooke, Grev. XIV. 20 (1885)	Gourlie's fomes
757	690	„ 5424	F. gryphæformis	Cooke, Grev. XIV. 19 (1885)	Shell-shaped fomes
758	692	„ 5449	F. hemileucus	Cooke, Grev. XIV. 19 (1885)	Whitish fomes
759	711	„ 5497	F. hemitephrus	Cooke, Grev. XIV. 21 (1885)	Half-ashy fomes
760	719	„ 5519	F. homalopilus	Cooke, Grev. XIV. 21 (1885)	Smooth-piled fomes ...
761	687	„ 5412	F. igniarius	Cooke, Grev. XIV. 18 (1885)	Tinder fomes
762	720	„ 5523	F. incrassatus	Cooke, Grev. XIV. 21 (1885)	Thickened fomes
763	699	„ 5468	F. inflexibilis	Cooke, Grev. XIV. 20 (1885)	Inflexible fomes
764	697	„ 5465	F. lineato-scaber ...	Cooke, Grev. XV. 51 (1886)	Rough-lined fomes
765	700	„ 5470	F. linteus	Cooke, Grev. XIV. 20 (1885)	Linty fomes ...
766	673	„ 5305	F. lucidus	Cooke, Grev. XIII. 118 (1885)	Varnished fomes
767	722	„ 5528	F. luridus	Cooke, Grev. XIV. 21 (1885)	Lurid fomes ...
768	677	„ 5352	F. marginatus	Cooke, Grev. XIV. 17 (1885)	Margined fomes
769	668	„ 5272	F. nigripes	Cooke, Grev. XIII. 117 (1885)	Black-stalked fomes ...
770	685	„ 5401	F. nigro-laccatus ...	Cooke, Grev. XIV. 18 (1885)	Black-lacquered fomes
771	715	„ 5507	F. oblinitus	Berk., Grev. XV. 22 (1886)	Variegated fomes
772	721	„ 5527	F. obliquus	Cooke, Grev. XIV. 21 (1885)	Oblique fomes
773	684	„ 5400	F. orbiformis	Cooke, Grev. XIV. 18 (1885)	Orbicular fomes
774	712	„ 5520	F. Palliseri	Cooke, Grev. XIV. 21 (1885)	Palliser's fomes
775	701	„ 5469	F. pectinatus	Cooke, Grev. XIV. 20 (1885)	Comb-like fomes
776	707	„ 5484	F. ponderosus	Cooke, Grev. XIV. 20 (1885)	Weighty fomes
777	670	„ 5282	F. pullatus	Cooke, Grev. XIII. 117 (1885)	Mourning fomes
778	696	„ 5461	F. pullus	Cooke, Grev. XIV. 19 (1885)	Russet-brown fomes ...
779	688	„ 5415	F. rimosus	Cooke, Grev. XIV. 18 (1885)	Cracked fomes
780	702	„ 5473	F. rubiginosus	Cooke, Grev. XIV. 20 (1885)	Rusty fomes...
781	669	„ 5281	F. rudis ...	Cooke, Grev. XIII. 117 (1885)	Rough fomes
782	671	„ 5283	F. rugosus	Cooke, Grev. XIII. 117 (1885)	Wrinkled fomes
783	691	„ 5429	F. salicinus	Cooke, Grev. XIV. 19 (1885)	Willow fomes
784	676	„ 5342	F. scansilis	Cooke, Grev. XIII. 119 (1885)	Climbing fomes
785	714	„ 5505	F. scopulosus	Cooke, Grev. XIV. 21 (1885)	Craggy fomes
786	675	„ 5335	F. senex ...	Cooke, Grev. XIII. 118 (1885)	Old fomes ...
787	698	„ 5466	F. spadiceus	Cooke, Grev. XIV. 20 (1885)	Bright-brown fomes ...
788	706	„ 5480	F. strigatus	Cooke, Grev. XIV. 20 (1885)	Stiff-haired fomes
789	674	„ 5315	F. superpositus ...	Cooke, Grev. XIII. 118 (1885)	Superposed fomes
790	695	„ 5452	F. tasmanicus	Cooke, Grev. XIV. 19 (1885)	Tasmanian fomes

OF AUSTRALIAN FUNGI—*continued*.

Agaricus, Polyporus, Boletus, Trametes, Ganoderma—*continued*.

Number	W.A	S.A	T.	V.	N.S.W.	Q.	B.	Occurrence	General Characters.
753	...	...	...	...	N.S.W.	Q.	...	Wood ...	Hard like iron, corky. Caps of current year fawn coloured; velvety of previous year, banded with brown.
754	...	...	...	...	N.S.W.	...	B.	Stumps ...	Hoof shaped, cushion like, thick, sooty brown, becoming hoary, throwing off a snuff-like powder.
755	W.A.	S.A	...	V.	N.S.W.	Q	B.	Trunks ...	Woody to corky, very hard at first; hairy, brown, then hoary.
756	...	...	T.	...	...	...	...	Bark ...	Semicircular, convex, sparingly zoned, velvety like tow, umber.
757	W.A.	...	...	...	...	...	...	Trunks ...	Very hard, hemispherical, shell shaped, cinnamon; margin rather thin, bay brown.
758	...	...	...	...	...	Q.	...	Stems ...	Corky, thick, rigid, semicircular, delicately downy, white.
759	...	...	...	V.	...	...	...	Trunks ...	Hoof shaped, concentrically furrowed, purple brown to ashy, very delicately downy.
760	...	...	...	...	...	Q.	...	Trunks ...	Semicircular, sessile, leathery to corky, rigid, thin, reddish brown.
761	W.A.	S.A.	T.	V.	N.S.W.	Q.	B.	Trunks; very common on *Casuarinas*	Hoof shaped. with thin hoary cuticle, rusty brown, becoming blackish, substance zoned and very hard. Common.
762	...	...	...	...	...	Q	...	Trunks ...	Hard, kidney shaped, at first thin, without zones, at length thickened, repeatedly zoned, coffee colour.
763	...	...	...	...	...	Q	...	Trunks ...	Hoof shaped, brown, crested, furrowed, quite hard.
764	...	...	...	...	...	Q.	...	Trunks ...	Semicircular, rigid, brown; margin frequently zoned, lineately radiate, rough.
765	...	...	...	...	...	Q.	...	Bark ...	Hard, heavy, semicircular, furrowed, radiately cracked, brown, but rendered pale by lint-white down.
766	...	...	T.	...	...	Q.	B.	Base of stumps	Cap corky to woody, fan shaped, rough, shining as if lacquered, as well as lateral stem. yellowish red or brown. Very common.
767	...	...	...	...	N.S.W.	Q.	...	Branches	Spread out, closely adherent to the wood; pores white, then livid or lurid.
768	...	...	...	V.	N.S.W.	...	...	Trunks ...	Corky to woody, hoof shaped. somewhat flattened, incrusted, concentrically furrowed, covered with greyish bloom.
769	...	...	...	...	N.S.W.	...	...	Trunks ...	Corky to woody. Cap convex, zoneless, opaque, rusty brown. Stem rooting, shining as if lacquered, black.
770	...	...	...	...	...	Q.	...	Wood ...	Fan shaped, corky or woody, rough; margin wavy, chestnut brown to black, shining as if lacquered.
771	...	...	...	...	N.S.W.	...	...	Trunks ...	Corky to woody, convex to flattened, kidney shaped, variegated with faint concentric zones, red brown.
772	...	S.A.	...	V.	N.S.W.	Q.	...	Trunks, *Eucalyptus*, &c.	A magnificent fungus. Thick, casting off the bark, pale to bay brown, then blackish.
773	...	...	...	V.	...	...	...	Trunks ...	Very hard, convex, circular, concentrically furrowed, crustaceous.
774	...	...	...	V.	N.S.W.	Q.	...	Trunks ...	Fleshy to tough and leathery, shell shaped, slightly silky, white behind, ashy brown in front.
775	...	...	...	...	...	Q.	B.	Trunks ...	Corky to woody. hard, concentrically lamellately folded, scurfy, rusty brown.
776	...	...	...	...	...	Q.	...	Trunks ...	Remarkable for hardness and weight. Woody, semicircular, sessile, imbricate, shell shaped, lurid grey, at base brownish.
777	...	...	...	V.	...	Q.	...	Ground ...	Circular, rough, with furrowed zones, at first glaucous, delicately velvety, at length brown.
778	...	...	...	...	...	Q.	...	Branches of *Jasminum racemosum*	Small, somewhat imbricate, laterally confluent, hard, semicircular, shell shaped, bay brown.
779	W.A.	S.A.	T.	...	N.S.W.	...	...	Gum-tree trunks ...	Woody, very hard, hoof shaped, at length cracked, deeply furrowed, dark umber, and nearly black when old.
780	...	...	T.	...	...	...	...	Rotten wood	Horizontal, solid, sessile, thin, zoned, rough, minutely velvety when young, rusty, when old tinged with brown.
781	...	...	T.	...	...	...	...	Rotten wood ...	Cap circular, brown, rough, covered with a bloom or fine velvet. Stem nearly central, rooting in wood, brown, covered with tawny bloom, shiny.
782	...	...	...	V.	N.S.W.	Q.	...	Ground ...	Leathery, rigid, concentrically furrowed, bay brown, turning black.
783	...	...	...	...	...	Q.	B.	Trunks of Willows, &c.	Woody, quite hard, wavy, smooth, cinnamon brown or rusty; scent of aniseed.
784	...	...	...	...	...	Q.	...	Trunks ...	Cushion shaped, brown, repeatedly deeply furrowed and ribbed, coffee colour
785	...	...	...	...	...	Q.	...	Wood ...	Woody, hard, fan shaped, fixed by the vertex, whitish, zoned, rough.
786	...	...	...	V.	N.S.W.	Q.	...	Trunks ...	Large, nearly plane, corky, chestnut brown.
787	...	...	...	...	...	Q.	...	Trunks ...	Hard, leathery or corky, thin, bright brown, minutely velvety, closely zoned.
788	...	...	...	...	...	Q.	...	Trunks ...	Rigid, thin, semicircular, brown, zoned, with small scattered stiff hairs.
789	...	...	...	...	N.S.W.	...	...	Trunks ...	Cap shell shaped, imbricate, arising from a common lateral cylindrical stem, pale, covered with bloom.
790	...	...	T.	...	...	...	...	Rotten wood ...	Narrow, furrowed, pale brown, downy.

Number	Cooke's Number	Saccardo's Number	Scientific Name.	Authority for Name.	English Name.

62. POLYSTICTUS.—Fries, Nov. Symb. 54 (1851).—

Number	Cooke's Number	Saccardo's Number	Scientific Name.	Authority for Name.	English Name.
791	764	VI. 5702	P. acutus	Cooke, Grev. XIV. 82 (1886) ...	Acute polystictus ...
792	735	„ 5572	P. adami	Cooke, Grev. XIV. 78 (1886) ...	Adam's Peak polystictus ...
793	740	„ 5584	P. affinis	Cooke, Grev. XIV. 78 (1886) ...	Allied polystictus ...
794	799	„ 5869	P. aratus	Cooke, Grev. XIV. 86 (1886)	Furrowed polystictus...
795	751	„ 5125	P. Beckleri	Cooke, Handb. Austr. Fungi 142 (1892)...	Beckler's polystictus ...
796	759	„ 5683	P. biformis	Cooke, Grev. XIV. 81 (1886) ...	Two-shaped polystictus
797	808	„ 5921	P. bireflexus	Cooke, Grev. XIV. 87 (1886) ...	Bireflexed polystictus
798	807	„ 5917	P. Braunii	Cooke, Grev. XIV. 87 (1886) ...	Braun's polystictus ...
799	806	„ 5909	P. breviporus ...	Cooke, Grev. XIV. 87 (1886) ...	Short-pored polystictus
800	810	„ 5931	P. Broomei ...	Cooke, Grev. XIV. 87 (1886) ...	Broome's polystictus ...
801	785	„ 5785	P. brunneo-albus	Cooke, Grev. XIV. 83 (1886)	Brownish-white polystictus ...
802	743	„ 5616	P. brunneolus	Cooke, Grev. XIV. 79 (1886) ...	Brown polystictus ... ...
803	729	„ 5546	P. bulbipes	Cooke, Grev. XIV. 77 (1886) ...	Bulbous-stalked polystictus ...
804	796	„ 5846	P. byrsinus	Cooke, Grev. XIV. 85 (1886)	Leathery polystictus ...
805	805	„ 5887	P. caperatus	Cooke, Grev. XIV. 86 (1886)	Wrinkled polystictus ...
806	738	„ 5581	P. carneo-niger ...	Cooke, Grev. XIV. 78 (1886)	Fleshy black polystictus ...
807	803	„ 5878	P. cichoraceus	Cooke, Grev. XIV. 86 (1886)	Chicory-coloured polystictus ...
808	770	„ 5711	P. cinnabarinus ...	Cooke, Grev. XIV. 82 (1886) ...	Vermilion polystictus... ...
809	727	„ 5542	P. cinnamomeus ...	Sacc., Syll. VI. 210 (1888) ...	Cinnamon polystictus... ...
810	798	„ 5866	P. citreus ...	Cooke, Grev. XIV. 85 (1886) ...	Lemon-yellow polystictus ...
811	763	„ 5700	P. cristatus	Cooke, Grev. XIV. 81 (1886) ...	Crested polystictus ...
812	792	„ 5836	P. cupreo-roseus ...	Cooke, Grev. XIV. 85 (1886)	Copper-rose polystictus ...
813	749	„ 5782	P. dispar	Cooke, Handb. Austr. Fungi 142 (1892)...	Unequal polystictus ...
814	750	„ 5640	P. elongatus	Cooke, Grev. XIV. 80 (1886) ...	Elongated polystictus ...
814A	750	„ 5640	P. elongatus, var. Hodgkinsoniae	Kalch., Grev. X. 96 (1882) ...	Hodgkinson's polystictus ...
815	809	„ 5922	P. eriophorus	Cooke, Grev. XIV. 87 (1886) ...	Cottony polystictus ... ...
816	769	„ 5709	P. Eucalypti	Cooke, Grev. XIV. 82 (1886) ...	Eucalyptus polystictus ...
817	765	„ 5703	P. extensus	Cooke, Grev. XIV. 82 (1886) ...	Extended polystictus ...
818	768	„ 5706	P. Feei	Cooke, Grev. XIV. 82 (1886) ...	Fee's polystictus ...
819	733	„ 5569	P. flabelliformis ...	Cooke, Grev. XIV. 78 (1886) ...	Fan-shaped polystictus ...
820	773	„ 5733	P. floridanus	Cooke, Grev. XIV. 82 (1886) ...	Florida polystictus ...
821	754	„ 5646	P. Friesii ...	Cooke, Grev. XIV. 80 (1886) ...	Fries' polystictus ...
822	758	„ 5665	P. funalis ...	Cooke, Grev. XIV. 80 (1886)	Cord-like polystictus ...
823	756	„ 5656	P. gallo-pavonis	Cooke, Grev. XIV. 80 (1886)	Peacock polystictus ...
824	779	„ 5770	P. gausapatus ...	Cooke, Grev. XIV. 83 (1886)	Friezed polystictus ... ...
825	780	„ 5773	P. glirinus	Cooke, Grev. XIV. 83 (1886)	Dormouse polystictus ...

OF AUSTRALIAN FUNGI—continued.

Trametes, Polyporus, Boletus.

Number	Habitat						B.	Occurrence.	General Characters.
	W.A.	S.A	T.	V.	N.S.W.	Q.			
791					N.S.W.	Q.		Branches	Spread out, reflexed, thin, pale ochrey, flexible, margin very acute.
792					N.S.W.	Q.		Rotten wood	Cap lateral, spoon shaped, thin, leathery, zoned, bay brown. Stem long, mealy, yellow.
793					N.S.W.	Q.		Branches	Cap papery, fan shaped, rigid, chestnut brown with darker zones. Stem lateral, thin, chestnut brown.
794					N.S.W.	Q.		Trunks ...	Yellowish olive, semicircular, flexible, concentrically furrowed, shining, smooth.
795					N.S.W.			Wood	Thin, semicircular, whitish or ochrey, roughened with very minute elevations.
796				V.				Trunks ...	Overlapping, cap reflexed, tow-like, leathery, soft, downy, white, not zoned.
797						Q.		Trunks ...	Spread out, attached; margin reflexed, incurved, pale, faintly linearly zoned, downy, then smooth.
798				V.				Trunks ...	Sessile, semicircular and overlapping in tufts or circular; soft when moist, hard when dry, chestnut brown.
799						Q.		Trunks ...	Rust coloured, membranous, rigid, radiately rough, shell shaped; margin crisped.
800						Q.		Old wood	Caps very numerous, minute, finally sessile, closely overlapping like scales, whitish yellow, brown when dry.
801			T.					Trunks ...	Overlapping, semicircular, somewhat zoned, thin, leathery, rusty to blackish brown, velvety.
802						Q.		Trunks ...	Corky to leathery, thin, kidney or fan shaped, pale fawn colour, thickly zoned, silky, and shining.
803	W.A.		T.	V.				Ground ...	Cinnamon. Cap leathery to membranous, fibrously radiate. Stem cylindrical, velvety, bulbous.
804						Q.		Bark	Broad, of one colour, leathery, soft, thin, elastic, reflexed, semicircular, downy, concentrically furrowed.
805					N.S.W.	Q.		Wood	Solid, woody to leathery, thin, zoned, variegated with brown, velvety at first, then naked, often shining.
806						Q.		Wood	Cap kidney shaped, thin, black, radiately rough, obscurely zoned. Stem of same colour, velvety, brown downwards.
807				V.		Q.		Trunks ...	Overlapping, thin, leathery, rigid, circular or kidney shaped, lobed, plaited, zoned, brown, silky.
808	W.A.	S.A.	T.	V.	N.S.W.	Q.		Trunks, &c.	Corky, somewhat zoned, rough, downy, becoming smooth, vermilion.
809		S.A.					B.	Woods, amongst moss	Bright cinnamon without and within. Cap zoned. Stem velvety.
810								Rotten wood	Narrow, spoon shaped, or split and fan shaped, lemon yellow, zoned a little.
811						Q.		Trunks ...	Widely extended with margins broadly reflexed, thin, bright ochre, flexible, fringed with nearly erect hairs.
812								Wood	Thin, leathery, copper coloured, silky, shining, radiately rough, thickly zoned.
813				V.				Trunks ...	Overlapping, confluent, base wedge shaped, sessile, fan shaped, lobed, slightly zoned, tan coloured.
814				V.		Q.		Dead leaves	Wedge shaped, rounded in front and lobed, tapering behind, thin, leathery, downy, pale ochre.
814A					N.S.W.	Q.		Wood ...	Cap rigid, spoon shaped, densely streaked, silky at first, then minutely warted, stem short, disc shaped.
815					N.S.W.	Q.		Branches	White, attached, cottony; margin slightly reflexed.
816		S.A.		V.				Trunks of *Eucalyptus*	Fleshy to corky (deformed), velvety, soft to the touch, zoneless, opaque, varying in colour from umber or bay to violet.
817						Q.		Trunks ...	Leathery, spread out, reflexed, radiately hispid, slightly concentrically zoned, thin, ochrey olive.
818	W.A.		T.	V.	N.S.W.	Q.		Trunks ...	Corky to leathery, with long hairs, zoned, brown, becoming hoary.
819				V.	N.S.W.	Q.		Wood	Cap leathery to membranous, zoned, covered with dingy evanescent down, somewhat bay brown. Stem very short, lateral.
820						Q.		Trunks ...	Pale bay, somewhat fan shaped, laterally growing together, thin, leathery, zoned, downy.
821			T.					Trunks ...	Leathery, thin, fan shaped, expanded from narrow base, densely concentrically streaked, silky, yellowish tan.
822						Q.		Trunks ...	Fibrously spongy, sessile, shell shaped, rusty, entirely resolved into rigid, much branched, cord-like fibres.
823						Q.		Trunks ...	Leathery, thin, rigid, flattened, shell shaped, slightly downy, greyish fawn colour, linearly zoned.
824					N.S.W.			Trunks ...	Fleshy to leathery, rigid, semicircular, sessile, shell shaped, zoned, hispid and rough to the touch, fawn to brownish.
825						Q.		Trunks ...	Semicircular or somewhat reniform, shell shaped, delicately downy, zoned, mouse or olive coloured.

Number.	Cooke's Number.	Saccardo's Number.	Scientific Name.	Authority for Name.	English Name.
				62. POLYSTICTUS.—Fries, Nov. Symb. 54 (1851).—	
826	801	VI. 5875	P. Hasskarlii	Cooke, Grev. XIV. 86 (1886)	Hasskarl's polystictus ...
827	777	„ 5760	P. hirsutus	Cooke, Grev. XIV. 83 (1886)	Hirsute polystictus ...
828	760	„ 5689	P. hololeucus ...	Cooke, Grev XIV. 81 (1886)	Entirely white polystictus ...
829	784	„ 5784	P. hypothejus ...	Cooke, Grev. XIV. 83 (1886)	Under-yellow polystictus ...
830	775	„ 5755	P. illotus ...	Cooke, Grev. XIV. 83 (1886) ...	Dirty polystictus ...
831	742	„ 5591	P. intonsus	Cooke, Grev. XIV. 79 (1886) ...	Unshorn polystictus ...
832	748	„ 5639	P. laceratus	Cooke, Grev. XIV. 80 (1886) ...	Torn polystictus ...
833	811	„ 5933	P. latus ...	Cooke, Grev. XIV. 87 (1886) ...	Broad polystictus ...
834	757	„ 5663	P. leonotis	Cooke, Grev. XIV. 80 (1886)	Leonine polystictus ...
835	745	„ 5630	P. libum	Cooke, Grev. XIV. 79 (1886) ...	Cake polystictus ...
836	767	„ 5708	P. lilacino-gilvus ...	Cooke, Grev. XIV. 82 (1886)	Lilac-yellow polystictus ...
837	781	„ 5777	P. limbatus	Cooke, Grev. XIV. 83 (1886)	Bordered polystictus ...
838	725	„ 5538	P. luteo-nitidus ...	Cooke, Grev. XIV. 77 (1886) ...	Bright-yellow polystictus ...
839	800	„ 5870	P. luteo-olivaceus ...	Cooke, Grev. XIV. 86 (1886)	Olive-yellow polystictus ...
840	737	„ 5577	P. luteus ...	Cooke, Grev. XIV. 78 (1886) ...	Yellow polystictus ...
841	753	„ 5644	P. multilobus ...	Cooke, Grev. XIV. 80 (1886)	Many-lobed polystictus ...
842	736	„ 5574	P. mutabilis	Cooke, Grev. XIV. 78 (1886)	Changeable polystictus ...
843	739	„ 5582	P. nephridius ...	Cooke, Grev. XIV. 78 (1886)	Kidney-shaped polystictus ...
844	728	„ 5545	P. oblectans ...	Cooke, Grev. XIV. 77 (1886)	Alluring polystictus ...
845	776	„ 5758	P. obstinatus ...	Cooke, Grev. XIV. 83 (1986)	Hard polystictus ...
846	794	„ 5843	P. occidentalis ...	Cooke, Grev. XIV. 85 (1886)	Western polystictus ...
847	752	„ 5642	P. ornithorhyuchj ...	Cooke, Grev. XIV. 80 (1886)	Ornithorhynchus polystictus ...
848	730	„ 5548	P. parvulus ...	Cooke, Grev. XIV. 77 (1886)	Very small polystictus
849	789	„ 5820	P. peradeniæ	Cooke, Grev. XIV. 84 (1886) ...	Peradenia polystictus
850	726	„ 5543	P. perennis ...	Cooke, Grev. XIV. 77 (1886)	Perennial polystictus ...
851	—	„ 5695	P. pergameus ...	Cooke, Grev. XIV. 81 (1886) ...	Parchment polystictus ...
852	744	„ 5623	P. peroxydatus ...	Cooke, Grev. XIV. 79 (1886) ...	Peroxide polystictus ...
853	791	„ 5832	P. Persoonii ...	Cooke, Grev. XIV. 85 (1886) ...	Persoon's polystictus ...
854	783	„ 5781	P. pinsitus ...	Cooke, Grev. XIV. 83 (1886) ...	Crushed polystictus ...
855	734	„ 5571	P. porphyrites ...	Cooke, Grev. XIV. 78 (1886)	Purple-coloured polystictus ...
856	761	„ 5693	P. proteiformis ...	Cooke, Grev. XIV. 81 (1886) ...	Proteus-like polystictus ...
857	731	IX. 730	P. quadrans ...	Cooke, Grev. XIV. 78 (1886) ...	Quadrate polystictus ...
858	771	VI. 5717	P. radiatus ...	Cooke, Grev. XIV. 82 (1886)	Radiate polystictus ...
859	755	„ 5648	P. radiato-rugosus ...	Cooke, Grev. XIV. 80 (1886)	Radiately-rough polystictus
860	747	„ 5634	P. rasipes ...	Cooke, Grev. XIV. 79 (1886)	Rough-stalked polystictus
861	795	„ 5841	P. rigens	Sacc. and Cub., Syll. Fung. VI. (1888) ...	Stiff polystictus ...

OF AUSTRALIAN FUNGI—continued.

Tramates, Polyporus, Boletus—continued.

Number	W.A.	S.A.	T.	V.	N.S.W.	Q.	D.	Occurrence	General Characters
826	...	...	...	...	...	Q.		Trunks ...	Corky, overlapping, zoned, velvety, bay brown.
827	...	...	...	V.	N.S.W.	Q.	B.	Trunks ...	Corky to leathery, hirsute, with rigid hairs, zoned with concentric furrows, whitish to tawny. Common.
828	...	...	...	V.	...	...		Trunks ...	Wholly white, loosely corky, semicircular, sessile, concentrically furrowed, flesh tinder-like.
829	...	...	...	...	N.S.W.	...		Trunks ...	Thin, leathery, shell shaped, narrow at base, with silky hairs, white, zoned. Pores becoming bright yellow.
830	...	...	...	V.	...	...		Trunks ...	Leathery to membranous, softly velvety, concentrically furrowed, dingy grey, turning brownish.
831	...	...	T.	...	...	...		Rotten wood	Small. Cap fan shaped, thin, velvety, brown. Stem excentric, short.
832	...	...	...	...	N.S.W.	Q.		Branches	Spread out and reflexed, thin, zoned, rough, streaked, wood colour.
833	...	...	T.	...	...	...		Branches	Inverted, obscurely zoned, corky to leathery, dingy brown, with velvety bloom.
834	...	...	...	...	...	...		Trunks ...	Spongy to fleshy, overlapping and grown together, rough haired, zoneless, dark rusty brown.
835	...	...	...	...	N.S.W.	Q.		Wood ...	Leathery, polished, fixed behind by a disc, lobed, somewhat zoned, white; margin yellowish.
836	W.A.	S.A.	T.	V.	...	Q.		Rotten wood	Somewhat overlapping, corky to leathery, delicate, rough, more or less zoned, brown when old.
837	...	...	...	V.	...	...		Trunks ...	Leathery, thin, somewhat velvety, concentrically furrow-zoned, becoming rusty.
838	...	...	...	...	...	Q.		Ground ...	Cap rough, yellow, silky, shining, irregularly lobed, thickly zoned, rather velvety at first. Stem deformed, with spongy coating.
839	...	...	...	...	...	Q.		Trunks ...	Woody, rigid, sessile, thin, downy, deeply and concentrically zoned, warted; substance olive yellow.
840	...	...	...	...	N.S.W.	Q.		Wood ...	Cap thin, rigid, leathery, fan shaped, yellowish. Stem marginal, dilated at base, yellowish.
841	...	...	...	...	N.S.W.	...		Trunks ...	Leathery, thin, rigid, somewhat kidney shaped, narrowed into lateral and very short stem, white to tan; margin lobed.
842	...	...	...	...	...	Q.		Wood ...	Cap leathery, rigid, fan or spoon shaped, zoned, yellowish white, narrowed down into stem, which is lateral and variable in length.
843	...	...	...	...	...	Q.		Branches	Small, thin. Cap veined, smooth, kidney shaped, bay brown. Stem very short, black.
844	W.A.	S.A.	T.	V.	...	Q.		Sandy soil	Cap thin, leathery, cut, zoned about centre, shining, bright cinnamon. Stem velvety, reddish brown.
845	...	...	...	...	...	Q.		Trunks ...	Leathery to woody, hardening, thin, somewhat ash coloured, velvety, variegated with narrow zones.
846	...	S.A.	...	V.	N.S.W.	Q.		Trunks ...	Corky to leathery, spread out and reflexed, concentrically furrowed, yellowish tan, becoming pale.
847	...	...	...	...	N.S.W.	...		Trunks ...	Somewhat tufted, thin, leathery, wedge shaped, zoneless, hairy to downy, rusty umber, tapering into short or obsolete stem of same colour.
848	...	S.A.	...	V.	...	...		Ground ...	Cap leathery to membranous, obsoletely silky, zoned, bay brown. Stem thin, tuberous, velvety.
849	...	...	...	...	...	Q.		Wood ...	Semicircular, laterally running together, somewhat zoned, silky, membranous, olive.
850	...	...	...	...	...	Q.	B.	Ground ...	Cap leathery, funnel shaped, velvety, zoned, cinnamon to bay brown. Stem firm, thickened downwards, velvety.
851	...	...	...	...	...	Q.		Trunks ...	Leathery to membranous, rigid, downy, furrowed concentrically, white.
852	...	...	...	...	N.S.W.	...		Trunks ...	Thin, nearly circular, rusty, powdery. Cap slightly zoned, rough. Stem short, thick.
853	...	...	...	V.	N.S.W.	Q.		Wood ...	Leathery, flattened, obsoletely zoned, dark blood red, becoming pale.
854	...	...	...	...	...	Q.		Wood ...	Leathery to membranous, tough, hairy, concentrically furrowed, ash coloured.
855	...	...	...	...	...	Q.		Rotting branches ...	Thin, leathery. Cap fan shaped, with ochrey zones, shining brown, becoming purplish. Stem short, of same colour.
856	...	...	...	V.	...	...		Trunks ...	Spread out and reflexed, tow-like, leathery, white within, slightly concentrically furrowed. Very variable at different stages.
857	...	...	...	...	...	Q.		Wood ...	Cap rigid, smooth, thin, furrowed, zoned, dark ochre. Stem short, excentric, of same colour.
858	...	...	...	V.	...	Q.	B.	Trunks ...	Corky, leathery, rigid, radiately rough, velvety at first, brown, then rusty brown.
859	...	...	T.	...	...	...		Trunks ...	Thickly overlapping, thin, radiately rough, dingy white or grey.
860	...	...	...	...	...	Q.		Trunks ...	Cap fan shaped, silky lineate, somewhat velvety, zoned, reddish brown when dry. Stem short, flattened, rather hispid.
861	...	...	...	...	N.S.W.	Q.		Trunks ...	Spread out, shortly reflexed, often run together, leathery, rigid, more or less concentrically furrowed, velvety, pale tawny to wood colour.

E

Number.	Cooke's Number.	Saccardo's Number.	Scientific Name.	Authority for Name.	English Name.
			62. POLYSTICTUS.—Fries, Nov. Symb. 54 (1851).—		
862	...	VI. 5831	P. rigidus	Cooke, Grev. XIV. 85 (1886)	Rigid polystictus
863	746	„ 5631	P. sanguineus	Cooke. Grev. XIV. 79 (1886)	Blood-red polystictus
864	788	„ 5811	P. scortens	Cooke, Grev. XIV. 84 (1886)	Leathery polystictus
865	762	„ 5696	P. seriatus	Cooke, Grev. XIV. 81 (1886)	Seriate polystictus
866	741	„ 5585	P. stereinus	Berk. and Curt., Linn. Journ. X. 308 (1869)	Solid polystictus
867	787	„ 5808	P. stereoides	Cooke, Grev. XIV. 78 (1886)	Stereum-like polystictus
868	802	„ 5876	P. tabacinus	Cooke, Grev. XIV. 86 (1886)	Tobacco-coloured polystictus
869	797	„ 5847	P. tephroleucus	Cooke, Grev. XIV. 85 (1886)	Pale-ashy polystictus
870	724	„ 5535	P. tomentosus	Cooke, Grev. XIV. 77 (1886)	Downy polystictus
871	786	„ 5787	P. trizonatus	Cooke, Grev. XIV. 84 (1886)	Three-zoned polystictus
872	782	„ 5779	P. vellerens	Cooke, Grev. XIV. 83 (1886)	Woolly polystictus
873	778	„ 5763	P. velutinus	Cooke, Grev. XIV. 83 (1886)	Velvety polystictus
874	772	„ 5732	P. venustus	Cooke. Grev. XIV. 82 (1886)	Graceful polystictus
875	790	„ 5827	P. vernicifluus	Cooke, Grev. XIV. 84 (1886)	Varnished polystictus
876	766	„ 5704	P. versatilis	Cooke, Grev. XIV. 82 (1886)	Changing polystictus
877	774	„ 5741	P. versicolor	Cooke, Grev. XIV. 83 (1886)	Variously-coloured polystictus
878	793	„ 5838	P. vinosus	Cooke. Grev. XV. 51 (1886)	Vinous polystictus
879	782	„ 5565	P. xanthopus	Cooke, Grev. XIV. 79 (1886)	Yellow-stalked polystictus
880	804	„ 5885	P. xerampelinus	Cooke, Grev. XIV. 86 (1886)	Purplish-umber polystictus
881	...	„ 5771	P. zonatus	Cooke, Grev. XIV. 83 (1886)	Zoned polystictus
			63. PORIA.—Pers. Syn. 512 (1801).—		
882	851	VI. 6062	P. aprica	Cooke, Grev. XIV. 112 (1886)	Exposed poria
883	840	„ 6153	P. Archeri	Cooke, Grev. XIV. 115 (1886)	Archer's poria
884	826	„ 5995	P. atro-vinosa	Cooke, Grev. XIV. 110 (1886)	Dark-vinous poria
885	817	„ 5948	P. calcea	Cooke, Grev. XIV. 109 (1886)	Chalky-white poria
886	820	„ 5961	P. callosa	Cooke, Grev. XIV. 110 (1886)	Thick-skinned poria
887	836	„ 6126	P. contigua	Cooke. Grev. XIV. 114 (1886)	Contiguous poria
888	833	„ 6083	P. corticola	Cooke, Grev. XIV. 113 (1886)	Bark-growing poria
889	838	„ 6131	P. dictyopora	Cooke, Grev. XII. 17 (1884)	Net-pored poria
890	823	„ 5982	P. epilintea	Cooke, Grev. XIV. 110 (1886)	Fibrous poria
891	839	„ 6138	P. fatiscens	Berk. and Rav., Grev. I. 65 (1872)	Cracking poria
892	835	„ 6123	P. ferruginosa	Cooke, Grev. XIV. 114 (1886)	Rusty poria
893	814	„ 5939	P. hyalina	Cooke, Grev. XIV. 109 (1886)	Hyaline poria
894	824	„ 5984	P. hyposclera	Cooke, Grev. XIV. 110 (1886)	Hard poria
895	828	„ 6004	P. livida	Cooke, Grev. X. 131 (1882)	Livid poria
896	816	„ 5947	P. medulla-panis	Cooke, Grev. XIV. 109 (1886)	Pith-of-bread poria
897	830	„ 6057	P. membranicincta	Berk. Grev. XV. 26 (1886)	Membrane-girt poria
898	832	„ 6068	P. merulina	Cooke, Grev. XIV. 112 (1886)	Merulius poria
899	813	„ 5936	P. mollusca	Cooke, Grev. XIV. 109 (1886)	Soft poria
900	818	„ 5949	P. nipholdes	Cooke, Grev. XIV. 109 (1886)	Snow-like poria
901	847	„ 6130	P. orbicularis	Cooke, Grev. XIV. 114 (1886)	Circular poria
902	821	„ 5969	P. parilis	Cooke, Grev. XIV. 110 (1886)	Equal poria
903	825	„ 5994	P. rufa	Cooke, Grev. XIV. 110 (1886)	Red poria

OF AUSTRALIAN FUNGI—continued.

Number.	W.A.	S.A.	T.	V.	N.S.W.	Q.	B.	Occurrence	General Characters.

Trametes, Polyporus, Boletus—continued.

Number.	W.A.	S.A.	T.	V.	N.S.W.	Q.	B.	Occurrence	General Characters.
862	...	...	...	...	...	Q.	...	Trunks ...	Leathery to corky, wedge shaped, light, lurid, brown zoned.
863	W.A.	S.A.	T.	V.	N.S.W.	Q.	...	Trunks ...	Vermilion. Cap leathery, thin, kidney shaped, shining, obsoletely concentrically furrowed. Stem lateral, short, circularly dilated at base.
864	...	...	...	...	N.S.W.	...	...	Trunks ...	Leathery, flattened on both sides, concentrically furrowed and zoned, shaggy haired, verdigris to grey.
865	...	...	...	V.	...	...	...	Trunks ...	Tow-like, semicircular, rigid, mostly run together in serial order, concentrically furrowed, velvety, brick red or yellowish tan.
866	...	...	...	...	N.S.W.	Q.	...	Trunks ...	Fan shaped, rigid, spread out and inflexed. Cap thin, many zoned, chestnut red. Stem disc shaped.
867	...	...	...	V.	N.S.W.	Q.	...	Trunks ...	Leathery, thin, rigid, spread out and reflexed, downy, then smooth, grey brown.
868	...	...	...	...	N.S.W.	...	...	Trunks ...	Overlapping, rusty to bay, leathery, thin, rigid, spread out and reflexed, shell shaped, downy, concentrically zoned.
869	...	...	...	V.	...	...	...	Wood ...	Semicircular, rigid, leathery, white, zoned, velvety, with rough hairs.
870	...	...	...	V.	...	Q.	...	Ground ...	Corky, hard, deformed, zoneless, persistently downy, as well as the unequal stem, brown.
871	...	...	...	V.		Q.	...	Trunks ...	Leathery, thin, rigid, ochrey, flattened, silky, variegated with three elevated darker zones.
872	...	...	...	...		Q.	...	Trunks ...	Semicircular, leathery, soft, thin, white, densely silky haired, zones darker, very narrow.
873	...	...	...	V.		Q.	B.	Trunks ...	Corky to leathery, velvety, soft, slightly zoned, white, at length yellowish.
874	W.A.	...	...	...		Q.		Trunks ...	Overlapping, forming elongated tufts, reflexed, leathery, zoned, whitish, downy, tufts in front.
875			T.	...		Q.	...	Rotten wood	Semicircular, somewhat fan shaped, thin, varnished and polished, reddish brown.
876	...	...	...	...	N.S.W.	Q.	...	Rotten wood	Very long, spread out, leathery to membranous; margin broadly reflexed.
877	...	...	T.	V.	N.S.W.	Q.	B.	Trunks ...	Leathery, thin, rigid, flattened, velvety, shining, variegated with other coloured zones.
878	...	...	...	...	...	Q.	...	Rotten wood	Kidney shaped, thin, zoned, delicately velvety, turning smooth, dark wine colour.
879	...	...	...	V.	N.S.W.	Q.	...	Branches	Cap leathery, papery, funnel shaped, zoned, shining, bay brown. Stem short, shining, yellowish.
880	...	...	...	...		Q.	...	Trunks ...	Corky to leathery, densely overlapping, shell shaped, shaggy, becoming smooth, with concentric furrowed zones, purplish umber.
881	...	...	...	...		Q.	...	Trunks ...	Corky to leathery, convex, tubercled and bulging behind, somewhat zoned, shaggy.

Polyporus, Boletus, Corticium.

Number.	W.A.	S.A.	T.	V.	N.S.W.	Q.	B.	Occurrence	General Characters.
882	...	...	T.	...	...	...	...	Wood ...	Inverted, spread out, loosely adhering, pale fawn.
883	...	...	T.	...	...	...	...	Rotten wood	Inverted, spread out, leathery to membranous; margin downy.
884	...	...	...	V.	N.S.W.	...	...	Trunks ...	Spread out, undefined, dark vinous purple; margin powdery to downy.
885	...	...	...	...		Q.	...	Wood ...	Chalky white. Inverted, spread out; margin very thin and membranous.
886	...	...	...	...		Q.	B.	Wood ...	Widely expanded, tough, entire, separable like soft leather, white.
887	...	S.A.	...	V.	N.S.W.	Q.	B.	Old wood	Spread out, thick, firm, cinnamon when young, margin at first shaggy.
888	...	S.A.	...	V.	...	Q.	...	Rotting bark	Widely expanded, firm, white.
889	...	...	...	V.	...	Q.	...	Burnt wood	Spread out, undefined, thin, white, incrusting.
890	...	...	...	...	N.S.W.	...	...	Trunks ...	Inverted, spread out; margin and substance fibrous.
891	...	...	...	...	...	Q.	...	Wood ...	Inverted, very thin, powdery, white.
892	W.A.	...	...	V.	N.S.W.	Q.	...	Wood, on posts	Spread out, thick, firm, unequal, tawny, then rusty bay.
893	...	...	T.	...	...	Q.	...	Wood ...	Inverted, white, hyaline, thin, margin downy.
894	...	...	...	V.	...	Q.	...	Trunks ...	Spread out, rather thick, margin thin, and pale ochrey or tinged with flesh colour.
895	...	...	...	...	N.S.W.	...	...	Bark ...	Spread out, defined, crustaceous, livid to sooty brown.
896	...	...	...	...	...	Q.	B.	Old wood	Spread out, defined, somewhat wavy, firm, smooth, white.
897	...	...	T.	...	...	...	...	Wood ...	Spread out, thin, pale ochrey, seated on a paler membranous layer.
898	...	...	T.	...	...	...	...	Wood ...	Inverted, expanded, orange.
899	...	S.A.	...	V.	...	...	B.	Dead bark of *Eucalyptus obliqua*	Expanded, thin, soft, white, circumference of finely filamentous texture.
900	...	...	...	...	N.S.W.	...	...	Wool ...	Inverted, snow white, margin very narrow.
901	...	...	T.	...	...	...	...	Living bark	Exactly circular, dark brown, margin membranous with rough down.
902	W.A.	...	...	...	...	...	...	Bark ...	Longitudinally expanded, closely attached, run together, dry, yellow, becoming pale.
903	...	...	...	V.	...	...	B.	Branches	Expanded, leathery, thin, attached, smooth, definite outline, blood red.

Number.	Cooke's Number.	Saccardo's Number.	Scientific Name.	Authority for Name.	English Name.
					63. PORIA.—Pers. Syn. 542 (1801).—
904	826 bis	Vl. 6029	P. rufitincta	Berk. and Curt., Grev. XV. 25 (1886)	Reddish poria
905	834	„ 6095	P. sinuosa	Cooke, Grev. XIV. 113 (1886)	Wavy poria
906	815	„ 5942	P. subvincta	Cooke, Grev. XIV. 109 (1886)	Somewhat-bound poria
907	819	„ 5952	P. tarda	Cooke, Grev. XIV. 109 (1886)	Slowly-developing poria
908	829	., 6035	P. vaporaria	Cooke, Grev. XIV. 111 (1886)	Sweating poria
909	827	„ 6003	P. victoriæ	Cooke, Grev. XIV. 111 (1886)	Victorian poria
910	822	., 5983	P. vincta	Cooke, Grev. XIV. 110 (1886)	Bound poria
911	812	„ 5935	P. vulgaris	Fries, S.M. I. 381 (1832)	Common poria
					64. TRAMETES.—Fries,
912	859	Vl. 6273	T. Curreyi	Cooke, Trans., Bot. Soc. Ed. 157 (1878)	Currey's trametes
913	848	„ 6197	T. devexa	Berk., Linn. Journ. XIII. 165 (1873)	Sloping trametes
914	854	„ 6240	T. epitephra	Berk., Linn. Journ. XIII. 165 (1873)	Ash-coloured trametes
915	852	., 6220	T. fibrosa	Fries, Epicr. 490 (1838)	Fibrous trametes
916	861	...	T. gausapata	Cooke, Grev. XV. 55 (1896)	Friezed trametes
917	846	,. 6188	T. heteromalla	Cooke, Grev. X. 132 (1882)	Shaggy trametes
918	853	„ 6235	T. hispidula	Berk. and Curt., Linn. Journ. X. 319 (1869)	Rough trametes
919	849	„ 6204	T. lactinea	Berk., Ann. Nat. Hist. X. 373 (1842)	Milk-white trametes
920	850	„ 6205	T. lævis	Berk., Hook., Lond. Journ. VI. 507 (1847)	Even trametes
921	858	„ 6264	T. mollis	Fries, Hym. Eur. 585 (1874)	Soft trametes
922	842	„ 6181	T. Muelleri	Berk., Linn. Journ. X. 320 (1869)	Mueller's trametes
923	856	„ 6247	T. ochroflava	Cooke, Grev. IX. 12 (1880)	Ochrey-yellow trametes
924	847	„ 5236	T. ochroleuca	Cooke, Grev. XIX. 99 (1891)	Ochrey-white trametes
925	841	„ 6170	T. phellina	Berk., Linn. Journ. XIII. 164 (1873)	Corky trametes
926	843	IX.p. 198	T. picta	Berk. and Br., Linn. Trans. II. 61 (1883)	Ornamented trametes
927	851	VI. 6213	T. Pini	Fries, S.M. I. 336 (1821)	Pine trametes
928	855	„ 6241	T. pyrrhocreas	Berk., Linn. Journ. XIII. 164 (1873)	Fleshy trametes
929	857	„ 6249	T. scrobiculata	Berk., Grev. VI. 70 (1877)	Pitted trametes
930	862	„ 6267	T. serpens	Fries, Hym. Eur. 586 (1874)	Spreading trametes
931	845	„ 6185	T. Sprucei	Berk., Hook., Journ. 236 (1856)	Spruce's trametes
932	860	., 5322	T. ungulata	Berk., Linn. Journ. XIII. 165 (1873)	Hoof-shaped trametes
933	844	,. 6183	T. versiformis	Berk., Linn. Journ. XIV. 56 (1875)	Variously-shaped trametes
					65. SCLERODEPSIS.—Cooke,
934	863	VI. 6237 (IX.p. 194)	S. colliculosa	Cooke, Grev. XIX. 49 (1890)	Hillocky sclerodepsis
					66. HEXAGONIA.—Fries,
935	883	VI. 627-	H. crinigera	Fries, Epicr. 496 (1838)	Hair-bearing hexagonia
936	894	„ 6333	H. decipiens	Berk., Linn. Journ. XIII. 166 (1873)	Deceptive hexagonia
937	890	„ 6320	H. discolor	Fries, Nov. Symb. 102 (1851)	Discoloured hexagonia
938	884	„ 6290	H. durissima	Berk. and Br., Linn. Journ. XIV. 57 (1875)	Very hard hexagonia

OF AUSTRALIAN FUNGI—continued.

Polyporus, Boletus, Corticium—continued.

Number	W.A.	S.A.	T.	V.	N.S.W.	Q.	B.	Occurrence	General Characters
904	...	...	...	...	...	...	...	Wood ...	Expanded, rather thick, firm, tawny, rusty bay when old, margin woolly; golden brown.
905	...	...	...	...	N.S.W.	...	...	Wood and bark	Broadly expanded, attached, dry, springing from temporary mycelium, white to yellowish. Pores wavy.
906	...	...	T.	...	...	...	...	Wood ...	Rather thick, widely expanded, white, separable.
907	W.A.	...	...	V.	...	...	...	Wood ...	White, then ochrey, mycelium waxy, like *Corticium*; margin narrow, downy. Pores slowly developed.
908	W.A.	...	T.	V.	...	Q.	B.	Creeping upon rotten wood, &c.	Expanded, mycelium creeping in the wood, woolly, brown or white. Very common.
909	...	...	...	V.	...	Q.	...	Trunks ...	Smoky colour, expanded, thin.
910	...	...	...	V.	...	Q.	...	Rotten wood	Inverted, thick in centre; margin thin, tinged above with red.
911	...	S.A.	...	V.	...	Q.	B.	Wood and branches	Broadly expanded, thin, closely attached and difficult to remove, dry, even, whitish. Very common.

Epicr. 488 (1838).—Polyporus, Dædalea, Boletus.

Number	W.A.	S.A.	T.	V.	N.S.W.	Q.	B.	Occurrence	General Characters
912	...	...	...	...	N.S.W.	Q.	...	Trunks ...	Expanded, reflexed, lobed, membranous to leathery, rusty umber.
913	...	...	...	...	N.S.W.	Q.	...	Trunks ...	Woolly, somewhat hoof-shaped, sloping behind, cap velvety, somewhat tawny.
914	...	S.A.	...	...	...	...	...	Trunks ...	Hoof-like, sloping behind, zoned, brown, somewhat rough; margin white.
915	...	S.A.	...	...	...	...	...	Trunks ...	Corky, thin, somewhat wavy, zoned, dark-brown, rough, with thickly grown branched fibres.
916	...	S.A.	...	...	...	Q.	...	Trunks ...	Spread out behind, reflexed, velvety, zoned, bright umber, leathery.
917	...	...	...	...	N.S.W.	...	...	Trunks ...	Corky, softish, shaggy, concentrically furrowed, whitish, then somewhat ashy.
918	...	S.A.	...	...	...	...	...	Wood ...	Small, hoof-shaped, rusty umber, rough behind, margin somewhat downy.
919	...	S.A.	...	...	N.S.W.	Q.	...	Wood	Sessile, irregular, thickish, hard, rigid, zoneless, corky, velvety, warty, milk white.
920	...	...	...	...	N.S.W.	Q.	...	Roots	Quite even, thick, hoof-shaped, pale-wood colour, delicately downy, somewhat zoned.
921	...	...	...	...	N.S.W.	...	B.	Branches	Inverted, distinct, somewhat membranous, pale wood-colour, finally brownish.
922	...	S.A	...	V.	N.S.W.	Q.	...	Wood	Semicircular, corky, delicately downy, white, rough, margin lobed, concentrically furrowed.
923	...	...	...	...	...	Q.	...	Trunks ...	Entirely ochrey-yellow, often overlapping, corky, compact, tuberculose, concentrically zoned.
924	W.A	...	T.	V.	N.S.W.	Q.	...	Bark ...	Hoof-shaped, corky, few-zoned, ochrey white, delicately downy at first; soon smooth and shining.
925	...	...	...	...	N.S.W.	...	...	Rotten wood	Corky, attached by circular disc, becoming whitish, rough, zoneless.
926	...	...	...	...	N.S.W.	Q.	...	Wood ...	Semicircular, corky, hard, smooth, pale, with darker concentric bands.
927	W.A.	...	...	...	...	...	B.	Trunks ...	Corky to woolly, very hard, concentrically furrowed, cracked, rusty brown, turning blackish, odour slight.
928	...	...	...	...	N.S.W.	Q.	...	Trunks ...	Thickish, zoned, umber brown behind, at length velvety; substance compact but soft, tawny.
929	...	...	...	V.	...	...	...	Trunks ...	Ochrey, semicircular, slightly furrowed, pitted and dotted, substance corky.
930	...	...	...	...	...	Q.	B.	Bark ...	Dry, at first breaking through, tubercular, circular, white; margin distinct, downy.
931	...	...	...	V.	N.S.W.	Q.	...	Wood ...	Thickish, humped, obtuse, becoming white.
932	...	S.A.	...	...	...	...	...	Trunks ...	Hard, whitish, hoof-shaped, sloping behind, delicately downy.
933	...	...	...	...	...	Q.	...	Wood ...	White, reflexed, lobed, radiately rough, opaque, varying extremely in thickness.

Grev. XIX. 49 (1890).—Trametes.

Number	W.A.	S.A.	T.	V.	N.S.W.	Q.	B.	Occurrence	General Characters
934	...	...	...	...	N.S.W.	Q.	...	Wood ...	Sessile, semicircular, thin, hard, leathery, somewhat silky, zoned, ochrey.

Epicr. 496 (1838).—Favolus, Polyporus.

Number	W.A.	S.A.	T.	V.	N.S.W.	Q.	B.	Occurrence	General Characters
935	...	...	...	...	...	Q.	...	Trunks ...	Corky to leathery, bristly, zoneless, brown, turning blackish.
936	W.A.	S.A.	...	V.	...	Q.	...	Trunks of *Casuarina*	Semicircular, zoned, umber, velvety to rough, furrowed. Species most distinct.
937	W.A	...	...	...	...	...	...	Bark ...	Expanded, reflexed, unpolished, pale umber.
938	...	S.A.	...	V.	...	...	...	Wood and trunks	Hoof-shaped, rough, zoned with red and brown; substance rather fleshy.

Number	Cooke's Number	Sacardo's Number	Scientific Name	Authority for Name.	English Name.
					66. HEXAGONIA.—Fries.
939	887	VI. 6310	H. Gunnii	... Berk., Hook., Lond. Journ. IV. 57 (1845)	... Gunn's hexagonia ... ...
940	885	„ 6300	H. Muelleri	... Berk., Linn. Journ. XIII. 166 (1873)	... Mueller's hexagonia ...
941	892	., 6325	H. polygramma ...	... Cooke, Grev. XV. 58 (1886) ...	... Many-lined hexagonia...
942	888	„ 6315	H. rigida	Berk., Linn. Journ. XVI. 54 (1878)	... Rigid hexagonia ...
943	855	„ 6306	H. sericea	Fries, Epicr. 497 (1838) ...	... Silky hexagonia
944	893	., 6328	H. similis	... Berk., Hook., Lond. Journ. V. 4 (1846) ...	... Similar hexagonia
945	891	„ 6324	H. tenuis ...	... Cooke, Grev. XV. 60 (1886) ...	... Thin hexagonia
945A	891	...	H. tenuis, var. sub-tenuis ...	Cooke, Grev. XIX. 103 (1891)	... Thinnish hexagonia
946	889	VI. 6319	H. umbrinella	Fries, Fung. Nat. 17 (1848)	... Umber hexagonia
947	892	„ 6274	H. Wightii	... Cooke, Grev. XV. 60 (1886)	Wight's hexagonia
					67. DÆDALEA.—Pers.
948	866	VI. 6184	D. ambigua	Berk., Hook., Lond. Journ. IV. 305 (1845)	... Ambiguous dædalea ...
949	874	.. 6374	D. aulacophylla	Berk., Linn. Journ. XIII. 166 (1873) ...	... Furrow-leaved dædalea
950	880	„ 6409	D. Bowmanni	Berk., Linn. Journ. XIII. 166 (1873)	Bowman's dædalea
951	884	„ 6359	D. glabrescens	Berk., Linn. Journ. XVI. 39 (1878)	Smooth dædalea
952	876	„ 6382	D. Hobsoni	Berk., Linn. Journ. XIII. 165 (1873) . .	... Hobson's dædalea
953	...	...	D. illudens	Cooke and Mass , Grev. XXI.. 37 1892)...	... Illusive dædalea
954	869	VI. 6363	D. incompta	Berk., Linn. Trans. II. 41 (1883)	... Unadorned dædalea ...
955	865	„ 6361	D. intermedia	... Berk., Linn. Journ. XVIII. 385 (1881) ...	... Intermediate dædalea ...
956	877	„ 6434	D. latissima	... Fries. S M. I. 340 (1821)	... Very broad dædalea ...
957	868	...	D. Muelleri	Berk., Grev. XIX. 93 (1891) ...	Mueller's dædalea ...
958	870	VI. 6364	D. scalaris ...	Berk. and Br., Linn. Trans. II. 61 (1883) ..	... Ladder-like dædalea ...
959	872	„ 6368	D. Schomburgkii ...	Berk. in Cooke, Austr Fung. 27 (1883) .	... Schomburgk's dædalea
960	878	.. 6106	D. sinulosa	Klotzsch, Linn. VIII. 482 (1833)	Flexuous dædalea
961	873	.. 6370	D. Sprucei	Berk., Hook., Journ. 236 (1856)	Spruce's dædalea
962	867	IX. 5 0	D. subcongener	Berk., Grev. XIX. 93 (1891) ...	Congeneric dædalea
963	879	VI. 6408	D. tasmanica	Sacc. Syll. VI. 364 (1888) ...	Tasmanian dædalea
964	871	.. 6367	D. tenuis ...	Berk., Hook., Lond. Journ. I. 151 (1842)	Thin dædalea
965	875	„ 6376	D. unicolor	Fries, S.M. I. 336 (1821)	One-coloured dædalea ...
					68. CERIOMYCES.—Battarr.
966	881	IX. 811	C. incomptus	... Sacc. Bull. Soc. Myc. Fr. V. 115 (1889)	Una lorned ceriomyces ...
					69. FAVOLUS.—Fries.
967	896	VI. 6437	F. Boucheanus	Cooke, Grev. XV. 53 (1886)	... Boucheanus favolus ...
968	900	.. 6465	F. hispidulus	Berk. and Curt., Linn. Journ. XIII. 167 (1873) ...	Hispid favolus ...
969	898	.. 6449	F. pusillus	Fries. Linn. V. 511 (1830)	Small favolus
970	899	„ 6460	F. Rhipidium	Cooke, Grev. XV. 54 (1886)	Fan-like favolus
971	897	„ 6439	F. scaber ...	Berk. and Br., Linn. Journ. XIV. 57 (1875)	... Rough favolus
972	895	„ 6430	F. squamiger	Berk., Linn. Journ. XIII. 166 (1873) ...	Scale-bearing favolus ...

OF AUSTRALIAN FUNGI—*continued.*

Number.	W.A.	S.A.	T	V.	N.S.W.	Q.	B.	Occurrence.	General Characters.

Epicr. 496 (1838).—Favolus. Polyporus—*continued.*

Number.	W.A.	S.A.	T	V.	N.S.W.	Q.	B.	Occurrence.	General Characters.
939	W.A.	...	T.	V.	...	...	...	Bark, &c.	Sessile, somewhat hoof shaped, overlapping, wood coloured, slightly zoned towards the margin.
940	...	...	...	...	N.S.W.	Q.	...	Trunks of *Eucalyptus*	Thin, rigid, attached by circular disc, many zoned, lobed, ochrey, becoming brownish.
941	...	S.A.	...	...	...	Q.	...	Trunks ...	Leathery, thin, kidney shaped, shining, pale wood colour to brownish, zoned with concentric furrows.
942	...	...	...	...	N.S.W.	Q.	...	Trunks ...	Semicircular, sloping behind, rigid, umber, concentrically zoned and furrowed, radiately rough, finally smooth.
943	...	...	...	...	...	Q.	...	Trunks of *Hormogyne cotinifolia*	Sessile, leathery to membranous, entire, somewhat bell shaped, silky hairy, pale.
944	S A.	...	...	...	N.S.W.	Q.	...	Rotten wood	Corky to leathery, thin, pale wood colour to tawny, zoned, silky, shining.
945	...	S.A.	...	...	...	Q.	...	Trunks ...	Leathery, kidney shaped, rigid, becoming hoary, closely concentrically furrowed; margin thin, brown.
945a	...	S.A.	...	...	...	Q.	...	Branches ...	Only slightly different.
946	...	...	...	...	...	Q.	...	Trunks ...	Corky to leathery, kidney shaped, closely concentrically smooth, umber, opaque.
947	...	...	...	V.	...	Q.	...	Trunks ...	Corky to leathery, fibrous, bristly, zoneless, brown.

Syn. 499 (1801).—Trametes, Polyporus.

Number.	W.A.	S.A.	T	V.	N.S.W.	Q.	B.	Occurrence.	General Characters.
948	...	...	...	...	N.S.W.	...	...	Trunks ...	Corky, thick, convex, zoneless, becoming white.
949	...	S A.	...	...	...	Q.	...	Trunks ...	Kidney shaped, sometimes with short stalk, whitish, downy, rather rough, slightly zoned.
950	...	...	...	...	...	Q.	...	Trunks ...	Narrowly reflexed, downy, pale.
951	...	...	...	...	N.S.W.	...	...	Trunks ...	Cushion shaped, thick, at first rough and downy, then smooth, zoned, pale.
952	...	S.A.	...	...	...	...	...	Trunks ...	Ochrey white, flabby, somewhat membranous.
953	...	...	...	V.	...	...	...	Trunks, &c.	Leathery, thin, running down behind, velvety, grey, with darker linear concentric zones.
954	...	...	...	...	...	Q	...	Trunks ...	Overlapping, pale, variegated with dingy brown spots, split, hard, rough, zoned.
955	...	...	...	...	...	...	...	Trunks ...	Semicircular, pale, zoned in front, radiately rough.
956	...	...	...	...	...	Q.	B.	Old trunks	Expanded, often spreading for a foot or more in a continuous sheet, corky to woody, thick, wavy, pale wood colour, zoned within.
957	...	...	...	V.	...	...	...	Trunks ...	Corky, thickish, narrow behind, zoneless, rough, whitish.
958	...	...	...	...	...	Q.	...	Trunks ...	White, overlapping, thick, bleached above.
959	...	S.A.	...	...	...	...	...	Trunks ...	Pale ochrey, corky, thin, flattened, semicircular, zoned, roughish, shortly velvety.
960	...	...	...	...	...	Q.	...	Trunks ...	Expanded, corky to leathery, thin, pale wood colour. Pores flexuous.
961	...	...	...	...	...	Q.	...	Trunks ...	Corky, dirty umber, thin, roughish. Margin zoned.
962	...	...	...	...	...	Q.	...	Trunks ...	Corky, semicircular, flattened, velvety, pale wood colour, concentrically furrowed.
963	...	...	T.	...	...	...	...	Rotten wood	Inverted, somewhat circular, thin, brown.
964	...	S.A.	...	...	...	Q	...	Stumps ...	Corky, umber to wood colour, semicircular, thin, zoned, rough, becoming almost smooth.
965	...	...	...	V	...	Q	B.	Trunks ...	Usually overlapping, leathery, velvety, grey, zones of same colour.

Hist. 62 (1759).

Number.	W.A.	S.A.	T	V.	N.S.W.	Q.	B.	Occurrence.	General Characters.
966	...	S A.	...	...	...	...	...	Rotten wood	Somewhat globose, corky to woolly, sessile, externally pale, internally sooty brown.

Syst. Myc. 342 (1821).—Hexagonia. Laschia.

Number.	W.A.	S.A.	T	V.	N.S.W.	Q.	B.	Occurrence.	General Characters.
967	...	...	...	V.	...	Q.	...	Trunks ...	Fleshy, tough, even, then scaly, yellowish tan.
968	...	S.A.	...	...	...	Q.	...	Stems ...	Cap thin, kidney shaped, hispid, netted, delicately downy. Stem short, cylindrical, rather rough.
969	...	...	T.	...	...	...	...	Trunks ...	Cap rather membranous, kidney shaped, smooth, tawny. Stem very short, blackening.
970	...	...	...	V.	N.S.W.	Q.	...	Branches, wood, &c.	Cap leathery, kidney shaped, concentrically furrowed, pale tan. Stem lateral, short, dilated above.
971	...	...	...	...	...	Q	...	Wood ...	Cap white, then smoky, rough. Stem very short arising from orbicular disc.
972	...	...	...	...	N.S.W.	...	...	Trunks	Cap variegated with scales. Stem short, scaly, dilated upwards.

SYSTEMATIC ARRANGEMENT

Number.	Cooke's Number.	Saccardo's Number.	Scientific Name.	Authority for Name.	English Name.
					70. LASCHIA.—Fries,
973	901	VI. 6507	L. cæspitosa	*Sacc.* Syll. VI. 407 (1888)	Tufted laschia
974	905	„ 6518	L. micropus	Berk., Linn. Journ. XIII. 170 (1873)	Small-stalked laschia
975	904	„ 6516	L. pustulata	Berk. and Br., Linn. Journ. XIV. 58 (1875)	Pustulate laschia
976	903	„ 6508	L. Thwaitesii	Berk. and Br., Linn. Journ. XIV. 58 (1875)	Thwaites' laschia
977	902	„ 6504	L. tremellosa	*Fries,* S.V. 325 (1819)	Gelatinous laschia
					71. CAMPBELLIA.—Cooke and Mass.,
978	906	VI. 6523 IX. p. 205	C. infundibuliformis	Cooke and Mass., Grev. XVIII. 87 (1890)	Funnel-shaped campbellia
					72. MERULIUS.—Hall.
979	910	VI. 6542	M. aureus	Fries, Elench. 62 (1828)	Golden merulius
980	909	„ 6538	M. Baileyi	Berk. and Br., Linn. Trans. II. 62 (1883)	Bailey's merulius
981	907	„ 6532	M. corium	Fries, Elench. 58 (1828)	Leathery merulius
982	911	„ 6563	M. lacrymans	Fries, S.M. I. 328 (1821)	Weeping merulius (Dry Rot)
983	913	„ 6559	M. pallens	Berk., Outl. 296 (1860)	Pale merulius
984	908	...	M. pelliculosus	Grev. XIX. 109 (1891)	Pellicle-like merulius
985	912	VI. 6543	M. serpens	Tode., Abh. Hall. I. 355 (1790)	Spreading merulius
986	911	„ 6550	M. tenuissimus	Berk. and Br., Linn. Trans. II. 62 (1883)	Very thin merulius
987	...		Xylostroma giganteum	Fries.	Gigantic xylostroma

OF AUSTRALIAN FUNGI—*continued*.

Number.	W.A.	S.A.	T.	V.	N.S.W.	Q	B.	Occurrence.	General Characters.

in Linn. V. 533 (1830).

Number.	W.A.	S.A.	T.	V.	N.S.W.	Q	B.	Occurrence.	General Characters.
973	...	...	...	...	N.S.W.	Q.	...	Branches ...	Densely tufted. Caps conical. Stems smooth, united at base.
974	...	...	...	V.	...	...	...	Trunks ...	Tawny yellow, minute, peziza-like, shortly stalked.
975	...	...	...	...	...	Q.	...	Rotten wood	Cap ochrey, then rusty, turberculose, circular. Stem oblique, of same colour.
976	...	...	...	...	...	Q.	...	Wood ...	Tufted and gregarious. Cap tremelloid, often oblique, orange. Stem slender, white.
977	W.A.	...	...	V.	N.S.W.	Q.	...	Rotten wood	Bell-shaped, attached behind, membranous, gelatinous, entirely dark red, rather thick, leathery when dry.

Grev. XVIII. 87 (1890).

Number.	W.A.	S.A.	T.	V.	N.S.W.	Q	B.	Occurrence.	General Characters.
978	...	...	...	V.	...	...	...	Wood	Gelatinous, large, stalked. Cap deeply funnel shaped, thick. Stem short, thick, expanded disc at base.

Helv. 150 (1768).—Boletus.

Number.	W.A.	S.A.	T.	V.	N.S.W.	Q	B.	Occurrence.	General Characters.
979	...	...	...	...	...	Q.	...	Wood ...	Expanded, thin, membranous, adherent, golden yellow.
980	...	...	...	...	...	Q.	...	Trunks ...	Fan shaped, viscid, smooth, almost orange coloured when fresh. Margin notched, rough, inflexed, flesh yellow.
981	W.A.	...	T.	V.	...	Q.	B.	Trunks and branches	Expanded, soft, rather papery, shaggy beneath, white.
982	W.A.	...	...	...	...	Q.	B.	Rotten wood chiefly of dwellings, in cellars	Large, spongy to fleshy, rusty yellow, web-like or velvety below.
983	...	...	T.	V.	...	...	B.	Trunks ...	Attached, fleshy, somewhat gelatinous, thin, slightly downy.
984	...	...	...	V.	...	...	...	Branches of *Acacia*	Broadly expanded, membranous, white, hyaline when dry, like a thin pellicle.
985	...	...	...	...	...	Q.	B.	Rotten wood ...	Crustaceous, attached, thin, pale, then reddish, spreading in wavy manner.
986	...	...	...	...	...	Q	...	Parasitic on *Hymeno-chæte*	Papery, forming very thin irregular yellowish-brown patches, nearly white at margin.
987	...	...	...	V.	N.S.W.	Q.	...	Heart wood of various Eucalypts	This is the sterile state of a wood-destroying fungus, probably belonging to Polyporaceæ, and consisting of thick dense leathery sheets like chamois leather.

73. Hydnum, Linn.
74. Sistotrema, Pers.
75. Irpex, Fries.

76. Radulum, Fries.
77. Phlebia, Fries.

Number.	Cooke's Number.	Saccardo's Number.	Scientific Name.	Authority for Name.	English Name.
					ORDER III.—HYDNACEÆ,
					73. HYDNUM.—Linn.,
988	934	VI. 6761	H. alutaceum	... Fries, S.M. I. 417 (1821) ...	Wash-leather hydnum
989	920	„ 6639	H. ambustum	Cooke and Mass. Grev XVI 32 (1887)	Scorched hydnum
990	...		H. calcareum	... Cooke and Mass., Grev. XXI. 38 (1892) ...	Chalky-white hydnum
991	938	VI. 6809	H. cervinum	Berk., Fl. Tasm. II. 256 (1860) ...	Fawn hydnum
992	...	„ 6696	H. cirrhatum	Pers., Syn. 558 (1801) ...	Curled hydnum
993	925	„ 6677	H. corallobies	Scop. Carn. 472 (1772)	Coralline hydnum
994	921	IX. 840	H. croceidens	Cooke, Grev. XIX. 45 (1890) ...	Saffron-coloured hydnum
99·	924	VI. 6664	H. cyathiforme	Schaeff., Fl. Dan. 1020	Cup-shaped hydnum ...
996	930	„ 6731	H. delicatulum	... Klotzsch., Ann. Nat. Hist. III. 395 (1839)	Delicate hydnum
997	933	„ 6757	H. dispersum	... Berk., Hook., Lond. Jonrn. IV. 58 (1845)	Scattered hydnum
998	940	„ 6824	H. filicicola	Berk., Fl. Tasm. II. 256 (1860) ...	Fern-growing hydnum
999	927	„ 6722	H. flavum	Berk., Ann. Nat. Hist. X. 389 (1842)	Yellow hydnum ...
1000	931	„ 6733	H. gilvum	Berk., Hook, Journ. 16· (1851)	Yellowish-tan hydnum
1001	923	„ 6660	H. graveolens	Delast. in Fries, Hym. Eur. 605 (1874) ...	Strong-smelling hydnum
1002	936	„ 6779	H. investiens	Berk., Hook., Lond. Journ. IV. 57 (1845)	Lining hydnum
1003	941	„ 6836	H. isidioides	Berk., Hook., Lond. Journ. IV. 58 (1845)	Isidium-like hydnum ...
1004	918	„ 6624	H. lævigatum ...	... Swartz., Vet. Akad. Handl. 243 (1810) ...	Smooth hydnum
1005	932	„ 6738	H. membranaceum	Bull., Champ 481 (1798)	Membranous hydnum ...
1006	926	„ 6705	H. merulioides	Berk., Linn. Trans. II. 63 (1883)	Merulius-like hydnum
1007	939	„ 6812	H. mucidum	Pers., Syn. 561 (1801) ...	Hoary hydnum
1008	929	„ 6727	H. Muelleri	Berk., Linn. Journ. XIII. 167 (1873)	Mueller's hydnum
1009	922	„ 6657	H. nigrum	Fries, S.M. I. 404 (1821)	Black hydnum
1010	928	„ 6725	H. ochraceum	Pers., Syn. 559 (1801)	Ochrey hydnum
1011	919	„ 6633	H. repandum	Linn., Sp. Pl. 1178 (1753)	Repand hydnum
1012	937	„ 6795	H. udum ..	... Fries, S.M. I. 422 (1821)	Moist hydnum
1013	935	„ 6778	H. xanthum ...	Berk. and Curt., Grev. I. 98 (1872) ...	Yellow hydnum
1013A	„	„	H. xanthum, var. teretiblens	Cooke., Handb. Aust. Fung. 172 (1892) ...	Cylindrical hydnum
					74. SISTOTREMA.—Pers.
1014	943	VI. 6572	S. irpicinum	Berk. and Br., Linn. Trans. II. 62 (1883) ...	Irpex-like sistotrema ...
					75. IRPEX.—Fries,
1015	949	VI. 6925	I. Archeri	Berk., Fl. Tasm. II. 257 (1860) ...	Archer's irpex
1016	947	„ 6895	I. flavus ...	... Klotzsch, Linn. VIII. 188 (1833)	Yellow irpex
1017	944	„ 6876	I. hexagonoides	Kalch., Grev. IX. 1 (1880) ...	Hexagonia-like irpex

ARRANGEMENT OF GENERA (9).

78. Grandinia, Fries.
79. Porothelium, Fries.

80. Odontia, Pers.
81. Kneiffia, Fries.

Number.	Habitat.						B.	Occurrence.	General Characters.
	W.A.	S.A.	T.	V.	N.S.W.	Q.			

FRIES, PL. HOM. 80 (1825).

Sp. Pl. 1178 (1753).

988	...	...	...	V.			...	B.	Rotten wood	...	Spore-bearing surface expanded longitudinally, crustose, smooth, pale ochre. Spines acute.
989	...	...	...	V.			...	...	Sandy soil	...	Cap fleshy to membranous, brick red, turning black. Stem erect, slender, paler than cap.
990	...	...	...	V.			...	...	Bark ...	...	Chalky white, opaque, widely expanded. Spore-bearing surface crustose, smooth, mealy. Has a scorched appearance when dry.
991	...	...	T.	...			...	...	Rotten wood	...	Inverted, expanded, indistinct margin, pale vinous, at first delicately downy.
992	...	...	...	V.			...	B.	Trunks ...	...	Fleshy, expanded, colour variable, the upper surface with long curled abortive spines.
993	...	...	...	...		Q.	B.	Rotten wood, trunks	Much branched, creamy, like cauliflower at first, then a mass of coralline branches.
994	...	...	...	V.			...	...	Ground ...	...	Cap fleshy, thin, ochrey yellow. Stem central, slender, smooth when dry.
995	...	...	...	...	...	Q.	H.	Woods	...	Small, commonly grown together. Cap leathery, thin, funnel shaped. zoned, ashy pale. Stem smooth, slender, disc rather downy.
996	...	...	...	...	N.S.W.		...	...	Trunks ...	...	Cap expanded, reflexed, leathery, thin, margin yellowish.
997	W.A.	...	...	...			...	...	Rotten wood	...	In long patches. Spore-bearing surface, thin, waxy, at length disappearing.
998	...	...	T.	...			...	...	Fern stems	...	Expanded, indistinct margin, white, thin.
999	...					Q.	...	Branches ...	...	Sessile, nearly circular, thin, pale yellow, smooth.
1000	...	...	...	...		Q.	...	Rotten wood	...	Fan shaped, overlapping, thin, ochrey to yellowish tan, clothed with cartilaginous radiating hairs.
1001	...	...	...	V.		Q.	B.	Woods	...	Cap leathery, thin, soft, zoneless, rough, dark brown, margin whitish. Stem slender. Odour of melilot.
1002	W.A.	...	...	...			...	In cavities of trunks of Xanthorrhœa	Widely expanded, spore-bearing surface at first downy, then compact, smooth.
1003	W.A.	...	...	...			...	Hymenium of Polyporus gryphæformis	Spore-bearing surface crustaceous, white, margin somewhat fringed, separating in chips.
1004	...	...	T.	...		Q.	B.	Solitary in wood	...	Cap fleshy, firm, smooth, umber. Stem thick, even.
1005	...	...	...	V.		Q.	B.	Rotten wood	...	Spore-bearing surface expanded, waxy, membranous, smooth, tawny to rusty.
1006	...	...	...	...		Q.	...	Wood ...	...	Semicircular, thick, smooth, pale, rough with prominent lines.
1007	...	...	...	...		Q.	...	Trunks ...	...	White. Spore-bearing surface very broad, membranous, soft, evanescent.
1008	...	...	...	...	N.S.W.	Q.	...	Wood	...	Inverted, then reflexed, sometimes growing together, velvety, somewhat zoned.
1009	...	...	...	...	N.S.W.		B.	Woods ...	...	Cap corky, rigid, downy, zoneless, bluish black. Stem stout, black.
1010	...	...	...	V.		Q.	B.	Trunks and sticks	...	Small. Expanded and reflexed, leathery, thin, zoned, ochrey.
1011	...	...	T.	...			B.	Woods ...	...	Gregarious. Cap fleshy, fragile, repand and wavy, deformed, pale, creamy buff. Stem rather short. Very common. Edible.
1012	...	...	T.	...	N.S.W.	Q.	B.	Rotten wood	...	Spore-bearing surface expanded, thin, somewhat gelatinous, flesh colour to watery yellow.
1013	...	...	...	...	N.S.W.		...	Trunks ...	...	Inverted, thin, margin delicately downy, then waxy. Spines compressed.
1013a	...	...	...	...	N.S.W.		...	Trunks ...	...	Spines not compressed and hanging from the brighter pale-orange cap.

Tent. Disp. 28 (1797).

| 1014 | ... | | | | | Q. | ... | Dead branches | ... | Somewhat cuticular, thick, delicately downy, pale, descending deeply behind. |

Pl. Homon. 81 (1825).—Polyporus.

1015	...	...	T.	...			...	Rotten wood	...	White, inverted. Margin web-like, downy.
1016	...	S.A	...	V.	N.S.W.	Q.	...	Trunks, &c.	Expanded, spongy soft, yellow, margin shortly reflexed, downy.
1017	...	...	...	...	N.S.W.		...	Trunks ...	...	Entirely white. Cap corky to leathery, running down behind, faintly zoned, teeth disposed in honeycomb manner.

Number.	Cooke's Number.	Saccardo's Number.	Scientific Name.	Authority for Name.	English Name.

75. IRPEX.—Fries,

1018	946	VI. 6891	I. maximus	Mont. Ann. Sci. Nat. 2 Ser. VIII. 364 (1837)	... Largest irpex ... ...
1019	950	„ 6883	I. sinuosus, var. cervicolor ...	Berk. and Br., Linn. Journ. XIV. 60 (1875)	... Fawn-coloured irpex ... ...
1020	948	„ 6902	I. tabacinus	Berk. and Curt., Grev. I. 102 (1872) ...	... Tobacco-coloured irpex ...
1021	945	„ 6888	I. zonatus	Berk., Hook., Journ. 168 (1854) ...	... Zoned irpex ... ...

76. RADULUM.—Fries,

| 1022 | 951 | VI. 6931 | R. molare ... | Fries, Elench. 151 (1828) | ... Tubercled radulum ... ... |

77. PHLEBIA.—Fries,

1023	956	VI. 6965	P. coriacea ...	Berk., Linn. Journ. XVIII. 385 (1881)	... Leathery phlebia ... ...
1024	954	„ 6960	P. hispidula ...	Berk. Linn., Journ XIII. 167 (1873)	... Rough phlebia ... ...
1025	952	„ 6950	P. merismoides ...	Fries, S.M. I. 427 (1821) ...	... Merisma-like phlebia ... ...
1026	953	„ 6951	P. radiata ...	Fries, S.M. I. 427 (1821) ...	... Radiate phlebia ...
1027	955	„ 6964	P. reflexa ...	Berk., Hook., Journ. 168 (1851) ...	... Reflexed phlebia ...

78. GRANDINIA.—Fries,

1028	960	VI. 6980	G. australis ...	Berk., Fl. Tasm. II. 257 (1860) ...	... Southern grandinia ... ...
1029	959	„ 6976	G. crustosa ...	Fries, Epicr. 527 (1838) ...	... Crustaceous grandinia .. ...
1030	961	IX. 865	G. glauca ...	Cooke, Grev. XVII. 55 (1889) ...	... Glaucous grandinia ... ...
1031	957	VI. 6969	G. granulosa ...	Fries, Epicr. 527 (1838) ...	... Granular grandinia ... ...

79. POROTHELIUM.—Fries,

| 1032 | 915 | VI. 6576 | P. subtile... | Fries. S.M. I. 506 (1821) | ... Delicate porothelium ... ... |

80. ODONTIA.—Pers.

| 1033 | 962 | VI. 7018 | O. secernibilis | ... Berk., Fl. Tasm. II. 257 (1860) ... | ... Separable odontia ... |

81. KNEIFFIA.—Fries,

| 1034 | 963 | VI. 7022 | K. Muelleri | ... Berk., Linn. Journ. XIII. 167 (1873) | ... Mueller's kneiffia ... ... |
| 1035 | 963bis. | „ 7020 | K. setigera | ... Fries, Epicr. 529 (1838) ... | ... Bristle-bearing kneiffia ... |

OF AUSTRALIAN FUNGI—*continued.*

Number	W.A.	S.A.	T.	V.	N.S.W.	Q.	B.	Occurrence.	General Characters.

Pl. Homon. 81 (1825).—Polyporus—*continued.*

1018	...	...	...	...	...	Q.	...	Trunks ...	Leathery, thin, kidney shaped, at first downy, then naked, concentrically furrowed.
1019	...	...	...	...	...	Q.	...	Wood ...	Expanded, shortly reflexed, thin, soft, zoneless, entirely fawn colour.
1020	...	...	...	...	...	Q.	...	Trunks ...	Slightly reflexed, somewhat zoned, bay brown, downy.
1021	...	...	...	V.	N.S.W.	Q.	...	Dead wood	Wood colour. Caps overlapping, somewhat fan shaped, leathery, zoned, delicately downy, becoming smooth.

Pl. Homon. 81 (1825).—Hydnum, Sistotrema.

| 1022 | ... | ... | ... | ... | ... | Q. | ... | Trunks, and on peach trees | Broadly expanded, crustaceous, smooth, somewhat yellow. Tubercles deformed, short. |

S.M. I. 426 (1821).—Auricularia, Thelephora.

1023	...	...	...	...		Q.	...	Ground ...	Spoon shaped, leathery, brown. Folds irregular.
1024	...	S.A.	...	...				Trunks ...	Inverted, reflexed, reddish, zoned, velvety and hispid.
1025	...	...	...	...		Q.	B.	Trunks, &c.	Expanded, even or incrusting, branched, flesh colour, white. Margin orange.
1026	...		...			Q.	B.	Bark	Roundish, smooth, red flesh colour. Margin radiately toothed.
1027	...		...	V.	N.S.W.	Q.	...	Wood	Inverted, semicircular, reflexed, cracked, clothed with spongy down, zoned, vinous brown.

Epicr. 527 (1838).—Hydnum, Thelephora.

1028	...	...	T.	V.		...	...	Rotten wood	Inverted, expanded, indistinct margin, pale, cracked, snow white within.
1029	...	...	...	V.		...	B.	Bark	Mealy, expanded irregularly, crustaceous, white.
1030	...	...	...	...		Q.	...	Wool ...	Waxy, broadly expanded, glaucous, margin distinct.
1031	...	...	T.	...		Q.	B.	Old wood ...	Waxy, broadly expanded, tan coloured. Margin distinct. Granules hemispherical, crowded.

Obs. II. 272 (1818).—Boletus.

| 1032 | ... | ... | T. | ... | | | ... | Bark | Irregularly expanded, membranous, snow white, porous warts distinct. |

Tent. Disp. 30 (1797).

| 1033 | ... | ... | T. | ... | | | ... | Rotten wood | Inverted, separable, membranous, white. |

Epicr. 529 (1838).

| 1034 | ... | S.A. | ... | ... | | ... | ... | Rotten wood | Resupinate, thin, mealy when young, cracking here and there. |
| 1035 | ... | ... | ... | ... | | ... | B. | Wood ... | Whitish, pale buff when dry, soft, forming a layer or sometimes fleshy. Bristles rigid, very minute, scattered. |

62

ORDER IV.—THELEPHORACEÆ PERS.

82. Craterellus. Fries.
83. Thelephora, Ehrh.
84. Cladoderris, Pers

85. Stereum, Pers.
86. Hymenochæte. Lev.
87. Corticium, Fries.

Number.	Cooke's Number.	Saccardo's Number.	Scientific Name.		Authority for Name.		English Name.	

ORDER IV.—THELEPHORACEÆ,

82. CRATERELLUS.—Fries, Epicr. 531 (1838).—

1036	965	VI. 7046	C. confluens	...	Berk. and Curt., Linn. Journ. IX. 423 (1867)	...	Confluent craterellus ...	...
1037	964	„ 7042	C. cornucopioides ...	...	Pers. Myc., Eur. II. 5 (1822) ...	...	Trumpet-shaped craterellos	...
1038	967	IX. 882	C. multiplex	...	Cooke and Mass., Grev. XVIII. 25 (1889)	...	Multiplex craterellus ...	
1039	966		C. pusio ...	..	Berk., Fl. Tasm. II. 258 (1860)	...	Little craterellus	...

83. THELEPHORA.—Ehrh. in Roth.

1040	977	VI. 7116	T. Archeri	...	Berk., Fl. Tasm. II. 258 (1860)	...	Archer's thelephora	...
1041	...	„ 7146	T. atra ...	...	Weinm., Ross. 380 (1836)	...	Black thelephora	...
1042	975	„ 7098	T. caryophyllea	...	Pers., Syn. 565 (1801) ...	...	Clove thelephora	
1043	974	„ 7087	T. concrescens		Fries, Pl. Preiss. II. 136 (1846)	...	Concrescent thelephora	
1044	976	„ 7107	T. congesta	...	Berk., Linn. Journ. XVI. 168 (1878)	...	Congested thelephora ...	
1045	984	„ 7159	T. cristata	...	Fries, S.M. I. 434 (1832)	...	Crested thelephora	...
1046	987	„ 7192	T. exsculpta	...	Berk., Linn. Journ. XIII. 168 (1873)	...	Carved thelephora	...
1047	980	„ 7144	T. intybacea	...	Pers., Syn. 567 (1801) ...	...	Endive thelephora	...
1048	982	„ 7147	T. laciniata	...	Pers., Syn. 567 (1801) ...	...	Cut thelephora	...
1049	979	„ 7142	T. multipartita	...	Schwein., in Fries. Epicr. 536 (1838)	...	Multipartite thelephora	...
1050	978	„ 7129	T. myriomera	...	Fries, Pl. Preiss. II. 137 (1846) ...	...	Myriad-partite thelephora	...
1051	...	„ 7103	T. palmata	...	Fries, S.M. I. 432 (1832)	...	Palmate thelephora	...
1052	986	„ 7188	T. pedicellata	...	Schwein., Syn. Car. 108 (1822) ...	...	Elevated thelephora	...
1053	985	„ 7173	T. riccioidea	...	Berk. Fl. Tasm. II. 258 (1860) ...	...	Riccia-like thelephora...	...
1054	...	„ 7088	T. spongipes	...	Berk., Linn. Trans. II. 63 (1883)	...	Spongy-stalked thelephora	...
1055	983	IX. 887	T. stereoides	...	Cooke and Mass., Grev. XVIII. 5 (1889)	...	Stereum-like thelephora	...
1056	981	VI. 7145	T. terrestris	...	Ehrh., Crypt. 179 (1788)	...	Ground thelephora	...

84. CLADODERRIS.—Pers. in Freye.

1057	989	VI. 7211	C. australica	...	Berk. in Grev. XI. 28 (1882)	...	Australian cladoderris	...
1058	988	„ 7210	C. australis	...	Kalch., in Thum. Syn. Myc. Aust. II. (1878)	...	Southern cladoderris ...	
1059	990	„ 7215	C. dendritica	...	Pers., Freye. Voy. (1826)	...	Tree-like cladoderris ...	...
1060	991	„ 7207	C. spongiosa	...	Fries, Fung. Nat. 20 (1848)	...	Spongy cladoderris	...

85. STEREUM.—Pers.

1061	1029	VI. 7418	S. acerinum	...	Fries, Hym. Eur. 645 (1874)	...	...	Maple stereum
1062	1027	„ 7375	S. amœnum	...	Mass., Mon. Thel. Linn. Journ. XXVII. 193 (1891)	Charming stereum
1063	992	„ 7074	S. caperatum	...	Mass., Mon. Thel. Linn. Journ. XXVII. 161 (1891)	Wrinkled stereum
1063A	„	„ 7075	S.caperatum, var. lamellatum	Cooke, Handb. Aust. Fung. 182 (1892) ...	Plaited stereum	...	...
1063B	„	„ 7088	S. caperatum, var. spongipes	Cooke, Handb. Aust. Fung. 182 (1892) ...	Spongy-stalked stereum	...
1064	1019	„ 7371	S. complicatum ...	...	Fries, Epicr. 548 (1838)	...	Crisped stereum	...

Arrangement of Genera (12).

<table>
<tr><td>88. Peniophora, Cooke.</td><td>91. Cyphella, Fries.</td></tr>
<tr><td>89. Coniophora, D. C.</td><td>92. Solenia, Hoffm.</td></tr>
<tr><td>90. Aleurodiscus, Rabh.</td><td>93. Lachnocladium, Lev.</td></tr>
</table>

Number.	Habitat.						B.	Occurrence.	General Characters.
	W.A.	S.A.	T.	V.	N.S.W.	Q.			

PERS. MYC. EUR. I. 109 (1822).

Cantharellus, Merulius, Peziza, Helvella.

1036	...	...	...	V.	...	Q.	...	Ground	Orange, deeply funnel shaped, margin incurved. Stem divided above into numerous caps.
1037	...	...	...	...	...	Q.	B.	Wood	Somewhat membranaceous, trumpet shaped, sooty brown, becoming black. Stem hollow, smooth, black. Common. *Edible.*
1038	...	...	T.	...				Ground ...	Kidney shaped, and attached at base to stem in series of five or six superimposed, ochrey. Stem slender, erect, wrinkled.
1039	...	...	T.	...				Among moss	Orange. Cap lateral, convex. Stem thickened upward.

Fl. Germ. I. 538 (1788).—Helvella, Clavaria, Merisma.

1010	...	...	T.	V.		Q.	...	Ground ...	Forked, branches compressed and dilated above, ochrey.
1011	...	...	...	V.		...	...	Ground ...	Growing in tufts. Expanded, soft, tufts arising from common tuber.
1012	W.A.	...	...	...		Q.	B.	Among grass	Purplish brown. Cap somewhat leathery, depressed, margin divided. Stem short and smooth. Inodorous.
1013	W.A.	...	...	...		...	...	Moist places	Tufted and growing together. Cap funnel shaped, brown as well as stem.
1014	...	...	...	V.	N.S.W.	Q.	...	Moist places	Small, gregarious, dark purple, sparingly branched in a forked manner.
1015	...	...	...	...		Q.	...	On moss, grass, &c.	Incrusting, tough, pale, spreading out into branches or fringes.
1016	...	...	...	V.		...	...	Bark	Circular, grey, tinged with pink, radiating with teeth towards the margin, powdery, dark purple beneath.
1047	...	...	...		N.S.W.	...	B.	Woods ...	Tufted, soft, whitish to rusty, then sooty brown.
1048	...	...	...	V.		...	B.	On trunks and ground	Incrusting, soft to leathery, rusty brown. Very common.
1049	...	...	...			Q.	...	Ground ...	Leathery, ashy brown. Cap smooth, much divided down to stem in a branching manner.
1050	W.A.	...	...	...			...	Ground ...	Tufted, papery, flattened and much branched, growing together.
1051	...	...	...	...		Q.	B.	Ground in woods	Fœtid, densely clustered. Brownish purple, soft, stem-like base with wedge-shaped branches.
1052	W.A.	...	...	V.		Q.	...	Branches	Spread out, finely filamentous, compact, cinnamon brown, rooting beneath with bundles of fibres.
1053	...	...	T.			...	...	Bare soil ...	Pale, closely attached and radiately branched.
1054	...	...	...		N.S.W.	Q.	...	Old wood	Funnel shaped, spongy and downy, radiately folded and wrinkled. Stem spongy.
1055				V.		...	...	Bark	Leathery, spread out, reflexed, downy, rust colour, crisped at the margin.
1056				V.		...	B.	Ground ...	Tufted, soft brown, turning black. Caps overlapping, flattened, running down into sub-lateral stem.

Voy. Bot. 176 (1826).—Thelephora.

1057	...	...	...	...	N.S.W.	...		Wood	Funnel shaped, oblique, fan shaped or semicircular. Stem excentric, short, of an umber colour like the cap.
1058	...	...	...	...	N.S.W.	...	...	Trunks ...	Cap somewhat excentric, funnel shaped, oblique, margin with fringed lobes. Stem woody and brown.
1059	...	...	...	V.	N.S.W.	Q.	...	Wood	Leathery to soft, yellowish tan. Cap kidney shaped, entire. Stem lateral, firm.
1060	...	...	...	V.			...	Trunks ...	Broadly funnel shaped, spongy, elastic, tan colour. Stem central, very short, woody.

Obs. Myc. I. 35 (1796).—Thelephora, Auricularia, Corticium, Elvella.

1061	...	...	T.	V.		...	B.	Bark	Crustaceous, smooth, white, thin. Surface generally covered with minute particles of lime.
1062	...	S.A.	...	...			...	Fallen branches	Gregarious. Leathery, membranaceous, hairy, white.
1063	...	...	...	V.	N.S.W.	Q.	...	Trunks ...	Leathery to membranaceous, wrinkled, folded, ochrey, hairy in centre. Stem central, thick, downy.
1063a	...	...	...	...		Q.	...	Wood	Ochrey. Cap wrinkled, plaited. Stem elongated, velvety.
1063b	...	...	...	...		Q.	...	Wood	Spongy and downy, both in cap and stem.
1064	...	...	...	V.		Q.	...	Branches ...	Papery, furrowed, brownish or ochrey, much crisped and lobed.

SYSTEMATIC ARRANGEMENT

Number.	Cooke's Number.	Saccardo's Number.	Scientific Name.	Authority for Name.	English Name.

85. STEREUM.—Pers. Obs. Myc. I. 35 (1796).—

1065	...	VI. 7299	S. concolor	Berk., Fl. Tasm. II. 259 (1860)	One-coloured stereum ...
1066	1000	„ 7255	S. crucibuliforme ...	Mass., Mon. Thel. Linn. Journ. XXVII. 168 (1891)	Crucible-shaped stereum
1067	993	„ 7223	S. cyathiforme ...	Fries, Epicr. 245 (1838)	Cup-shaped stereum ...
1068	994	„ 7233	S. elegans ...	Fries, Epicr. 545 (1838)	Elegant stereum ...
1069	1011	„ 7271	S. fasciatum ...	Fries, Epicr. 546 (1838)	Banded stereum ... ...
1070	1012	„ 7270	S. gausapatum	Fries, Hym. Eur. 638 (1874)	Rough-coated stereum... ...
1071	1014	„ 7228	S. hirsutum	Fries, Epicr. 549 (1838)	Hairy stereum
1071a	...	...	S. hirsutum, var. tenellum ..	Sacc., Notes Myc. 5 (1890)	Tender stereum ...
1071b	...	...	S. hirsutum, var. glaucellum	Sacc., Notes Myc. 5 (1890)	Glaucous stereum ...
1072	1015	VI. 7329	S. illudens ...	Berk., Hook., Lond. Journ. Bot. IV. 59 (1845) ...	Deceptive stereum ...
1073	1009	„ 7272	S. involutum ...	Klotzsch, Linnæa VII. 499 (1832) ...	Involute stereum
1074	...	„ 7310	S. Kalchbrenneri ...	Sacc. Syll. VI. 568 (1888)	Kalchbrenner's stereum ...
1075	1007	„ 7267	S. Leichardtianum	Sacc. Syll. VI. 559 (1888)	Leichardt's stereum ...
1076	1008	„ 7311	S. lobatum ...	Fries, Epicr. 547 (1838)	Lobed stereum ..
1077	1028	„ 7360	S. molle ...	Sacc. Syll. VI. 577 (1889) ...	Soft stereum ...
1078	998	„ 7247	S. Moselei	Berk., Linn. Journ. XVI. 48 (1878) ...	Moseley's stereum ...
1079	995	„ 7229	S. nitidulum	Berk., Hook., Lond. Journ. Bot. II. 638 (1843) ...	Shining stereum ...
1080	1017	„ 7283	S. ochroleucum ...	Fries, Hym. Eur. 639 (1874) ...	Yellowish-white stereum ...
1081	999	„ 7253	S. prolificans ...	Berk., Linn. Journ. XVI. 41 (1878)	Prolific stereum ...
1082	1018	„ 7284	S. purpureum	Pers. Obs. Myc. II. 92 (1796)	Purple stereum ...
1083	1005	„ 7263	S. pusillum	Berk., Ann. Nat. Hist., X. 381 (1842)	Small stereum ...
1084	1001	„ 7282	S. radiatofissum ...	Berk. and Broome, Linn. Trans. II. 63 (1883)	Radiately divided stereum
1085	1020	., 7520	S. radicale	Mass., Mon. Thel. Linn. Journ. XXVII. 187 (1891)	Root-growing stereum
1086	1019	„ 7388	S. retirugum	Cooke, Proc., Roy. Soc. Ed. XI. 456 (1882) ...	Net-veined stereum ...
1087	1023	„ 7336	S. rugosum ...	Fries, Epicr. 552 (1838)	Wrinkled stereum
1088	1010	„ 7278	S. semilugens	Kalch., Grev. IX. 1 (1880) ...	Gloomy stereum ...
1089	1021	„ 7310	S. simulans	Berk. and Broome, Linn. Trans. II. 64 (1883) ...	Simulating stereum ...
1090	996	., 7070	S. Sowerbei ...	Mass., Mon. Thel. Linn. Journ. XXVII. 164 (1891)	Sowerby's stereum ...
1091	1022	., 7289	S. spadiceum	Fries, Epicr. 549 (1838) ..	Bright-brown stereum ...
1092	1032	„ 7387	S. sparsum ...	Berk., Linn. Journ. XIII. 169 (1873)	Scattered stereum ...
1093	1002	., 7257	S. spathulatum ...	Berk., Hook., Journ. VIII. 274 (1856) ...	Spoon-shaped stereum ...
1094	1006	„ 7295	S. striatum	Fries, Hym. Eur. 641 (1874)	Streaked stereum ...
1095	1030 &	„ 7410	S. strumosum	Fries, Nov. Symb. Myc. 111 (1851) ...	Swollen stereum ...
1096	1031 1024	„ 7300	S. sulphuratum	Berk. and Rav., Linn. Journ. X. 331 (1869) ...	Sulphur-coloured stereum ...
1097			S. thelephoroides ... ... (substituted for S. pannosum)	McAlpine, Syst. Arr. Austr. Fung. (1894) Cooke and Mass., Grev. XXI. 38 (1892)	Thelephora-like stereum ...
1098	997	VI. 7251	S. Thozetii ...	Berk., Linn. Journ. XVIII. 385 (1881)	Thozet's stereum

OF AUSTRALIAN FUNGI—*continued.*

Thelephora, Auricularia, Corticium, Elvella—*continued.*

Number.	W.A.	S.A.	T.	V.	N.S.W.	Q.	B.	Occurrence.	General Characters.
					Habitat.				
1065	...	...	T.	...		...	...	Dead branches ...	Dirty white, semicircular, decurrent behind, soft, downy, contracted when dry.
1066	...	...	...	V.	...	...	...	Wood	Hairy, margin inflexed. Resembling *Crucibulum vulgare.*
1067	...	S.A.	...	V.	...	...	...	Wood and ground ...	Leathery, whitish. Cap broad ($1\frac{1}{4}$–$2\frac{1}{2}$ inches), wine-glass shaped, bristly. Stem about $\frac{1}{2}$ inch long, central, smooth.
1068	...	S.A.	...	V.	N.S.W.	Q.	...	Ground	Cartilaginous to leathery. Cap funnel shaped usually, chestnut, shining. Stem short and slender.
1069	...	S.A.	...	V.		Q.	...	Trunks of *Eucalypts*	Leathery, tufted, thin, ash grey, zoned with shining bay-brown bands, and tapering towards base.
1070	...	...	...	...	...	Q.	...	Trunks	Tufted, sessile. Caps soft, corky, shell shaped, rough haired, tawny, becoming pale.
1071	W.A.	S.A.	T.	V.	N.S.W.	Q.	B.	Trunks and branches	Leathery, rigid. Caps overlapping, spread out, rough haired, somewhat zoned, dirty ochre, becoming pale. Common.
1071A	...	S.A.	...	...	...	...	...	Rotten wood of *Acacia*	Membranous. Caps reflexed, whitish, zoned.
1071B	...	S.A.	...	...	...	...	...	Trunks	Cap velvety, discoloured and zoned. Hymenium greyish, glaucous.
1072	W.A.	S.A.	T.	V.	N.S.W.	Q.	...	Branches, &c. ...	Leathery and rather rigid. Caps confluent, zoned, radiately plaited, bay brown.
1073	...	...	...	...		Q.	...	Trunks	Leathery, tufted. Caps overlapping, grown together, ear shaped, longitudinally furrowed, tapering at base into lateral black stem.
1074	...	...	...	...	...	...	...	...	Leathery to membranous, sessile, rough haired, whitish, zoned, whitish, finally tawny.
1075	...	S.A.	...	...	N.S.W.	Q.	...	Trunks	Leathery, flattened, tapering behind, zoned, velvety, tawny.
1076	...	...	T.	V.	N.S.W.	Q.	...	Trunks	Sessile, leathery. Cap 3–5 inches across, more or less lobed, downy, zoned, reddish cinnamon or brownish.
1077	...	...	...	...		Q.	...	Trunks	Sessile, leathery, flexible, tapering behind, zoned, soft and spongy, pale brown.
1078	...	...	...	V.		...	...	Sticks ...	Gregarious. Cap funnel to fan shaped, delicately velvety, somewhat zoned, fawn. Stem thin.
1079	...	...	...	V.		Q.	...	Among grass	Cap funnel shaped, thin, rather membranous, smooth, shining, zoned, brown. Stem central, thin.
1080	...	...	T.	...	...	Q.	B.	Bark and dead branches	Leathery, thickish, expanded, silky, zoned, hoary.
1081	...	...	...	V.	N.S.W.	Q.	...	Wood, &c. ...	Gregarious. Cap funnel shaped, zoned and furrowed, velvety, bay brown. Stem very short.
1082	W.A.	S.A.	T.	V.	N.S.W.	Q.	B.	Trunks, branches, &c.	Tough, soft. Cap spread out, zoned, silky and downy. Hymenium more or less purple.
1083	...	...	T.	...		...	...	Wood	Cartilaginous to leathery. Cap fan shaped, shining, reddish umber, silky streaked. Stem short, $\frac{1}{2}$ inch high.
1084	...	...	...	...		Q.	...	Wood ...	Crowded, thin, fan shaped, silky, shining, bay brown, many zoned, usually split into numerous lobes.
1085	W.A.	...	...	...		...	...	Base of living shrubs	Cap thick, white within, spread out, rough haired, whitish tawny.
1086	...	...	...	...	...	...	...	Wood	Leathery to membranous, mouse coloured. Cap spread out, cup shaped, then flattened, margin fringed. Hymenium net veined.
1087	W.A.	...	...	...	N.S.W.	...	B.	Trunks ...	Corky, rigid, becoming red when bruised. Cap spread out and partly reflexed, bay brown.
1088	...	...	...	...		Q.	...	Trunks ...	Membranous, somewhat tufted, sessile, flattened, semicircular, zoned, rust coloured to umber.
1089	...	...	...	...		Q.	...	Branches ...	Cap circular, rigid, wrinkled, yellowish tan, zoned and slightly reflexed.
1090	...	S.A.	T.	V.	N.S.W.	Q.	B.	Ground ...	Snow white, funnel shaped, soon discoloured, rough, with radiating processes projecting from surface.
1091	...	S.A.	T.	V.	N.S.W.	Q.	B.	Trunks, &c. ...	Leathery, spread out or reflexed, hairy, rusty, blood red where scratched, with dark-coloured hymenium.
1092	...	...	...	V.		...	...	Bark ...	White or pale ochrey, in hard new and then confluent pustules. Evidently immature condition of some species.
1093	...	...	...	...		Q.	...	Wood	Cap spoon shaped or somewhat fan shaped, bristly behind, smooth in front. Stem yellowish, velvety, lateral.
1094	...	...	...	...	N.S.W.	...	...	Wood ...	Leathery, spread out, flattened, way, roughly striated, rusty brown.
1095	...	...	...	...	...	Q.	...	Wood ...	Thick, firm, isolated patches, ochrey or lemon yellow. Including *S. sulphureum*, Fries.
1096	...	S.A.	...	V.		...	...	Dead wood and branches	Reflexed, lobed, crisped, sulphur coloured. rough haired, somewhat spongy.
1097	...	...	...	V.		...	...	Among moss and on ground	Tufted, growing together, sessile, soft, spongy, flexible, reflexed, pale umber, velvety, concentrically zoned. Species' name pre-occupied by Cooke himself.
1098	...	S.A.	...	V.		Q.	...	Trunks ...	Cap funnel shaped, downy, somewhat zoned, pale. Stem tapering downwards.

Number.	Cooke's Number.	Saccardo's Number.	Scientific Name.	Authority for Name.	English Name.
				85. STEREUM.— Pers., Obs. Myc. I. 35 (1796).—	
1099		VI. 7316	S. umbrinum	Fries, Pl. Preiss. II. 137 (1846)...	Umber stereum
1100	1004	„ 7367	S. vellereum	Berk., Fl. N. Zeal. 183 (1855)	Woolly stereum
1101	1003	., 7276	S. versicolor	Fries, Epicr. 547 (1838)	Variously coloured stereum
1102	1026	„ 7376	S. versiforme	Berk. and Curt., Grev. I. 164 (1873)	Variously shaped stereum
1103	...	„ 7338	S. vittæforme	Fries, Pl. Preiss. II. 137 (1846)	Striped stereum
1104	1025	., 7286	S. vorticosum	Fries, Obs. II. 275 (1818)	Obscurely zoned stereum
				86. HYMENOCHÆTE.—Lev., Ann. Sci. Nat.	
1105	...	VI. 7476	H. Archeri	Cooke, Grev. VIII. 149 (1880)	Archer's hymenochæte
1106	1035	„ 7438	H cacao ...	Berk., Linn. Trans. I. 403 (1879)	Chocolate-brown hymenochæte
1107	1044	.. 7461	H. crassa ...	Berk., Grev. VIII. 148 (1888)	Thick-margined hymenochæte
1108	1040	., 7473	H. innata	Cooke and Mass., Grev. XV. 99 (1887)	Innate hymenochæte
1109	1048	IX. 927	H. Kalchbrenneri ...	Mass., Mon. Thel. Linn. Journ. XXVII. 116 (1891)	Kalchbrenner's hymenochæte
1110	1041	VI. 7449	H. Mougeotii	Cooke, Grev. VIII. 1147 (1880)...	Mougeot's hymenochæte
1111	1047	„ 7464	H. olivacea	Cooke, Grev. XIV. 11 (1885)	Dark-olive hymenochæte
1112	1034	„ 7441	H. phœa ...	Cooke, Grev. VIII. 116 (1880)	Dusky hymenochæte
1113	1046	„ 7462	H. purpurea	Mass., Mon. Thel. Linn. Journ. XXVII. 115 (1891)	Purple-gilled hymenochæte
1114	...	„ 7440	H. rigidula	Berk. and Curt., Linn. Journ. X. 334 (1869)	Rigid hymenochæte
1115	1043	„ 7467	H. rhabarbarina	Cooke, Grev. VIII. 148 (1880)	Rhubarb-gilled hymenochæte
1116	1033	„ 7427	H. rubiginosa	Lev., Ann. Sci. Nat. Ser. 3, V. (1846)	Rusty hymenochæte
1117	1045	., 7312	H. Schomburgkii ...	Mass., Mon. Thel. Linn. Journ. XXVII. 115 (1891)	Schomburgk's hymenochæte
1118	1037	„ 7444	H. spadicea	Berk. and Br., Linn. Journ. XIV. 68 (1875)	Bay-brown hymenochæte
1119	1036	„ 7436	H. strigosa	Berk. and Br., Linn. Journ. XIV. 68 (1875)	Strigose hymenochæte
1120	1042	„ 742	H. tabacina	Lev., Ann. Sci. Nat. Ser. 3, V. 151 (1846)	Dark-brown hymenochæte
1121	1039	IX. 923	H. tasmanica	Mass., Mon. Thel. Linn. Journ. XXVII. 105 (1891)	Tasmanian hymenochæte
1122	1038	VI. 7443	H. tenuissima	Berk., Linn. Journ. XIV. 67 (1875)	Very thin hymenochæte
				87. CORTICIUM.—Fries, Epicr. 556 (1838).—Thelephora. Hypochnus,	
1123	1076	VI. 7786	C. anthochroum	Fries, Hym. Eur. 661 (1874)	Bright-coloured corticium
1124	1070	.. 7528	C. arachnoideum ...	Berk., Ann. Nat. Hist. 345 (1844)	Arachnoid corticium
1125	1069	., 7668	C. Archeri	Berk., Fl. Tasm. II. 260 (1860)	Archer's corticium
1126	1079	„ 7510	C. atrovirens	Fries, Hym. Eur. 651 (1874)	Dark-green corticium
1127	1068	., 7552	C. Auberianum	Mont., Crypt. Cuba 372 (1842)	Auber's corticium
1128	1064	„ 7596	C. calceum	Fries, Hym. Eur. 652 (1874)	Chalky corticium
1129	1074	IX. 946	C. cinnabarinum ...	Mass., Mon. Thel. Linn. Journ. XXVII. 140 (1891)	Vermilion-red corticium
1130	1078	VI. 7539	C. cæruleum	Fries, Hym. Eur. 651 (1874)	Blue corticium
1131	1080	„ 7616	C. comedens	Fries, Hym. Eur. 656 (1874)	Wasting corticium
1132	1067	„ 7527	C. lacteum	Fries, Hym. Eur. 649 (1874)	Milk-white corticium

Number	Habitat						B.	Occurrence.	General Characters.
	W.A.	S.A.	T.	V.	N.S.W.	Q.			

Thelephora, Auricularia, Corticium, Elvella—*continued.*

1099	W.A.	...	...	...	...	...	...	Bark of *Banksia* ...	Sessile, hemispherical, leathery, rather flaccid, fan shaped, undulating radiations from base towards margin, umber.
1100	...	...	...	V.	...	...	...	Branches and dead twigs	Spreading, with broad lobed zoned margin, dirty white above, and clothed behind with coarse tow-like fibres.
1101	...	...	...	V.	...	...	...	Trunks ...	Spread out, reflexed, fan shaped, thin, rigid, with raised concentric zones, variously coloured, or whitish to brown.
1102	...	...	...	V.	...	...	...	Dead branches, &c.	Small, at first circular, often running together, thickish, bright brown. Margin thin, upraised.
1103	W.A.	...	...	...	...	...	...	Bark of *Acacia* ...	Entire, leathery, rigid, somewhat bell shaped, with elevated concentric lines, becoming mud colour.
1104	...	...	...	...	...	Q.	B.	Bark and wood ...	Leathery, spread out, reflexed, obscurely zoned, rough haired, pale.

Ser. 3, V. 151 (1846).—Thelephora, Stereum, Corticium.

1105	...	...	T.	...	...	...	...	Rotten wood ...	Spreading, without distinct margin, soon detached, wine colour to brown, bristly, umber within.
1106	...	...	...	...	...	Q.	...	Wood	Dense dark-brown circular patches closely overlapping, cap fan shaped, deeply lobed and folded, furrowed with a few zones, velvety.
1107	...	...	...	...	N.S.W.	...	B.	Trunks ...	Leathery, velvety, pale reddish brown, margin thick, at length free.
1108	...	...	...	...	...	Q.	...	Wood ...	Thin, innate, scarcely distinct from underlying matrix, externally fawn colour, internally brick red.
1109	...	...	...	V.	...	Q.	...	Dead trunks of *Eucalypts*	Brown, rather thickly membranous, broadly spread out, loosely adherent to matrix.
1110	...	...	T.	V.	N.S.W.	...	...	Trunks of *Pinus picea*, &c.	Broadly spread out, dry, attached, dark blood red.
1111	...	...	...	...	...	Q.	...	Branches...	Spread out, dark olive, rough, velvety. Margin thinner and paler.
1112	...	...	...	V.	N.S.W.	Q.	...	Bark and wood ...	Semicircular, sessile, thin, leathery, flexible, concentrically zoned, shortly hairy, somewhat velvety, bay brown.
1113	...	...	...	V.	N.S.W.	...	...	Wood ...	Broadly spread out, closely attached, texture soft and spongy.
1114	...	...	...	...	N.S.W.	...	...	Wood ...	Broadly spread out, reflexed, rigid, thickish, zoned, velvety, bay brown.
1115	...	...	...	...	...	Q.	...	Wood ...	Broadly spread out, closely attached. Hymenium velvety, rhubarb colour.
1116	W.A.	...	T.	...	...	Q.	B.	Hard wood, posts, &c.	Leathery, rigid, spread out, reflexed, velvety, reddish brown to bay brown.
1117	...	S.A.	...	...	...	Q.	...	Wood ...	Somewhat circular, then shell shaped, umber, somewhat zoned in front, velvety, and about 1 inch broad.
1118	...	...	...	...	N.S.W.	Q.	...	Wood	Thin, elastic when dry, semicircular or circular, attached behind, zoned, rusty to bay brown.
1119	...	...	...	...	...	...	...	Wood ...	Semicircular, thin, lobed, zoned, bay brown tinged with purple, strigose or rough haired, 1 to 3 inches across.
1120	...	...	...	V.	N.S.W.	...	B.	Trunks ...	Somewhat leathery, thin, flaccid, covering underside of fallen logs. Cap spread out, reflexed, silky, rusty. Margin golden yellow.
1121	W.A.	...	T.	V.	...	...	...	Wood ...	Broadly spread out, crustaceous, rather thick. Margin thinner and paler, and sometimes slightly curled.
1122	...	...	...	...	...	Q.	...	Wood and bark ...	Sessile, very thin and flexible, 1 inch or more long, rusty to tawny, zoned, clothed with coarse down.

Peniophora, Hymenochæte, Coniophora, Stereum, Auricularia.

1123	...	...	...	...	...	Q.	B.	Bark ...	Broadly spread out, membranous, brick red or rosy, turning pale.
1124	...	S.A.	T.	V.	...	Q.	B.	Wood, bark, &c. ...	Delicate, spread out, spider web-like patches, snow white, often remaining barren.
1125	...	...	T.	...	...	Q.	...	Charred wood ...	Pale red, white within, rather thick, cracking.
1126	...	...	...	...	...	Q.	B.	Rotten wood, sticks, &c.	Irregularly spread out, thin, blackish or verdigris green.
1127	...	...	...	V.	...	...	...	Branches...	Circular at first, thin, mealy, from white to pale ochre.
1128	...	...	T.	...	...	...	B.	Wood and branches	Broadly spread out, thin, waxy, smooth, white, hard, sometimes continuous, sometimes much cracked.
1129	...	...	...	...	N.S.W.	...	...	Wood	Spreading for several inches, rather thin, without distinct margin, Hymenium waxy, vermilion.
1130	...	...	...	...	N.S.W.	Q.	B.	Wood, branches, &c.	Irregularly spread out, attached, downy, bright blue. Said to be phosphorescent.
1131	W.A.	...	...	...	...	...	B.	Branches...	Spread out and developed beneath the bark, which eventually is ruptured, lilac then pale.
1132	...	...	...	V.	...	...	B.	Wood	Broadly spread out, somewhat membranous and usually broken up, whitish, ochrey or buff when dry.

Number	Cooke's Number	Saccardo's Number	Scientific Name	Authority for Name	English Name
			87. CORTICIUM.—Fries, Epicr. 556 (1838).—Thelephora, Hypochnus,		
1133	1072	VI. 7530	C. læve ...	Pers., Tent. Disp. 30 (1797) ...	Even corticium ...
1134	...	., 7597	C. lividum ...	Pers., Obs. I. 38 (1796)...	Livid corticium ...
1135	...,	„ 7643	C. Marescalchianum	Marc. and Sacc., Syll. VI. 633 (1888) ...	Marescalchi's corticium ...
1136	1075	„ 7676	C. miniatum ...	Cooke, Grev. IX. 2 (1880) ...	Vermilion corticium ...
1137	1073	„ 7609	C. nudum ...	Fries, Epicr. 564 (1838) ...	Naked corticium ...
1138	1071	...	C. penetrans ...	Cooke and Mass., Grev. XIX. 90 (1891)...	Penetrating corticium ...
1139	1065	VI. 7162	C. sebaceum ...	Mass., Mon. Thel. Linn. Journ. XXVII. 127 (1891)	Waxy corticium ...
1140	1066	„ 7593	C. simulans ...	Berk. and Broome, Linn. Journ. XIV. 72 (1875) ...	Simulating corticium ...
1141	1077		C. sulphurellum	Cooke and Mass., Grev. XX. 33 (1891) ...	Sulphur-yellow corticium ...
			88. PENIOPHORA.—Cooke, Grev. VIII. 20 (1879).—		
1142	1053	VI. 7158	P. albo-marginata ...	Mass., Mon. Thel. Linn. Journ. XXV. 114 (1890)...	White-margined peniophora ...
1143	1058	„ 7712	P. bambusicola	Sacc. Syll. VI. 647 (1888)	Bamboo-loving peniophora ...
1144	1059	„ 7697	P. carnea	Cooke, Grev. VIII. 21 (1879) ...	Flesh-coloured peniophora ...
1145	1051	„ 7694	P. cinerea	Cooke, Grev. VIII. 20 (1879) ...	Ash-coloured peniophora
1146	..	„ 7707	P. deglubens	Berk., Linn. Journ. XVIII. 383 (1881) ...	Peeling peniophora ...
1147	1057	„ 7605	P. incarnata ...	Mass., Mon. Thel. Linn. Journ. XXV. 117 (1890)...	Bright peniophora ...
1148	1049	„ 7689	P. papyrina	Cooke, Grev. VIII. 20 (1879)	Papery peniophora ...
1149	1050	IX. 969	P. puberula	Sacc. Syll. IX. 288 (1891) ...	Downy peniophora ...
1150	1056	VI. 7531	P. rosea ...	Mass., Mon. Thel. Linn. Journ. XXV. 146 (1890)...	Rosy peniophora ...
1151	1060	„ 7702	P. sparsa	Cooke, Grev. VIII. 21 (1879) ...	Scattered peniophora
1152	1052	„ 7695	P. tephra...	Cooke, Grev. VIII. 20 (1879) ...	Ashy peniophora ...
1153	1055	., 7477	P. vinosa ...	Mass., Mon. Thel. Linn. Journ. XXV. 115 (1890)	Vinous peniophora ...
			89. CONIOPHORA.—De Candolle, Fl. Fr. V. 34 (1815).—		
1154	1084	VI. 7719	C. luteo-cincta ...	Cooke, Grev. VIII. 89 (1880) ...	Yellow-margined coniophora ...
1155	1083	„ 7724	C. membranacea ...	D. C., Fl. Fr. V. 634 (1815) ...	Membranous coniophora ...
1156	1086	„ 7683	C. murina ...	Mass., Mon. Thel. Linn. Journ. XXV. 138 (1890) ...	Mouse-coloured coniophora ...
1157	1082	„ 7723	C. olivacea ...	Cooke, Grev. VIII. 89 (1880) ...	Olive-coloured coniophora ...
1158	1085	,. 7535	C. sulphurea ...	Mass., Mon. Thel. Linn. Journ. XXV. 132 (1890) ...	Sulphur-coloured coniophora ...
1159	1081	„ 7203	C. viridis...	Cooke, Grev. VIII. 89 (1879) ...	Green coniophora
			90. ALEURODISCUS.—Rabh.—		
1160	1062	IX. 930	A. albidus	Mass., Grev. XVII. 55 (1889) ...	White aleurodiscus ...
1161	1061	VI. 7506	A. amorphus	Rabh., Fung. Eur. No. 1824	Shapeless aleurodiscus ...
1162	1063	„ 7510	A. tabacinus	Cooke, Grev. XIV. 11 (1885)	Dark-brown aleurodiscus ...
			91. CYPHELLA.—Fries, Syst. Myc. II. 201 (1823).—		
1163	1087	VI. 7817	C. albo-violascens ...	Karst. Fung. Fenn. Exs. No. 715	Pale-violet cyphella ...
1164	1091		C. australiensis ...	Cooke, Grev. XX. 9 (1891) ...	Australian cyphella ...

OF AUSTRALIAN FUNGI—continued.

Number.	Habitat.						B.	Occurrence.	General Characters.
	W.A.	S.A.	T.	V.	N.S.W.	Q.			

Poniophora, Hymenochæte, Coniophora, Stereum, Auricularia—continued.

Number.	W.A.	S.A.	T.	V.	N.S.W.	Q.	B.	Occurrence.	General Characters.
1133	...	...	T.	V.	...	Q.	B.	Wood, bark, &c. ...	Spread out over fibrils, membranous, separating from matrix, pinkish or pale.
1134	...	...	...	...	...	Q.	B.	Rotten wood ...	Expanded, soft, waxy, thin, irregular, smooth, bluish grey tinged with purple.
1135	...	...	...	...	...	Q.	...	Wood ...	Hymenium powdery, rhubarb colour, broken into little elevations.
1136	...	...	...	...	N.S.W.	Q.	...	Bark ...	Spread out, vermilion, margin fringed, whitish. When dry hymenium resembles patches of dried blood (Cooke).
1137	...	...	...	...	...	Q.	B.	Bark of Citrus ...	Waxy, cracking, flesh colour to pale. Margin decided, smooth.
1138	...	...	...	V.	...	...	...	Rotten wood, &c. ...	White, spreading, encrusting, thick, soft, with profuse penetrating mycelium.
1139	...	...	T.	...	...	...	B.	On ground and running up stems of plants	Spread out, fleshy or waxy, turning hard, incrusting and variable in form, whitish.
1140	...	...	...	V.	...	...	...	Running over mosses and twigs	Soft, tawny, arising from a white membranous woolly mycelium.
1141	...	...	...	V.	...	...	...	Dead branches ...	Broadly spread out, usually forming a thin, powdery, bright sulphur-yellowish stratum.

Corticium, Stereum, Thelephora, Hymenochæte.

Number.	W.A.	S.A.	T.	V.	N.S.W.	Q.	B.	Occurrence.	General Characters.
1142	...	...	...	V.	N.S.W.	...	...	Bark and wood ...	Very broadly spread and confluent, umber, velvety in centre, and margin white and downy.
1143	...	...	...	...	...	Q.	...	Rotting bamboo ...	Roundish patches, yellowish tan, tough, thin, fringed, cracked in drying.
1144	...	...	...	...	...	...	...	On Pinus contorta ...	Broadly spread, ochrey to flesh colour. Margin white and with loose fibres.
1145	...	...	...	V.	...	...	B.	Branches...	Waxy, cracking, confluent, ash coloured or brownish.
1146	...	...	...	...	...	Q.	...	Trunks ...	Pale, spreading, thick, peeling, downy.
1147	W.A.	...	...	...	...	Q.	B.	Thin layer on wood and bark	Broadly spread, thin, waxy, radiating at margin, red or orange with pink bloom.
1148	...	...	...	V.	...	Q.	...	Bark ...	Very thin, leathery to papery, very broadly spread out and reflexed, rough haired, ash coloured, and margin tawny.
1149	...	...	...	...	...	Q.	...	Wood ...	Membranous, leathery, broadly spread out, ochrey yellow, downy. Margin reflexed.
1150	...	...	T.	...	...	...	B.	Wood and bark ...	Spread out, clear rose pink, turning white, rather fleshy, fringed. Beautiful species, sometimes in scattered patches.
1151	...	...	...	V.	...	Q	...	Bark ...	Minute, snowy white, rather circular scattered patches.
1152	...	S.A.	...	...	...	...	...	Bark ...	Spread out, margin reddish brown, notched, free, downy.
1153	W.A.	...	...	...	...	...	...	Wood and bark ...	Isolated round patches becoming confluent and widely extending, irregularly lobed, wine colour or dark brown.

Thelephora, Coniophora, Auricularia, Merulius, Corticium, Hypochnus.

Number.	W.A.	S.A.	T.	V.	N.S.W.	Q.	B.	Occurrence.	General Characters.
1154	...	...	...	V.	...	...	...	On ground and bark	Spreading. Hymenium brown, powdery, yellow at circumference.
1155	...	...	...	V.	...	...	B.	Walls, &c. ...	In thin patches, foot or more in diameter, fragile, yellowish.
1156	...	...	...	V.	...	...	...	Branches, &c. ...	Widely spreading, at length breaking up, mouse coloured.
1157	...	...	...	V.	...	Q.	B.	Decayed pine	Widely spreading, membranous. Margin whitish, fringed. Hymenium dark olive, powdery.
1158	...	...	T.	...	...	...	B.	Bark, wood, leaves, &c.	Spreading, often spongy, passing into radiating cord-like branching sulphur-coloured threads.
1159	...	...	T.	...	...	...	...	Rotten wood ...	Developed beneath bark, spreading, downy, greenish.

Corticium, Cyphella, Peziza.

Number.	W.A.	S.A.	T.	V.	N.S.W.	Q.	B.	Occurrence.	General Characters.
1160	...	...	...	...	...	Q.	...	Branches...	Pure white, at first scattered, becoming confluent and forming irregular patches, cup-shaped at first, then expanded and flattened.
1161	...	...	...	...	...	Q.	...	Fir trunks and branches	Waxy, tough, rather leathery, cup shaped, then expanded, confluent, white, downy.
1162	...	...	...	V.	N.S.W.	...	...	Wood ...	Gregarious, cup shaped, cap somewhat elliptical, wrinkled, crisped, brown, downy.

Peziza, Thelephora, Cantharellus.

Number.	W.A.	S.A.	T.	V.	N.S.W.	Q.	B.	Occurrence.	General Characters.
1163	...	S.A.	...	V.	...	...	B.	Bark and wood, branches of Vine	Somewhat corky, sessile or nearly so, spherical to hemispherical, white. Hymenium pale violet.
1164	...	...	...	V.	...	...	...	Bark ...	Gregarious, cup shaped, sessile, pale, with closely pressed silky hairs.

Number.	Cooke's Number.	Saccardo's Number.	Scientific Name.	Authority for Name.	English Name.
				91. CYPHELLA.—Fries, Syst. Myc. II. 201 (1823).—	
1165	1088	VI. 7856	C. capula...	Fries. Epicr. 568 (1838) ...	... Cup-like cyphella ... ...
1166	...	...	C. filicola	... Cooke, Grev. XIV. 129 (1886)	... Fern-growing cyphella ...
1167	...	...	C. longipes	... Cooke and Mass.. Grev. XXI. 38 (1892) ...	... Long-stalked cyphella ...
1168	1092	VI. 7889	C. muscigena	... Fries. Epicr. 567 (1838)	... Moss-growing cyphella ...
1169	...	„ 7906	C. parasitica	... Berk. and Br , Linn. Journ. XIV. 74 (1873)	... Parasitic cyphella ...
1170	1090	IX. 1006	C. polycephala	... Sacc., Hedw. 126 (1889) ...	... Many-headed cyphella ...
1171	...	„ 1010	C. Schneideri	... Berk. and Br., Linn. Trans. II. 220 (1887)	... Schneider's cyphella ...
1172	1089	VI. 7868	C. villosa...	... Karst., Myc. Fenn. III. 325 (1871) ...	... Villous cyphella ...
				92. SOLENIA.—Hoffm.,	
1173	916	VI. 5589	S. candida ...	... Pers., Tent. Disp. 36 (1797) ...	... White solenia ... ...
1174	917	„ 6594	S. anomala, Fries, var. ochracea (Mass.)	Brit , Fung. Fl. I. 144 (1892) ...	... Ochrey solenia ... ...
1175	917*	„ 6596	S. sulphurea ...	... Sacc. and Ellis., Mich. II. 564 (1882)	... Sulphur solenia ...
				93. LACHNOCLADIUM.—Lev., Orb. Dict. VIII. 487 (1819).—	
1176	969	VI. 8177	L. Brasiliense	... Sacc. Syll. VI. 738 (1888)	... Brazilian lachnocladium ...
1177	973	„ 8018	L. flagelliforme	... Cooke, Handb. Aust. Fung. 179 (1892) ...	... Whip-shaped lachnocladium ...
1178	968	„ 8175	L. furcellatum	... Sacc. Syll. VI. 738 (1888)	... Forked lachnocladium ...
1179	972	„ 8183	L. rameale	... Berk. and Broome. Linn. Journ. XIV. 67 (1875) ...	... Branch-growing lachnocladium
1180	970	„ 8180	L. semivestitum ...	... Berk. and Curt., Grev. I. 161 (1873) ...	... Half-clothed lachnocladium ...
1181	971	„ 8188	L. setulosum ...	... Sacc. Syll. VI. 740 (1888)	... Bristly lachnocladium ...
1182		IX. 1043	L. simulans	... Berk and Broome, Linn. Trans. II. 219 (1887) ...	... Simulating lachnocladium ...

OF AUSTRALIAN FUNGI—continued.

Peziza, Thelephora, Cantharellus—continued.

Number	W.A.	S.A.	T.	V.	N.S.W.	Q.	n.	Occurrence	General Characters.
1165	...	...	T.	...	...	Q.	B.	Herb stems	Membranous, obliquely bell shaped, running down into oblique stem, whitish.
1166	...	...	..	V.	...	...		Fronds of *Adiantum*	Whitish, somewhat discoid, concave, smooth. Margin entire or two cusped.
1167	...	...	...	...		Q.	...	Logs and stems of trees in wet scrubs	Gregarious, membranous, white, cap narrow, funnel shaped, tapering into long thin curved stem.
1168	...	...	T.	V.	...	...	B.	On larger mosses ...	Membranous, soft, flattened nearly semicircular, white.
1169	...	...	...	...	...	...	...	Parasitic on some sphæria	Minute, cup shaped, snow white, externally hairy. Margin inflexed.
1170	...	S.A.	...	...	N.S.W.	...	...	Herb stems, *Senecio hypoleucus*	Gregarious. Caps closely joined in common base, whitish brown, urn shaped.
1171	...	...	...	...		Q.	...	Wood	Gregarious, tubular, membranous, pale yellow.
1172	...	...	...	V.	...	Q.	B.	Herb stems and rotting stems of Castor Oil plant	Sessile, dry, spherical, white, covered with snow-white persistent villous down.

Bot. Tasch. 68 (1793).

Number	W.A.	S.A.	T.	V.	N.S.W.	Q.	n.	Occurrence	General Characters.
1173	...	...	...	...	...	Q.	...	Rotten wood	Scattered, cylindrical, smooth, white.
1174a	...	...	T.	...	...		B.	Rotten wood	Scattered, club shaped to cylindrical, downy, ochrey.
1175	...	...	...	V.	...	...		Branches	Thickly crowded, minute, cup-like, shortly stalked, sulphur coloured, with rough hairs.

Clavaria, Eriocladus.

Number	W.A.	S.A.	T.	V.	N.S.W.	Q.	n.	Occurrence	General Characters.
1176	...	...	...	...	...	Q.	...	Trunks	Very shortly stalked, much branched. Branches tapering, forked, ochrey white.
1177	...	...	...	...		Q.	...	Bare ground	Very much branched, divided to the base. Branches tufted, cylindrical, elongated, forked.
1178	...	...	...	...		Q.		Rotten wood	Ascending, somewhat rusty. Branches solid, repeatedly forked, distant, tough, velvety.
1179	...	...	...	...	...	Q.	...	Branches, &c.	Dark purple, thread-like, forked, encrusting fresh branches, leaf stalks and leaves.
1180	...	...	...	...	...	Q.	...	Ground ...	Delicate, repeatedly forked. Branches downy.
1181	W.A.	...	...	...	...		...	Ground ...	Small, ochrey. Stem short, irregularly divided. Branches compressed, forked, downy.
1182	...	...	...	...		Q.	...	Ground ...	Dark brown when dry, downy. Stem simple below, repeatedly branched above. Branches slender, tips forked.

Authority for Name.		English Name.

ORDER V.—CLAVARIACEÆ,

94. SPARASSIS.—Fries,

Fries, S.M. I. 465 (1821)	...	Crisped sparassis ... ...

95. CLAVARIA.—*Linn.*,

Pers., Comm. 46 (1797)	... ...	Fir clavaria ... ...
Berk., Fl. Tasm. II. 261 (1860)		Archer's clavaria ...
Fries, S.M. I. 482 (1821)		Clay-coloured clavaria
Cooke and Mass., Grev. XVI. 33 (1887) ...		Orange clavaria
Schaeff. 287 (1762) ...		Golden clavaria
Pers., Syn. 587 (1801) ...		Clustered clavaria
Bull, Champ. 354 (1791)		Ash-coloured clavaria ...
Berk., Fl. N. Zeal. 186 (1855)		Colenso's clavaria ...
Linn., Sp. Pl. 1182 (1753)		Coral-like clavaria
Fries, S.M. I. 470 (1821)		Curled clavaria ...
Pers., Syn. 591 (1801) ...		Crested clavaria
Pers., Ic. and Desc. 36 (1798)		Saffron clavaria
Linn., Sp. Pl. 1182 (1753)	...	Fastigiate clavaria
Schaeff. 175 (1762) ...		Yellow clavaria
		(Pollard fungus)
Pers., Ic. and Desc. 11 (1798) ...		Elegant clavaria
Holmsk. I. 7 1818 ...		Brittle clavaria ...
Sow., Fung. 224 1797		Spindle-shaped clavaria
Pers., Comm. 44 (1797)		Dingy-grey clavaria ...
Muell., Fl. Dan. 873 ...		Unequal clavaria
Fries, S.M. I. 473 (1821)		Rush-like clavaria
F. v. M., Linn. Soc. N.S.W. 105 1882 ...		Kalchbrenner's clavaria
Fries, Epicr. 572 1838		Krombholz's clavaria ...
Fries, S.M. I. 474 1821		Kunze's clavaria
Berk. and Br., Linn. Journ. XIV. 76 (1875) ...		Bright clavaria ...
Pers. in Linn. Journ. XVIII. 386 (1881) ...		Very bright clavaria ...
Berk., Linn. Journ. XIII. 169 1873		Crooked bush clavaria ...
Kalch., Linn. Soc. N.S.W. 105 1882		Lurid clavaria
Berk., Hk. Journ. 147 1852		Crimson clavaria
Pers., Comm. 2 (1797)		Musty clavaria

Arrangement of Genera (2).

95. Clavaria, *Linn.*

Number	Habitat						R.	Occurrence	General Characters.
	W.A.	S.A.	T.	V.	N.S.W.	Q.			

CORDA, IC. FUNG. II. 35 (1838).

S.M. I. 464 (1821).—Clavaria.

1183	...	...	...	...	N.S.W.	...	B.	Wood	Very much branched, whitish. Branches intricate, zoneless, serrate. *Edible.*

Sp. Pl. 1182 (1753).—Typhula.

1184	...	...	...	V.	...	...	B.	Firwood ...	Ochrey, very much branched, green where bruised. Trunk whitish, downy, thick. Branches crowded.
1185	...	...	T.	...	...	Q.	...	Ground ...	Tufted, short, orange, fan to club shaped, rather rough.
1186	...	...	...	V.	N.S.W.	Q.	B.	Ground ...	Tufted, fragile, clay coloured, pale. Clubs simple, variable. Stem shining, yellow.
1187	...	...	...	V.	...	...	...	Ground ...	Orange, quite simple, straight, thickened upwards into club, tapering downwards into stem.
1188	...	...	...	V.	N.S.W.	Q.	B.	Woods ...	Yellow. Trunk thick, divided into stout straight forked much-divided tapering branches.
1189	W.A.	...	T.	V.	N.S.W.	Q.	B.	Ground ...	Fragile, white. Trunk very thick, much branched. Branches swollen, and tips red.
1190	...	...	...	V.	...	...	B.	Woods ...	Fragile, stuffed, grey. Trunk rather thick, short, much branched, wrinkled.
1191	...	...	...	...	...	Q.	...	Bare ground and decayed wood	Small, branched from compressed base, branches erect, forked, brown when dry.
1192	...	...	...	V.	...	...	B.	Moist woods ...	Rather fragile, usually tufted, white, hollow inside. Trunk thick, repeatedly and irregularly branched.
1193	W.A.	...	...	...	...	...	B.	At base of trunks	Very much branched, tan to ochrey. Trunk thin, shaggy, and rooting. Branches flexuous, spreading.
1194	...	...	T.	V.	...	Q.	B.	Woods ...	Tough, stuffed, dingy white. Branches dilated above, and fringed.
1195	...	...	...	V.	...	...	B.	Woods ...	Minute, thin, saffron yellow. Stem naked, pale. Branches and branchlets somewhat forked.
1196	...	...	...	V.	N.S.W.	Q.	B.	Pastures ...	Yellow tufted, tough, much branched. Branches equal, short, spreading, branchlets twiggy.
1197	...	S.A.	T.	V.	N.S.W.	Q.	B.	Gravelly ground ...	Fragile. Trunk thick, fleshy, white, very much branched. Branches tapering, twiggy, yellow.
1198	...	...	...	V.	N.S.W.	Q.	B.	Woods ...	Trunk thick, whitish, very much branched. Branches rosy orange, branchlets yellowish.
1199	...	...	...	V.	...	...	B.	Ground ...	In bundles, very delicate, yellow above, white below, sometimes entirely white. Clubs hollow, variable.
1200	...	...	...	...	...	Q.	B.	Among grass	Tufted and run together, yellow, soon hollow. Clubs somewhat spindle shaped, simple and toothed.
1201	...	...	...	V.	...	...	B.	Ground in woods ...	Grey, firm, fragrant. Trunk thick, whitish. Branches tapering and dingy grey, as well as branchlets.
1202	...	...	T.	V.	N.S.W.	...	B.	Among sand	Tufted, yellow, fragile. Clubs various, simple or forked. Apex jagged.
1203	...	...	T.	V.	...	...	B.	Among dead leaves	Gregarious. Thin, thread-like, flabby, hollow, pale to reddish brown, with creeping base of fibrils.
1204	...	...	...	V.	...	...	...	Ground ...	Thin, pale, orange yellow, somewhat tufted. Trunk, thin, naked. Branches short, forked, or tufted.
1205	...	...	...	V.	...	...	B.	Ground ...	Fragile, tufted, white, sparingly branched. Branches rather compressed, obtuse.
1206	...	...	...	...	...	Q.	B.	Woods	Rather fragile, very much branched from the thin base, white. Branches elongated, crowded, repeatedly forked.
1207	...	...	...	V.	...	...	...	Red soil ...	Simple, tufted, acute, shining red, without evident stem.
1208	...	...	...	...	...	Q	...	Ground ...	Tufted, orange, repeatedly forked, compressed. Apices dilated, subdivided, tawny.
1209	...	...	...	V.	...	...	...	Ground ...	Pale umber. Branches straight. Apices shortly bifid, and rather acute.
1210	...	...	...	V.	...	...	...	Ground ...	Tufted, very much branched, dirty white. Trunks thin. Branches and branchlets crowded, elongated, tawny when dry.
1211	...	...	...	V.	...	Q.	...	Rotten wood ...	Gregarious, fragile. Clubs simple, acute, crimson.
1212	...	...	...	...	N.S.W.	Q.	...	Ground and musty wood	Gregarious, minute, simple or very sparingly branched, white, becoming yellowish, sometimes rosy.

Number.	Cooke's Number.	Saccardo's Number.	Scientific Name.	Authority for Name.	English Name.
					95. CLAVARIA.—*Linn.*,
1213	1136	...	C. Muelleri	Berk., Grev. XX. 10 (1891)	Mueller's clavaria
1214	1098	VI. 7938	C. muscoides	Linn., Sp. Pl. 1183 (1753)	Moss-like clavaria
1215	1130	„ 8096	C. paludicola	Libert., in Fries Hym. Eur. 678 (1874)	Marsh-growing clavaria
1216	1128	„ 8085	C. pistillaris	Linn., Sp. Pl. 1182 (1753)	Pestle-shaped clavaria
1217	1106	„ 7955	C. plebeja	Fries, Pl. Preiss. II. 137 (1846)	Plebeian clavaria
1218	1116	„ 7996	C. portentosa	Berk. and Br., Linn. Trans. II. 65 (1883)	Monstrous clavaria
1219	1107	„ 7957	C. pyxidata	Pers., Comm. (1797)	Box-like clavaria
1220	1132	„ 8112	C. rhizomorpha	Berk., Fl. Tasm. II. 261 (1860)	Root-shaped clavaria
1221	1121	„ 8063	C. rosea	Fries, S.M. I. 482 (1821)	Rosy clavaria
1221a	1121	...	C. rosea, var. attenuata	Fries, Obs. 2 (1814)	Attenuated clavaria
1222	1120	VI. 8062	C. rufa	Muell., Fl. Dan. 755	Reddish clavaria
1223	1103	„ 7947	C. rugosa	Bull, Champ. t. 448 (1798)	Wrinkled clavaria
1224	1114	„ 7948	C. stricta	Pers., Comm. 45 (1797)	Straight clavaria
1225	...	„ 7953	C. subtilis	Pers., Comm. (1797)	Slender clavaria
1226	1135	...	C. tasmanica	Berk., in Herb., Grev. XX. 10 (1891)	Tasmanian clavaria
1227	1126	VI. 8079	C. vermicularis	Scop., in Fries, S.M. I. 484 (1821)	Worm-like clavaria

OF AUSTRALIAN FUNGI—*continued*.

Sp. Pl. 1182 (1753).—Typhula—*continued*.

Number	W.A.	S.A	T.	V.	N.S.W.	Q.	D.	Occurrence	General Characters.
1213	...	...	...	V.	...	Q.	...	Ground	Simple, club shaped, white, slender, tapering below into thin cylindrical stem.
1214	...	...	...	V.	N.S.W.	...	D.	Pastures among moss	Tough, slender, yellow, two to three times forked. Stem thin. Branches curved, long, graceful.
1215	...	S.A.	...	...	...	...	...	Moist places among ferns	Small, simple, slightly compressed, rough, yellow, orange when dry. Clubs short, obtuse.
1216	...	...	...	V.	...	...	B.	Among grass ..	Simple, tall, fleshy, stuffed, yellow to reddish. Club large.
1217	W.A.	...	...	..	...	...	...	Sandy soil ...	Tough, white, becoming yellow. Trunk thickish. Branches and branchlets very much divided and crested at top.
1218	...	...	...	...	...	Q.	...	Among leaves ...	Whitish. Stem somewhat cylindrical, rough, repeatedly much branched. Apices elongated.
1219	...	...	...	V.	N S.W.	...	B.	Rotten wood ...	Pale tan to reddish. Trunk thin, branched. Branches and branchlets all excavated at the tips into little cups.
1220	...	...	T.	...	...	...	...	Dead bark ...	Erumpent, confluent, chestnut red, nearly simple.
1221	...	...	...	V.	...	...	B.	Ground among moss	In bundles, fragile, rosy. Clubs stuffed, at length yellowish at tips, tapering downwards, white.
1221A	...	...	...	V.	...	...	...	Ground among moss	Clubs tapering at apex.
1222	...	...	...	V.	...	Q.	B.	Among grass ...	Tufted, rufous. Clubs stuffed, thickened, sometimes bifid, acute.
1223	...	S.A.	...	V.	...	Q.	B.	Moist places ...	Tough, simple or sparingly branched, thickened upwards and wrinkled, white. Branches deformed.
1224	...	...	...	...	N.S.W.	Q.	B.	Trunks ...	Very much branched, pale yellow, turning brown when bruised. Trunk thickish. Branches and branchlets straight.
1225	...	...	...	V.	...	...	...	Ground in woods ...	Scattered, slender, somewhat tough, whitish, becoming pale, smooth at base.
1226	...	...	T.	...	...	...	...	Tree ferns, wood, &c.	Clubs simple, single or two or three together, sooty brown, base expanded in a white woolly web.
1227	...	...	..	V.	...	...	B.	Among grass ...	Tufted, fragile, white. Clubs tufted, simple, cylindrical, often flexuous or incurved.

8 of 252

ORDER VI.—TREMELLACEÆ, FRIES.

SUB-ORDER I.—Auriculariæ, Bref.—Basidia or spore-bearing

Genera (2)—
96. Auricularia.

SUB-ORDER II.—Tremelleæ, Bref.—Basidia globose or ovoid,

Genera (5)—
98. Exidia, Fries.
99. Ulocolla, Bref.
100. Tremella, Linn.

SUB-ORDER III.—Dacryomyceteæ, Bref.—Basidia

Genera (3)—
103. Dacryomyces, Nees.
104. Guepinia, Fries.

Number.	Cooke's Number.	Saccardo's Number.	Scientific Name.	Authority for Name.	English Name.

ORDER VI.—TREMELLACEÆ,

96. AURICULARIA.—Bull,

1228	1145	VI. 8302	A. albicans	Berk., Linn. Journ. XIII. 170 (1873)	Whitish auricularia
1229	1144	„ 8293	A. lobata	Somm., Mag. Nat. Vid. (1827)	Lobate auricularia
1230	1143	„ 8294	A. mesenterica	Fries, Epicr. 555 (1838)	Intestine-like auricularia
1231	1146	„ 8303	A. minuta	Berk., Hook., Lond. Journ. IV. 59 (1845)	Minute auricularia
1232	1147	„ 8305	A. pusio	Berk., Linn. Journ. XVIII. 386 (1881)	Small auricularia

97. HIRNEOLA.—Fries,

1233	1150	VI. 8312	H. auricula-judæ	Berk., Outl. 289 (1860)	Jew's ear hirneola
1234	1148	„ 8309	H. auriformis	Fries, Fung. Nat. 26 (1848)	Ear-shaped hirneola
1235	1151	„ 8319	H. fusco-succinea	Mont., Cuba 364 (1842)	Amber-brown hirneola
1236	1153	„ 8323	H. hispidula	Sacc. Syll. VI. 769 (1888)	Hispid hirneola
1237	1149	„ 8311	H. polytricha	Mont., in Bel. Voy. 154.	Many-haired hirneola
1238	1152	„ 8320	H. rufa	Fries, Fung. Nat. 27 (1848)	Reddish-brown hirneola
1239	1154	„ 8328	H. vitellina	Fries, Fung. Nat. 27 (1848)	Egg-yellow hirneola
1239a	..	„ „	H. vitellina, var. tasmanica	Berk., Fl. Tasm. II. 262 (1860)	Tasmanian hirneola

98. EXIDIA.—Fries,

| 1240 | 1156 | VI. 8352 | E. albida | Bref., Unters. VII. 94 (1888) | Whitish exidia |
| 1241 | 1155 | „ 8347 | E. glandulosa | Fries, S.M. II. 224 (1821) | Glandulous exidia |

99. ULOCOLLA.—Bref.,

| 1242 | 1157 | VI. 8367 | U. foliacea | Bref. Unters. VII. 95 (1888) | Leafy ulocolla |

100. TREMELLA.—Linn.,

1243	1158	VI. 8375	T. frondosa	Fries, S.M. II. 212 (1821)	Frondose tremella
1244	1160	„ 8384	T. fuciformis	Berk., Hook., Journ. 277 (1856)	Seaweed-like tremella
1245	1159	„ 8377	T. lutescens	Pers., Syn. 622 (1801)	Yellowish tremella

ARRANGEMENT OF GENERA (10).

bodies, elongated or fusoid, transversly divided.

97. Hirneola.

four partite in a cruciate manner when mature.

101. Seismosarca, Cooke.
102. Tremellodon, Pers.

cylindrical or club shaped, forked upwards.

105. Calocera, Fries.

Number.	Habitat.						B.	Occurrence.	General Characters.
	W.A.	S.A.	T.	V.	N.S.W.	Q.			

FRIES, SYST. MYC. II. 207 (1823).

Herb. Fr. I. 36 (1787).—Helvella, Thelephora.

1228	...	...	...	...	...	Q.	...	Trunks ...	... Circular, whitish, delicately downy beneath.
1229	...	...	...	...	N.S.W.	Q.	B.	Bark ...	... Expanded, reflexed, lobed, variegated with hispid zones, or velvety, or smooth, dusky to whitish.
1230	W.A.	...	...	...		Q.	B.	Trunks	... Reflexed, entire, shaggy, zoned, bownish to ashy. Hymenium ribbed and folded.
1231	W.A.	...	T.	...			...	...	Dead branches	... Gregarious, expanded behind. Caps minute, lobed, hispid, zoned, tawny umber.
1232	...	...	...	...		Q.	...	Trunks ...	... Cap attached behind, reflexed, white, downy, wrinkled. Margin lobed. Small but distinct species.

Pl. Homon. 93 (1825).—Tremella, Peziza, Auricularia, Exidia.

1233	...	...	T.	V.	N.S.W.	Q.	B.	Trunks ...	... Hollow ear-like cups, flexuous, thin, blackish, with vein-like folds on both sides, downy beneath.
1234	...	...	...	...		Q.	...	Trunks ...	... Tufted, stalked, glaucous brown. Cups semicircular, veined beneath. Stem short, twisted, lateral.
1235	W.A.	...	...	...	N.S.W.	Q.	...	Bark	... Broad, sessile, shell shaped, then flattened. Margin wavy, internally netted with veins, amber brown.
1236	...	...	...	V.	...	Q.	...	Wood	... Globose to bell shaped, oblique, sessile, internally dark brown, externally with short fawn woolly hairs.
1237		S.A.	...	V.	N.S.W.	Q.	B.	Trunks	... Cups hemispherical, expanded, ear shaped, shaggy, grey, produced into very short oblique stem.
1238	...	...	...	...		Q.	...	Trunks ...	... Cup shaped, somewhat lateral, sessile, beset with tufted short reddish-brown bristles.
1239	...	...	T.	...		...	...	Trunks ...	... Cup shaped, sessile, excavated. Hymenium egg yellow.
1239a	...	...	T.	...		...	...	Wood ...	... Pale, circular, wavy, small. Stem short, compressed.

S.M. II. 220 (1823).—Tremella.

1240	W.A.	S.A.	T.	V.	N.S.W.	Q.	B.	Branches and bark of dead logs	Ascending, tough, expanded, wavy, whitish, tawny when dry.
1241	W.A.	...	T.	...			B.	Trunks and woods	Expanded, rather flattened, thick, turning black, with conical pimples, ashy beneath and somewhat downy.

Unters. VII. 95 (1888).—Tremella.

1242	W.A.	...	T.	V.		Q.	B.	Old trunks	... Tufted, wavy, cinnamon to flesh colour, folded at the base.

Sp. Pl. 1157 (1753).—Elvella, Thelephora.

1243	...	...	T.	...			B.	Old trunks	... Tufted, large, yellow to pale, folded at base. Lobes folded and waved.
1244	...	...	...	...		Q.		Trunks ...	... White, tufted, repeatedly lobed or forked. Lobes dilated in fan-like manner.
1245	...	S.A.	T.	V.	N.S.W.	Q.	B.	Fallen branches	... Tufted, small, very soft, wavy, and folded, yellowish. Lobes entire, naked.

Number.	Cooke's Number.	Saccardo's Number.	Scientific Name.	Authority for Name.	English Name.
					100. TREMELLA.—*Linn*.,
1246	1161	VI. 8387	T. mesenterica ...	Retz., Vet. Ak. Handl. 249 (1769) ...	Contorted tremella ... ...
1247	...	IX. 1071	T. microscopica ...	Berk. and Br., Linn. Trans. II. 220 (1887)	Microscopic tremella ... ...
1248	1162	VI. 8397	T. olens ...	Berk., Fl. Tasm. II. 262 (1860) ...	Scented tremella ...
1249	...	„ 8444	T. sarcoides	*Fries*, S.M. II. 215 (1821)	Flesh-coloured tremella
1250	1163	„ 8402	T. viscosa	Berk., Outl. 288 (1860) ...	Sticky tremella ...
					101. SEISMOSARCA.—Cooke,
1251	1164	IX. 1082	S. hydrophora	Cooke, Grev. XVIII. 25 (1889)	Watery seismosarca ... ...
					102. TREMELLODON.—Pers.,
1252	942	VI. 6862	T. gelatinosum	*Pers.*, Myc. Eur. II. 172 (1822) ...	Gelatinous tremellodon ...
					103. DACRYOMYCES.—Nees,
1253	1167	VI. 6472	D. deliquescens	*Duby*, Bot. Gall. 729 (1822) ...	Deliquescent dacryomyces ..
1254	1165	„ 8469	D. miltinus	Berk., Fl. Tasm. II. 263 (1860) ...	Vermilion dacryomyces ..
1255	1166	„ 8471	D. rubrofuscus	Berk., Hook., Lond. Journ. IV. 61 (1845)	Reddish-brown dacryomyces ..
1256	1169	„ 8483	D. Sacchari	Berk. and Br., Linn. Trans. II. 65 (1883)	Sugar-cane dacryomyces ...
1257	1171	„ 8502	D. sclerotioides	Berk., Fl. Tasm. II. 263 (1860) ...	Sclerotium-like dacryomyces ...
1258	1170	„ 8488	D. seriatus	Berk., Fl. Tasm. II. 263 (1860) ...	Seriate dacryomyces ...
1259	1168	„ 8473	D. stillatus	Nees, Syst. 89 (1816) ...	Dripping dacryomyces ...
					104. GUEPINIA.—Fries,
1260	1172	VI. 8514	G. merulina	*Quel* / Quelq. Esp. II. 11 (1878)	Merulius-like guepinia ...
1261	1173	„ 8518	G. pezizæformis	Berk., Hook., Lond. Journ. IV. 60 (1845)	Cup-shaped guepinia ...
1262	1174	„ 8520	G. spathularia	*Fries*, Elench. II. 32 (1828) ...	Spoon-shaped guepinia
					105. CALOCERA.—Fries,
1263	1139	VI. 8158	C. cornea	Fries, S.M. I. 485 (1821)	Horny calocera ... ...
1264	1138	IX. 1042	C. digitata	Cooke and Mass., Grev XVII. 7 (1888) ...	Digitate calocera ...
1265	1142	VI. 8165	C. glossoides	*Fries*, S.M. I. 487 (1821)	Tongue-like calocera ... ...
1266	1137	„ 8154	C. guepinioides	Berk., Hook., Lond. Journ. IV. 61 (1845)	Guepinia like calocera ...
1267	1141	IX. 1041	C. nutans	Sacc. Hedw. 154 (1890)	Nodding calocera ...
1268	1140	VI. 8163	C. stricta ...	Fries, Epicr. 581 (1838)	Erect calocera ...

OF AUSTRALIAN FUNGI—*continued.*

Number	Habitat						B.	Occurrence.	General Characters.
	W.A.	S.A.	T.	V.	N.S.W.	Q.			

Sp. Pl. 1157 (1753).—Elvella, Thelephora—*continued.*

1246	W.A.	S.A.	...	V.	N..S.W	Q.	B.	Dead branches	... Toughish, variable in form, folded and wavy, bright orange.
1247	...	...	...	V.		Q.	...	Leaves ...	Minute, hemispherical, point-like, dark-green scattered spots on upper surface of leaf.
1248	...	...	T.	...			...	Rotten wood	... Irregular, gelatinous, pale, scented.
1249	...	...	...	V.			B.	Trunks ...	Tufted, soft, viscid, pale-flesh colour, at first club shaped, then compressed. Lobed and folded. Conidial stage of *Ombrophila sarcoides.*
1250	...	...	T.	...			B.	Old wood	... Expanded, flattened, wavy, rather viscid, white to hyaline, brown when dry.

Grev. XVIII. 25 (1889).

| 1251 | ... | ... | ... | ... | N.S.W. | ... | ... | Wood | ... Inflated, gelatinous, lobate, pale sooty brown, very soft and watery, covered with scattered coloured hairs. |

Myc. Eur. II. 172 (1825).—Hydnum.

| 1252 | ... | ... | ... | ... | | Q. | B. | Ground and trunks | Gelatinous, tremulous, semicircular, somewhat stalked, greyish green to brownish. Only tremelloid fungus with true spines. |

Syst. 89 (1816).—Tremella.

1253	...	...	T.	V.		...	B.	Rotten wood	... Nearly round, rooting, yellowish, contorted at length, hyaline.
1254	...	S.A.	T.	...	N.S.W.	...	...	Dry wood	... Small, vermilion, lobate and folded.
1255	W.A.	...	...	...		...	...	Rotten branches	... Small, reddish brown, black when dry, cracked and folded.
1256	...	...	...	...		Q.		Stems of *Saccharum*	Irregular, thin, gelatinous, orange red, seated on whitish layer, spreading over charred stems of sugar cane.
1257	...	...	T.	...		...		Bark ...	White, circular, depressed in centre, cup shaped.
1258	...	...	T.	...		...		Bark ...	Bursting through, arranged in a row, whitish, then yellowish.
1259	...	S.A.	...	V.	N.S.W.	...	B.	Rotten wood	... Nearly round, folded, yellow to orange, colour persistent.

Pl. Homon. 92 (1825).—Dacryomyces, Merulius, Tremella.

1260	...	S.A.	...	...		...	...	Rotten wood of *Melaleuca*	Jelly-like to tough, orange yellow, solitary or in tufts. Stem at first clubbed, then expanding into cup shape.
1261	W.A.	...	T.	V.		Q.	...	Wood ...	... Minute, velvety, red. Stem short, velvety. Hymenium obliquely cup shaped.
1262	...	...	...	V.	N.S.W.	Q.	...	Wood and fences ...	Tufted, somewhat erect, rooting. Cap semicircular, spoon shaped. Stem downy, glaucous.

Syst. Myc. I. 485 (1821).—Clavaria.

1263	...	...	...	V.		Q.	B.	Rotten wood	... Tufted, rooting, viscid, orange yellow. Clubs short, grown together at base.
1264	...	...	...	V.			...	Damp logs	... Branched, tough, pale. Trunk thin, twice or thrice forked. Branches expanded at apex like a spoon each bearing from three to five finger-like processes.
1265	...	...	T.	V.		...	B.	Trunks ...	... Simple, solitary, jelly-like, yellow. Clubs thickened, compressed. Stem tapering.
1266	W.A.	S.A.	T.	V.		Q.	...	Rotten wood	... Small, bursting through, reddish brown. Stem compressed, palmate above. Branches few.
1267	...	S.A.	...	V.		...	...	Trunks ...	... Scattered, tapering, compressed, honey yellow, curved. Stem very short, but distinct.
1268	...	...	...	V.		...	B.	Wood and dead fir leaves	Simple, solitary, elongated, linear, yellow. Clubs short, grown together at base.

GENERAL CLASSIFICATION OF GASTROMYCETES.

GROUP II.—GASTROMYCETES, WILLD.

ARRANGEMENT OF ORDERS (6).

Above ground—

7. PILACREACEÆ—Minute. Peridium eventually disintegrating. Intermediate between Hymenomycetes and Gastromycetes.
8. PHALLOIDEACEÆ—Fleshy to gelatinous. Receptacle and spore-bearing surface enclosed in universal volva.
9. NIDULARIACEÆ—Leathery. Spores never powdery.
10. LYCOPERDACEÆ—Membranous to leathery. Spores forming powdery mass when mature.
11. SCLERODERMACEÆ—Leathery. Peridium thick, sessile or stalked, opening at apex.

Subterranean—

12. HYMENOGASTRACEÆ—Fleshy to firm. Peridium indehiscent.

ORDER VII.—PILACREACEÆ, BREF.

Genus (1)—
106. Pilacre, Frios.

ORDER VIII.—PHALLOIDEACEÆ, FRIES.

Genera (8)—

107. Dictyophora, Desv.	109. Mutinus, Fries.	111. Colus, Cav. and Sec.	113. Anthurus, Kalch.
108. Ithyphallus, Fries.	110. Clathrus, Linn.	112. Lysurus, Fries.	114. Aseroë, La Bill.

ORDER IX.—NIDULARIACEÆ, FRIES.

Genera (3)—

115. Cyathus, Hall.	116. Crucibulum, Tul.	117. Sphærobolus, Tode.

ORDER X.—LYCOPERDACEÆ, EHR.

Genera (18)—

118. Secotium, Kunze.	123. Gymnoglossum, Mass.	127. Calostoma, Desv.	132. Areolaria, Forq.
119. Chainoderma, Mass.		128. Geaster, Scop.	133. Castoreum, Cooke and Mass.
120. Cycloderma, Klot.	124. Protoglossum, Mass.	129. Diploderma, Link.	
121. Mesophellia, Berk.	125. Tulostoma, Pers.	130. Bovista, Scop.	134. Xylopodium, Mont.
122. Podaxon, Fries.	126. Battarrea, Pers.	131. Lycoperdon, Linn.	135. Favillea, Fries.

ORDER XI.—SCLERODERMACEÆ, FRIES.

Genera (4)—

136. Scleroderma, Pers.	137. Polysaccum, D.C.	138. Arachnion, Schw.	139. Paurocotylis, Berk.

ORDER XII.—HYMENOGASTRACEÆ, VITT.

Genera (5)—

140. Octaviania, Vitt.	142. Hymenogaster, Vitt.	143. Hydnangium, Wallr.	144. Gautieria, Vitt.
141. Rhizopogon, Fries.			

G

GROUP II.—GASTROMYCETES.—

ORDER VII.—PÍLACREACEÆ,

Number.	Cooke's Number	Saccardo's Number.	Scientific Name.	Authority for Name.	English Name.

106. PILACRE.—Fries,

Number	Cooke	Sacc	Name		
1269	1989	IV. 2750	P. divisa...	.. Berk., Fl. N.Z. II. 197 (1855)	Divided pilacre
1270	1990	„ 2752	P. Petersii	... Berk. and Curt., Ann. Nat. Hist. III., 3rd Ser. 362 (1889)	Peters' pilacre

ORDER VIII.—PHALLOIDEACEÆ,

Number.	Cooke's Number	Saccardo's Number.	Scientific Name.	Authority for Name.	English Name.

107. DICTYOPHORA.—Desv., Journ.

1271	1179	VII. 13	D. merulina	Berk., Linn. Journ. XIII. 172 (1873)	... Merulius-like dictyophora
1272	1178	„ 11	D. multicolor	... Berk. and Br., Linn. Trans. II. 65 (1883)	... Many-coloured dictyophora ...
1273	1175	„ 2	D. phalloidea	... Desv., Journ. Bot. II. 88 (1809)	... Phallus-like dictyophora ...
1274	1177	„ 6	D. speciosa	... Meyen., Nov. Act, XIX. 239 (1843)	... Handsome dictyophora ...
1275	1176	„ 3	D. tahitensis	... Fisch., Sacc. Syll. VII. 4 (1888)...	... Tabitian dictyophora ...

108. ITHYPHALLUS.—Fries, S.M. II. 283 (1823).—

1276	1183	VII. 22	I. aurantiacus	... Fisch., Sacc. Syll. VII. 9 (1888)...	... Orange-coloured ithyphallus ...
1277	1182	„ 21	I. calyptratus	... Fisch., Sacc. Syll. VII. 9 (1888)...	... Capped ithyphallus
1278	1180	„ 18	I. impudicus	... Fisch., Sacc. Syll VII. 8 (1888)	... Impure ithyphallus ... (Stinkhorn)
1279	1181	„ 23	I. novæ hollandiæ	... Fisch., Sacc. Syll. VII. 10 (1888)	New Holland ithyphallus
1280	1181	„ 20	I. quadricolor	... Fisch., Sacc. Syll. VII. 9 (1888)	... Four-coloured ithyphallus
1281	1185	„ 27	I. retusus	... Fisch., Sacc. Syll. VII. 11 (1888)	... Blunt ithyphallus ...
1282	1186	„ 29	I. rubicundus	... Fisch., Sacc. Syll. VII. 11 (1888)	Rubicund ithyphallus ...

WILLD., BEMERK. FARR. (1802).

BREF. UNT. VII. (1888).

Number.	Habitat.						B.	Occurrence.	General Characters.
	W.A.	S.A.	T.	V.	N.S.W.	Q.			

Pl. Homon. 364 (1825).

| 1269 | ... | ... | T. | ... | ... | ... | ... | Bark | ... Head globose, clay coloured. Stem divided, brownish. |
| 1270 | ... | ... | ... | ... | N.S.W. | ... | B. | Trunks | ... Gregarious, often covering half-dead trunks for a considerable distance. Head relatively large. Stem short, whitish. |

FRIES, S.M. II. 281 (1823).

Number.	Habitat.						B.	Occurrence.	General Characters.
	W.A.	S.A.	T.	V.	N.S.W.	Q.			

Bot. II. 92 (1809).—Phallus.

1271	...	...	...	...	...	Q.	...	Ground ...	Gregarious. Cap bell shaped, ochrey, covered with fœtid brown spore-bearing mass. Stem distinct, white.
1272	...	...	...	...	N.S.W.	Q.	...	Ground ...	... About 7 inches high. Cap conical, orange, netted. Stem cream coloured, tapering to base.
1273	...	...	...	...		Q.	...	Sandy soil	... Cap thickened at apex, bell shaped, white, netted. Stem white and pitted.
1274	...	...	...	...	...	Q.	...	Ground ...	... Stem tapering upwards, white. Cap joined to stem at apex with short collar, bell shaped.
1275	...	...	...	...	N.S.W.	Q.	...	Ground ...	... Stem cylindrical, walls pitted. Cap at apex of stem without collar ovate, roughly netted.

Phallus, Omphalo-phallus, Cynophallus.

1276	...	...	...	...		Q.	...	Ground ...	Stem 6 to 8 inches long, ¾ inch thick, orange. Cap without collar or ring, thimble shaped, orange.
1277	...	...	...	...	...	Q.	...	Among grass	... Scarcely 2 inches high. Stem slightly tapering upwards. Cap somewhat hemispherical, orange.
1278	...	...	...	...		Q.	B.	Ground ...	... Up to 10 inches high. Stem tapering above and below, white. Cap conical, netted ; gelatinous mass of spores, dark olive.
1279	...	...	...	V.	N.S.W.	Q.	...	Ground ...	... Stem white, slender, tapering upwards. Cap narrow bell shaped, netted.
1280	...	...	...	...		Q.	...	Ground ...	... About 4½ inches high. Stem cylindrical, lemon colour, veil white; mycelium purple. Cap orange coloured.
1281	...	...	...	...	N.S.W.	...	...	Ground ...	... About 6 inches high. Stem 1 inch or more thick, cylindrical. Cap ovate, blunt above apex of stem, white.
1282		...	...	V.	...	...	...	Ground ...	... Stem between 5 and 6 inches high, spindle shaped, red. Cap conical to bell shaped, clad with brownish gluten.

Number.	Cooke's Number.	Saccardo's Number.	Scientific Name.	Authority for Name.	English Name.
					109. MUTINUS.—Fries, Summ. Veg.
1283		VII. 30	M. caninus	... Fries, S.V.S. II. (1849)	... Dog mutinus ... ...
1284	1188	„ 38	M. curtus	... Fisch., Sacc. Syll. VII. 13 (1888)	... Short mutinus ... ...
1285	1190	„ 40	M. discolor	... Fisch., Sacc. Syll. VII. 14 (1888)	... Discoloured mutinus ... ...
1286	1189	„ 39	M. papuasius	... Kalch., Grev. IV. 74 (1875)	Papuan mutinus ...
1287	1187	„ 37	M. Watsoni	Fisch., Sacc. Syll. VII. 13 (1888)	Watson's mutinus ...
					110. CLATHRUS.—Linn.,
1288	1194	VII. 61	C. albidus	Becker, Schles. Gesell. 81 (1874)	... Whitish clathrus ...
1289	1195	„ 59	C. cibarius	Fisch., Sacc. Syll. VII. 20 (1888)	... Edible clathrus ...
1290	1196	„ 60	C. crispus	Turp., Dict. Sci. Nat. (1822) ...	... Curled clathrus ...
1291	1193	„ 58	C. gracilis	Schl., Linn. 166 (1861)... ...	... Graceful clathrus ...
1292	1192	„ 55	C. pusillus	Berk., Hook., Lond. Journ. IV. 67 (1845)	... Small clathrus ... ...
1293	1191	„ 51	C. triscapus	Fries, S.M. II. 287 (1823)	... Three-branched clathrus
					111. COLUS.—Cav. and Sec.,
1294	1197	VII. 62	C. hirudinosus	... Cav. and Sec., Ann. Sci. Nat. 253 (1835)	... Leech colus ... ...
1295	...	IX. 1093	C. Muelleri	... Fisch., Unters. Phall. 61 (1890)	... Mueller's colus ...
					112. LYSURUS.—Fries,
1296	1198	IX. 1095	L. australiensis	Cooke and Mass., Grev. XVIII. 6 (1889)	... Australian lysurus ... ...
					113. ANTHURUS.—Kalch.,
1297	1200	VII. 71	A. Archeri	.. Fisch., Sacc. Syll. VII. 24 (1888)	Archer's anthurus ...
1298	1199	„ 69	A. Muellerianus ...	... Kalch., Grev. IX. 2 (1880) ...	... Mueller's anthurus
					114. ASEROË.—La Bill.,
1299	1202	VII. 78	A. lysuroides	... Fisch., Jahrb. Bot. Gart. IV. (1886)	Lysurus-like aseroë ... ...
1300	1201	„ 76	A. rubra ...	... La Bill. Voy. 44 (1798)	... Red aseroë ... ...
1300a	„	„	A. rubra, var. pentactina	... (Sacc.) Syll. VII. 26 (1883) ...	... Five-rayed aseroë .. ...

OF AUSTRALIAN FUNGI—*continued*.

Number.	Habitat.						B.	Occurrence.	General Characters.
	W.A.	S.A.	T.	V.	N.S.W.	Q.			

Scan. II. 434 (1849).—Phallus, Cynophallus.

1283	...	...	...	...	...	...	B.	Woods	..., Receptacle somewhat spindle shaped, white or rosy, spore-bearing portion short, red. Stem 3 to 4 inches high, scentless.
1284	W.A.	...	...	...	...	...		Ground	... About 1 inch; receptacle broadly truncate at apex. Stem yellow; very fœtid.
1285	...	...	...	...	Q.	...		Ground ...	... Stem cylindrical, orange; spore-bearing part one-sixth of receptacle; apex yellowish grey or turning black.
1286	...	...	...	...	...	Q.	...	Ground ...	... About 3 to 4 inches high; receptacle thin and slender; spore-bearing part pear shaped, black.
1287	...	...	...	...	...	Q.		Ground ...	... About 2½ inches high; spore-bearing part conical, minutely veined, red.

Sp. Pl. II. 1179 (1753).—Ileodictyon.

1288	...	...	...	V.	...	...	...	Ground ...	... Branches of receptacle with broad channel, white, then yellowish.
1289	...	...	...	V.	...	Q.	...	Ground ...	... Receptacle spherical or ovoid, white, interstices broad; gregarious and common. *Edible*.
1290	...	...	...	...	N.S.W.	Q.	...	Sandy soil	... Receptacle spherical or obovate, vermilion or salmon colour; interstices rounded or oval.
1291	W.A.	S.A.	T.	V.	N.S.W.	Q.	...	Ground ...	Veil globose, splitting into about four lobes; receptacle ovoid, white; interstices hexagonal.
1292	W.A.	...	...	...		Q.	...	Ground	... Small; veil nearly cylindrical; receptacle bright ruby red.
1293	...	...	...	...		Q.	...	Ground ...	... Receptacle of three vertical branches, slender, thin, straight, white below, vermilion about apex.

Ann. Sci. Nat. 2 Ser. III. 253 (1835).—Clathrus.

| 1294 | W.A. | ... | ... | ... | ... | ... | ... | Ground ... | ... Receptacle spindle shaped, white, reddening at apex. Stem inversely conical, with meshes at top. |
| 1295 | ... | ... | ... | V. | | ... | ... | Ground ... | ... Receptacle unequally perforated, the superior meshes of equal diameter, inferior greatly elongated. Stem short. |

S.M. II. 285 (1823).—Mutinus.

| 1296 | ... | ... | ... | ... | | Q. | ... | Ground ... | ... Receptacle tawny, mostly five lobed. Stem cylindrical, hollow, whitish. Veil globose, torn in lobes. |

Grev. IX. 2 (1880).—Lysurus.

| 1297 | ... | ... | T. | ... | ... | ... | ... | Ground | ..., Receptacle 3½ inches high, rosy. Stem very short, divided into five long erect lobes. |
| 1298 | ... | ... | ... | V. | N.S.W. | ... | ... | Ground | ... Receptacle yellow to reddish, cup shaped or funnel shaped above, dilated. |

Voy. 44 (1798).—Lysurus.

1299	...	...	T.	...	...	...	...	Ground ...	... Stem white, slender, long, disc carmine rose above, externally distinct from stem.
1300	...	S.A.	...	V.	N.S.W.	Q.	B.	Ground ...	... Stem becoming red, margin divided into five to eight teeth, vermilion above, forked.
1300A	...	...	...	...	N.S.W.	...	...	Ground ...	... With five bifid rays.

86

SYSTEMATIC ARRANGEMENT

Number.	Cooke's Number.	Saccardo's Number.	Scientific Name.	Authority for Name.	English Name.

ORDER IX.—NIDULARIACEÆ,

115. CYATHUS.—Hall,

1301		VII. 101	C. ambiguus	... Tul., Ann. Sci. Nat. 75 (1844) ...	... Ambiguous cyathus ... ...
1302			C. Baileyi	... Mass., Grev. XXI. 3 (1892)	... Bailey's cyathus ...'
1303	1214	VII. 127	C. Colensoi	... Berk., Fl. N. Zeal. II. 192 (1855) ...	... Colenso's cyathus ... ...
1304	1212	„ 113	C. dasypus	... Nees. Phys. Ber. 41 (1820)	... Hairy-stalked cyathus ...
1305	1206	„ 104	C. desertorum	... F. v. M., Linn. Journ. XVIII. 387 (1881)	... Desert cyathus ... ...
1306	...	...	C. dimorphus	... Cobb., Ag. Gaz. N.S.W. III. Pt. 12 (1892)	... Dimorphic cyathus ... ...
1307	1213	VII. 121	C. fimetarius	... D. C. Fl. Fr. V. 104 (1815)	... Dung cyathus ..
1308	1209	„ 116	C. fimicola	... Berk., Linn. Journ. XVIII. 387 (1881) ...	... Dung-borne cyathus ..
1309	1205	„ 98	C. intermedius	... Tul., Mon. Nid. Ann. Sci. Nat. 72 (1844)...	... Intermediate cyathus ...
1310	1207	„ 109	C. Lesueurii ...	... Tul., Mon. Nid. Ann. Sci. Nat. 79 (1844)...	... Lesueur's cyathus ...
1311	1204	„ 96	C. Montagnei ...	... Tul., Mon. Nid. Ann. Sci. Nat. 70 (1844)...	... Montagne's cyathus ...
1312	1203	„ 95	C. novæ-zealandiæ	Tul., Mon. Nid. Ann. Sci. Nat. 66 (1844)...	New Zealand cyathus ...
1313	1210	„ 117	C. pezizoides ...	Berk., Linn. Journ XVIII. 387 (1881) ...	Peziza-like cyathus ...
1314	1211	„ 118	C. pusio ...	... Berk., Linn. Journ. XVIII. 387 (1881) ...	... Small cyathus
1315	1208	„ 110	C. vernicosus	... D. C. Flor. Fr. II. 270 (1805)	... Varnished cyathus

116. CRUCIBULUM.—Tul. Mon. Nid. Ann. Sci.

| 1316 | 1216 | | C. simile ... | ... Mass., Grev. XIX. 94 (1891) | ... Similar crucibulum ...' |
| 1317 | 1215 | VII. 128 | C. vulgare | Tul., Mon. Nid. Ann. Sci. Nat. 90 (1844)... | Common crucibulum ... |

117. SPHÆROBOLUS.—Tode,

| 1318 | 1217 | VII 136 | S. stellatus | Tode, Meck. 43 (1790) . | ... Stellate sphærobolus ... ... |

OF AUSTRALIAN FUNGI—*continued.*

Number	Habitat					Q.	B.	Occurrence	General Characters
	W.A.	S.A.	T.	V.	N.S.W				

FRIES, S.M. III. 296.

Helv. III. 127 (1768).—Nidularia, Peziza.

1301	...	...	...	...	...	Q.	...	Garden soil	Obconical to cup shaped, cartilaginous to membranous, tawny, rusty, or tawny cinnamon, beautiful silky gloss.
1302	...	...	...	...	...	Q.	...	Dung	More or less gregarious, inversely conical or bell shaped, thin and cartilaginous, minutely downy, cinnamon
1303	...	S.A.	T.	...	N.S.W.	...	...	Ground ...	Densely crowded, cup shaped, thin, flexible, dirty umber, downy.
1304	...	...	...	V.	...	...	...	Bare ground	Bell shaped, somewhat cylindrical, pale ochre, and minutely downy.
1305	...	S.A.	...	V.	N.S.W.	...	...	Sandy soil	Pale, downy, smooth within and even.
1306	...	...	...	...	N.S.W.	...	...	Ground ...	Gregarious. Peridium ash colour, bell shaped, with obscure circle of small markings half way up.
1307	...	...	...	...	...	Q.	...	Dung ...	Hemispherical, brown to tawny, velvety.
1308	...	...	...	V.	...	Q.	...	Dung ...	Cup shaped, umber, becoming pale, minutely velvety.
1309	...	...	...	...	...	Q.	...	Rubbish ...	Cup shaped, inversely conical, rusty, becoming yellowish, hairy, slightly streaked.
1310	...	...	...	V.	N.S.W.	Q.	...	Rotten wood	Membranous, thin, grey, clad with somewhat star-shaped hairs or naked.
1311	W.A.	...	...	...	...	Q.	...	Chips	Crucible shaped, rusty, with a few woolly hairs, internally smooth, streaked and ciliate above.
1312	...	...	...	...	N.S.W.	...	...	Rotten wood	Elongated, narrow, brown, woolly outside, streaked and furrowed inside.
1313	...	...	...	...	...	Q.	...	Rotting herbs	Cup shaped, densely downy outside with flexuous hairs, umber, very small, smooth inside.
1314	...	...	...	...	...	Q.	...	Trunks of *Eucalyptus*	Wine-glass shaped, whitish, clad outside with a fine down, smooth inside.
1315	W.A.	...	...	V.	...	Q.	B.	Ground and twigs ...	Clustered, bell shaped, nearly sessile, pale ochre to ash colour, then dusky, downy, lead colour or brown within.

Nat. 3 Ser. I. 89 (1844).—Cyathus, Nidularia, Peziza.

| 1316 | ... | ... | ... | ... | ... | ... | ... | Bark and wood ... | Crowded or scattered, somewhat cylindrical, thin, flexible, externally densely covered with ochrey-brown shaggy down. |
| 1317 | ... | ... | ... | V. | ... | Q. | B. | Twigs, wood, &c. ... | Gregarious, cylindrical to bell shaped, ochrey, then rusty, downy when young, then smooth, inside shining yellow. |

Merr. I. 43 (1790).—Lycoperdon.

| 1318 | ... | ... | T. | V. | ... | ... | B. | Wood, chips, &c. ... | Nearly spherical, fleshy, dehiscing with five to eight acute teeth, yellow, interior whitish. |

88

SYSTEMATIC ARRANGEMENT

Number.	Cooke's Number.	Saccardo's Number.	Scientific Name.	Authority for Name.	English Name.

ORDER X.—LYCOPERDACEÆ,

118. SECOTIUM.—Kunze

1319	1218	VII. 146	S. acuminatum	... Mont., Fl. Alg. I. 371 (1846) ...	... Acuminate secotium ... ...
1320	1219	,, 150	S. coarctatum	... Berk., Hook., Lond. Journ. IV. 63 (1845)	Compact secotium ...
1321	1221	,, 152	S. erythrocephalum	... Tul., Ann. Sci. Nat. 115 (1844) ...	... Red-headed secotium ... ...
1322	1222	,, 156	S. Gunnii...	... Berk.. in Cooke's Handb. Austr. Fung. 221 (1892)	Gunn's secotium ... ...
1323	1220	,, 151	S. melanosporum ...	Berk., Hook., Lond. Journ. IV. 62 (1845)	... Dark-spored secotium .. ...
1324	1223		S. scabrosum	... Cooke and Mass., Grev. XX. 35 (1891) ...	... Scabrous secotium ...

119. CHAINODERMA.—Massee,

| 1325 | 1224 | | C. Drummondii | ... Mass., Grev. XIX. 46 (1890) | ... Drummond's chainoderma ... |

120. CYCLODERMA.—Klotzsch,

| 1326 | 1225 | VII. 1584 | C. platyspora | Cooke and Mass., Grev. XVI. 73 (1888) ... | ... Broad-spored cycloderma ... |

121. MESOPHELLIA.—Berk.,

| 1327 | 1226 | VII 162 | M. arenaria | Berk., Linn. Trans. XXII. 131 (1857) | ... Sandy mesophellia ... |
| 1328 | 1227 | ,, 163 | M. ingratissima | De Toni, Sacc. Syll. VII. 57 (1888) | ... Strong-smelling mesophellia ... |

122. PODAXON.—Fries,

1329	1231	VII. 170	P. axata ...	... Mass., Mon. Pod. Journ. Bot. (1890)	... Axate podaxon ...
1330	1229	,, 168	P. carcinomalis	Fries, S.M. III. 62 (1829) ...	Cancerous podaxon ...
1331	1230	,, 171	P. indica ...	Spreng., Syst. Veg. V. 518 (1828)	Indian podaxon ...

123. GYMNOGLOSSUM.—Massee,

| 1332 | 1232 | | G. stipitatum | ... Mass., Grev. XIX. 97 (1891) | ... Stalked gymnoglossum |

124. PROTOGLOSSUM.—Mass.,

| 1333 | 1233 | | P. luteum | ... Mass., Grev. XIX. 97 (1891) ... | ... Yellow protoglossum ... |

125. TULOSTOMA.—Pers.,

1334	1240	...	T. album ...	Mass., Grev. XIX. 95 (1891)	White tulostoma ...
1335	1238	VII. 185	T. fimbriatum	... Fries, S.M. III. 43 (1829) ...	... Fringed tulostoma ...
1336	1239	,, 193	T. granulosum	Lev., Demid. Voy. IV. 120 (1842)	... Granular tulostoma ...
1337	1235	,, 177	T. leprosum	... Kalch., Grev. IV 72 (1875)	... Leprous tulostoma ...
1338	1234	,, 175	T. mamanosum	... Fries, S.M. III. 42 (1829)	... Teat-like tulostoma ...
1339	1237	,, 184	T. maximum	Cooke and Mass., Grev. XV. 94 (1887)	... Maximum tulostoma ...
1340	1241	IX. 1113	T. pulchellum	... Sacc., Bull., Soc. Myc. V. 118 (1889)	Beautiful tulostoma ...
1341	1236	VII. 182	T. Wightii	... Berk., Hook., Lond. Journ. I. 157 (1842)	Wight's tulostoma ...

OF AUSTRALIAN FUNGI—*continued.*

Number.	Habitat.						R.	Occurrence.	General Characters.
	W.A.	S.A.	T.	V.	N.S.W.	Q.			

EHR. SYLV. BER. 14 (1818).

Fl. 321 (1840).

1319	W.A.	...	...	...	...	...	...	Ground ...	Solitary, ovoid, tan or ochrey, acuminate at apex, cuticle breaking up into scales. Stem short, twisted.
1320	W.A.	S.A.	...	...	...	...	...	Ground ...	Minute, strong scented, inversely ovate; margin bent inwards. Stem slender, cylindrical.
1321	...	...	T.	...	...	...	...	Ground ...	Gregarious, rather long stemmed. Stem erect, naked, white. Peridium smooth, carmine red.
1322	...	...	T.	...	...	...	...	Ground ...	Small. Stem slender, solid, pale brown. Peridium somewhat globose, pale brown, smooth.
1323	W.A.	...	...	...	...	...	...	Ground ...	Tufted. Cap somewhat globose, mealy. Margin obtuse and rounded. Stem solid. Spores dark chocolate brown.
1324	...			V.			...	Ground ...	Hemispherical, dingy olive or greyish, rugged.

Grev. XIX. 46 (1890),—Secotium.

| 1325 | W.A. | ... | ... | ... | | ... | ... | Ground ... | Club to spindle shaped, dingy brown, smooth. Mass of spores dingy brown. |

Linn. VII. 203 (1832).

| 1326 | ... | ... | V. | | | ... | ... | Ground | Ovate, external thick, flexible, ochrey, internal thin, shining. |

Fl. Tasm. II. 266 (1860).—Inoderma.

| 1327 | ... | ... | T. | V. | ... | ... | ... | Ground ... | Thick, elliptical, externally clad with whitish woolly threads. |
| 1328 | ... | ... | ... | V. | ... | ... | ... | Ground ... | Strong scented, crustaceous, very fragile, somewhat globose. |

S.M. III. 62 (1829).—Lycoperdon, Mitremyces.

1329	...		...	...	N.S.W.	...	...	Dry sandy places ...	Tuberous rooting, oblong. Stem hollow, substance woody. Peridium ovate.
1330	...	S.A.	...	...		Q.	...	Sandy places	Peridium ovate, oblong, whitish. Stem cylindrical, curved.
1331	...	...	...	V.		Q.	...	Ground ...	Stem corded, often twisted lengthwise. Peridium club shaped, invested with saffron-yellow membrane.

Grev. XIX. 97 (1891).

| 1332 | ... | ... | ... | ... | N.S.W. | ... | ... | Ground ... | Obtusely conical, pitted, pale brown. Stem solid, pale brown. |

Grev. XIX. 97 (1891).

| 1333 | ... | ... | ... | ... | V. | Q. | ... | In rich black mould | Cylindrical, growing vertically, with extreme apex above, orange yellow, underground portion yellowish. No stem. |

Tent. Disp. 6 (1797).—Lycoperdon, Schizostoma.

1334	W.A.	...	...	...		...	...	Ground ...	Stem ochrey, wrinkled lengthwise. Peridium globose, pure white shining.
1335	W.A.	...	...	V.		...		Sandy soil	Peridium almost naked, scales falling away, becoming tawny. Stem tawny ochre, mouth torn, fringed.
1336	W.A.	...	...	V.		...	...	Ground ...	Peridium globose, depressed, brown; mouth teat-like; margin torn and toothed. Stem thickish.
1337	...	...	...	...		Q.	...	Ground	Peridium clad with a lurid umber mealy scurf, at length falling away.
1338	...	...	...	V.		Q.	B.	Ground	Stem hollow, covered more or less with falling scales. Peridium globose, with minute prominent teat-like mouth.
1339	W.A.	...	...	...		...	...	Ground	Peridium smooth, ochrey, with rounded mouth. Stem elongated, of same colour.
1340	...	S.A.	...	V.		...	...	Branches ...	Minute, shortly stalked. Stem cylindrical, smooth, whitish. Peridium membranous, sub-globose.
1341	...	...	...	...		Q.	...	Ground ...	Peridium papery, egg shaped to globular. Stem somewhat scaly.

Number	Cooke's Number	Saccardo's Number	Scientific Name	Authority for Name	English Name
					126. BATTARREA.—Pers.,
1342	1244	VII. 199	B. Muelleri	Kalch., Grev. IX. 3 (1880)	Mueller's battarrea
1343	1242	„ 195	B. phalloides	Pers., Syn. 129 (1801)	Phallus-like battarrea
1344	1243	„ 196	B. Steveni	Fries, S.M. III. 7 (1829)	Steven's battarrea
1345	1245	IX. 1115	B. Tepperiana	Ludw., Bot. Centr. 337 (1889)	Tepper's battarrea
					127. CALOSTOMA.—Desv.,
1346	1249	...	C. æruginosa	Mass., Grev. XIX. 96 (1891)	Verdigris-green calostoma
1347	1247	VII. 206	C. fusca	Mass., Ann. Bot. II. 43 (1888)	Brown calostoma
1348	1246	„ 205	C. lurida	Mass., Ann. Bot. II. 43 (1888)	Lurid calostoma
1349	1248	„ 207	C. viridis	Mass., Ann. Bot. II. 40 (1888)	Green calostoma
					128. GEASTER.—Scop.,
1350	1270	VII. 1590	G. Archeri	Berk., Fl. Tasm. II. 264 (1860)	Archer's geaster
1351	1271	IX. 1123	G. argenteus	Cooke, Grev. XVII. 75 (1889)	Silvery geaster
1352	1259	VII. 1592	G. australis	Berk., Fl. Tasm. II 265 (1860)	Southern geaster
1353	...	IX. 1119	G. Berkeleyi	Mass., Ann. Bot. 79 (1889)	Berkeley's geaster
1354	1253	VII. 226	G. Drummondi	Berk., Hook., Lond. Journ. IV. 63 (1845)	Drummond's geaster
1355	1269	„ 261	G. dubius	Berk., Linn. Journ. XVI. 40 (1878)	Doubtful geaster
1356	1264	„ 248	G. floriformis	Vitt., Mon. Lycop. 167 (1842)	Flower-shaped geaster
1357	1258	„ 238	G. fimbriatus	Fries, S.M. III. 16 (1829)	Fringed geaster
1358	1268	„ 257	G. hygrometricus	Pers., Syn. 135 (1801)	Hygrometric geaster
1359	1262	„ 246	G. lageniformis	Vitt., Mon. Lycop. 160 (1842)	Flask-shaped geaster
1360	1256	„ 229	G. lignicola	Berk., Linn. Journ. XVIII. 386 (1881)	Wood-growing geaster
1361	1267	„ 255	G. lugubris	Kalch., Gast. 10 (1883)	Mourning geaster
1362	1257	„ 232	G. minimus	Schwein., Syn. Car. 327 (1822)	Least geaster
1363	1265	„ 249	G. pusillus	Fries, Pl. Preiss. II. 139 (1846)	Small geaster
1364	1254	„ 1591	G. Readeri	Cooke and Mass., Grev. XVI. 73 (1888)	Reader's geaster
1365	1266	„ 251	G. rufescens	Pers., Syn. 134 (1801)	Reddish geaster
1366	1261	„ 245	G. saccatus	Fries, S.M. III. 16 (1829)	Saccate geaster
1367	1263	„ 247	G. Spegazzinianus	De Ton., Rev. Geast. in Rev. Myc. 19 (1887)	Spegazzini's geaster
1368	1252	„ 224	G. striatulus	Kalch., Grev. IX. 3 (1880)	Furrowed geaster
1369	1251	„ 222	G. striatus	D. C., Fl. Fr. II. 267 (1805)	Streaked geaster
1370	1255	„ 228	G. subiculosus	Cooke and Mass., Grev. XV. 97 (1887)	Subiculous geaster
1371	1250	„ 218	G. tenuipes	Berk., Hook., Lond. Journ. VII. 576 (1848)	Slender-stalked geaster
1372	1260	„ 242	G. vittatus	Kalch., Grev. IX. 3 (1880)	Vittate geaster

OF AUSTRALIAN FUNGI—*continued.*

Number.	W.A.	S.A.	T.	V.	N.S.W.	Q.	B.	Occurrence.	General Characters.

Syn. 129 (1801).—Lycoperdon, Dendromyces.

1342	...	S.A.	...	...	...	...	...	Ground ...	Entirely white, at length rusty from the scattered spores. Peridium bell to mitre shaped, seated on solid very long stem.
1343	W.A.	...	...	V.		...	B.	Sandy soil	Veil ovate, whitish, with mucus. Stem cylindrical, tapering a little to each end. Peridium bell shaped, smooth below, powdery brown above.
1344	W.A.	...	...	...		...	...	Sandy soil	... Stem bellied, covered with scales, hollow. Peridium somewhat plane, leathery, thin.
1345	...	S.A.	...	V.		...	...	Sandy soil	... Stem very long, thickened upwards, woody, hollow, in upper part torn into large membranous scales.

Journ. Bot. II. 94 (1809).—Mitromyces.

1346	...	...	...	V.		...	...	Ground ...	... Outer peridium becoming broken up into small verdigris-green scales; inner peridium sub-globose, dingy green.
1347	W.A.	...	T.	V.		...	...	Ground ...	... Simple or tufted, outer peridium dark brown, dingy red within; inner peridium pale brown, sub-globose, mouth vermilion, teeth erect.
1348	W.A.			...		...	...	Sandy soil	... Outer peridium breaking up early into small blackish granules, adhering to ochrey inner peridium.
1349	...			V.		...	Q.	Ground and dead timber	Outer peridium in form of dingy-green irregular scales, adhering to pale-green inner peridium, stem-like, base stout, greenish.

Carn. II. 489 (1772).—Lycoperdon.

1350	...	...	T.	...		...	Q.	Ground ...	... Outer peridium cut to the middle into six to seven lobes; inner peridium globose, purplish umber.
1351	...	...	...	V.		...		Ground	... Outer peridium cut into eight to ten teeth, whitish and shining, internally dingy umber; inner peridium globose.
1352	W.A.	...	T.	...		...	...	Ground ...	... Outer peridium leathery, rigid, cut to the middle in eight to ten lobes; inner peridium sub-globose, pale umber.
1353	...	...	...	...		...	B.	Ground ...	... Outer peridium thinnish, split to centre into a number of segments; inner peridium short stalked, thick, pale brown.
1354	W.A.	S.A.	...	V.		...	...	Ground ...	... Outer peridium simple, rigid, flattened, many lobed; inner peridium globose, delicately rough.
1355	...	...	...	...	N.S.W.	Q.	...	Ground ...	... Outer peridium thick, globose, delicately powdery, fawn colour, seated on stem-like mycelium.
1356	...	S.A.	...	V.		Q.	...	Ground	... Outer peridium cut into five to eight lobes; inner peridium ovate-oblong, papery, greyish white, shining.
1357	...	S.A.	T.	V.		Q.	B.	Grassy spots	... Outer peridium simple, five to fifteen lobed, flattened, tawny brown; inner peridium sub-globose, whitish, yellow or umber.
1358	W.A.	...	...	...		Q.	B.	Ground ...	... Outer peridium cut to the base into seven to twenty lobes, rarely six; inner peridium compressed, brown or grey.
1359	...	...	...	...		Q.	B.	Ground ...	... Outer peridium cut nearly to middle in six to nine lobes; inner peridium nearly spherical, soft, membranous.
1360	...	...	...	...		Q.		Trunks ...	... Outer peridium downy to granular, pale, irregularly ruptured; inner peridium brown.
1361	W.A.	...	...	...		...	...	Ground ...	... Outer peridium cut into seven to eight narrow lance-shaped teeth, with thin continuous black layer; inner peridium clay colour to brownish.
1362	W.A.	S.A.		V.	N.S.W.	Q.	...	Moist clay soil	Outer peridium for most part seven to nine lobed; inner peridium shortly but distinctly stalked, size of pea, white.
1363	W.A.		...	...		...	...	Sandy soil	Outer peridium splitting into eight lobes; inner peridium globose, becoming whitish.
1364	...	...		V.	N.S.W.	...	...	Ground ...	... Outer peridium thin, cut into seven to nine lobes, umber within; inner peridium somewhat stalked, globose, ochrey umber.
1365	W.A.	...	...	...	N.S.W.	Q.	B.	Ground ...	... Outer peridium rigid, cut into about six lobes, reddish; inner peridium somewhat ovate, pale.
1366	W.A.	...	T.	...	N.S.W.	Q.	B.	Ground ...	... Outer peridium cut into six to nine lobes, thin, soft; inner peridium globose, collapsed.
1367	...	S.A.	...	V.		Q.	...	Ground ...	... Outer peridium split into eight to sixteen stellate fringes; inner peridium globose, tough, yellowish tan.
1368	...	S.A.		...		Q.	...	Ground ...	Small. Outer peridium with few lobes, mealy outside, smooth inside, umber or tawny; inner peridium conically globose.
1369	W.A.	...	...	V.		Q.	B.	Ground ...	... Outer peridium often multifid beyond the middle, brown within; inner peridium globose, umber.
1370	...	...	...	...		Q.		Rotten wood	... Gregarious, springing from expanded, white subiculum, or filamentous mass. Outer peridium mealy, wood colour; inner peridium darker, globose.
1371	...	...	T.	...		...	...	Ground ...	... Outer peridium soft, papery, pale umber, about seven lobed; inner peridium on long stalk, globose, dark brown.
1372	...	...	...	...		...	...	Ground ...	Outer peridium membranous to leathery, cut into about eight fringes, tan colour, cracked lengthwise as if channelled; inner peridium globose, tawny.

92

SYSTEMATIC ARRANGEMENT

Number.	Cooke's Number	Saccardo's Number.	Scientific Name.	Authority for Name.	English Name.
					129. DIPLODERMA.—Link,
1373	1275	IX. 1125	D. albnm ...	Cooke and Mass., Grev. XVI. 2 (1887)	White diploderma ... ...
1374	1276	„ 1126	D. fumosum	Cooke and Mass., Grev. XVI. 2 (1887)	Smoke-coloured spored diploderma
1375	1272	VII. 269	D. glaucum	Cooke and Mass., Grev. XV. 99 (1887) ...	Glaucous diploderma ... ...
1376	1277	...	D. melaspermum ...	Cooke and Mass., Grev. XX. 35 (1891) ...	Dark-spored diploderma
1377	1274	IX. 1127	D. pachythrix	Cooke and Mass., Grev. XVIII. 50 (1890)	Thick-fibred diploderma
1378	...	...	D. sabulosum	Cooke and Mass., Grev. XXI. 38 (1892)	Sandy diploderma
1379	1273	VII. 270	D. suberosum	Cooke and Mass., Grev. XV. 100 (1887)	Corky diploderma
					130. BOVISTA.—Scop.,
1380	1282	IX. 1130	B. anomala	Cooke and Mass., Grev. XVIII. 6 (1889)	Anomalous bovista
1381	...	VII. 283	B. brasiliensis	De Ton., Sacc. Syll. VII. 100 (1888)	Brazilian bovista
1382	1278	„ 283	B. brunnea	Berk.. Fl. N. Zeal. II, 189 (1855)	Brown bovista
1383	1284	„ 291	B. cervina	Berk., Ann. Nat. Hist. IX. 447 (1842)	Fawn-coloured bovista
1384	1303	„ 296	B. dermoxantha	De Ton., Sacc. Syll. VII. 100 (1888)	Yellow-skinned bovista
1385	1280	., 1603	B. hyalothrix	Cooke and Mass., Grev. XVI. 73 (1888) ...	Colourless-threaded bovista
1386	1281	...	B. hypogæa	Cooke and Mass., Grev. XX. 55 (1891) ...	Subterranean bovista ...
1387	1279	VII. 293	B. Muelleri	Berk., Linn. Journ. XIII. 171 (1873)	Mueller's bovista
1388	1307	„ 286	B. mundula	De Ton., Sacc. Syll. VII. 98 (1888) ...	Neat bovista ...
1389	1283	„ 1600	B. olivacea	Cooke and Mass., Grev. XVI. 77 (1888) ...	Olive bovista ...
1390	1306	., 325	B. pusilla	De Ton. Rev. Geast (1887)	Little bovista
					131. LYCOPERDON.—Linn.,
1391	1298	VII. 387	L. australe	Berk., Fl. Tasm. II. 266 (1860) ...	Southern lycoperdon (or puff-ball)
1392	1293	„ 324	L. Bovista	Linn., Sp. Pl. 1653 (1753)	Bovista lycoperdon ...
1393	1287	IX. 1133	L. bovistoides	Sacc., Bull., Soc. Myc. Fr. V. 118 (1889) ..	Bovista-like lycoperdon
1394	1295	VII. 352	L. cælatum	Bull., Champ. 130 (1812)	Embossed lycoperdon ...
1395	1296	„ 1615	L. Cookei	Mass., Mon. Lyc. Trans. R.M.S. 714 (1887)	Cooke's lycoperdon ...
1396	1301		L. coprophilum	Cooke and Mass., Grev.	Dung-loving lycoperdon
1397	1289	VII. 320	L. gemmatum	Batsch., Elen. 147 (1783)	Warty lycoperdon ... (Root fungus)
1398	1292	„ 386	L. glabrescens	Berk., Fl. Tasm. II. 226 (1860) ...	Smooth lycoperdon ...
1399	1309	„ 341	L. Gunnii	Berk., Fl. Tasm. II. 265 (1860) .	Gunn's lycoperdon
1400	1285	„ 403	L. lilacinum	Speg., Fung. Arg. 110 (1882) ...	Lilac lycoperdon

OF AUSTRALIAN FUNGI—continued.

Number	W.A.	S.A.	T.	V.	N.S.W.	Q.	n.	Occurrence.	General Characters.

Berl. Mag. VII. 41 (1816).

1373					N.S.W.			Ground	Somewhat globose. Outer peridium thin, persistent; inner peridium whitish, cartilaginous.
1374	...	...	...	V.	N.S.W.			Ground	Globose, depressed, white. Outer peridium fibrous; inner peridium pale, fragile.
1375	...	...	T.		...			Ground, amongst sand	Somewhat globose, greyish green. Outer peridium fragile, soon falling away; inner peridium thin, yellowish.
1376	...		V.		...			Ground ...	Somewhat globose. Outer peridium thin, persistent, densely velvety, grey; inner peridium cinnamon.
1377	...		...	V.				Underground	Somewhat globose, about 1 inch in diameter. Outer peridium thin, ashy; inner peridium pale. Sterile threads thick, fibrous.
1378	...	S.A.	...					Sandy soil	Nearly globose, pale. Outer peridium thick, somewhat gelatinous, collecting grains of sand; inner peridium membranous.
1379	...	S.A.	...	...		Q.		Ground ...	Somewhat globose. Outer peridium corky, ochrey; inner peridium cartilaginous, turning black.

Carn. II. 487 (1772).

1380	...	...	...	V.	...			Ground ...	Somewhat globose, depressed, whitish. Peridium thick, leathery, delicately velvety, dingy ochre.
1381		...				Q.		Trunks ...	Tufted. Peridium membranous, persistent, globose, rough pointed, rooting.
1382	...		...	...	N.S.W.			Ground	Globose, about 1 inch across, with minute apiculate rooting base. Peridium brownish umber, smooth, shining.
1383	...		...	V.		Q.		Ground ...	Small, globose. Peridium membranous, pale, fawn-coloured, cortex rather rigid.
1384	...		...	V.	...			Grassy places	Peridium very thin, sessile, irregularly globose, root rather long, slender, bright yellow becoming brownish.
1385	...		...	V.				Ground ...	Somewhat globose. Cortex very thick and fibrous, forming persistent base like acorn cup. Peridium minutely rugged.
1386	...		...	V.				Subterranean partly exposed	or Globose and depressed. Outer cortex persistent, thin, white, silky; inner layer thin, whitish, flexible.
1387			...	...		Q.		Ground ...	Somewhat globose, with short stout rooting base. Cortex soon broken up into minute warts. Peridium firm, brown.
1388	...		V.		...			Ground ...	Peridium fluffy, becoming smooth, white, size of hazel-nut.
1389	...		V.		...		B.	Ground ...	Globose, with short stout rooting base. Cortex very thin and evanescent. Peridium thick, soft and pliant like leather, pale ochrey.
1390	W.A.			V.	N.S.W.	Q.	B.	Ground	Among the smallest of puff-balls. Somewhat globose, slightly tapering below, with minute scurfy scales, becoming smooth, pale-olive ochre.

Sp. Pl. II. 1183 (1753).—Bovista.

1391	W.A.	S.A.	T.	V.	N.S.W.	Q.		Sand turfy meadows	Sessile, globose, and depressed, densely covered with small pointed warts. Root long tapering.
1392	...	...	...	...	...	Q.	B.	Ground ...	Peridium subglobose, sessile, fragile above, greyish white to yellowish white, then olive grey. Edible.
1393	...	S.A.	...	V.	...			Ground ...	Peridium nearly sessile, attached by a broad base, globose, then depressed, membranous, yellowish.
1394	...		T.	V.	N.S.W.	Q.	B.	Among grass	Solitary, sessile or stalked, large, somewhat globose or depressed. Cortex or bark pale-creamy ochre, broken into large angular patches.
1395	...	...	...	...	N.S.W.		B.	Ground ...	Hemispherical or globose, abruptly contracted into short thick stem-like base, smoky brown above, white below.
1396	...	...	...	...		Q.		Dung ..	Globose, sessile, whitish, covered with delicate persistent spines.
1397	W.A.	...	T.	...	N.S.W.	Q.	B.	Ground ...	Stalked, somewhat globose, depressed above, with prominent spiny warts, eventually falling off. Stem stout.
1398	...	...	T.	V.	...			Ground ...	Nearly hemispherical, at first covered with slender spines becoming smooth. Stem short, stout.
1399	...	...	T.	V.	...	Q.		Pastures ...	Globose, sessile, with very minute stellate warts.
1400	W.A.	...	T.	V.	N.S.W.	Q.		Ground ...	Broadly obovate, contracted below into stout stem-like base. Peridium thin; cortex white, polished.

Number.	Cooke's Number.	Saccardo's Number.	Scientific Name.	Authority for Name.	English Name.
					131. LYCOPERDON.—*Linn.*,
1401	1288	VII. 1610	L. natalense	Cooke and Mass., in Mon. Lyc. Trans. R.M.S. 709 (1887)	Natal lycoperdon
1402	1291	„ 359	L. pyriforme	Schaeff., Icon 185 (1762)	Pear-shaped lycoperdon
1403	1304	330	L. reticulatum	Berk., Fl. N. Zeal. II. 190 (1855)	Reticulated lycoperdon
1404	1299	„ 333	L. stellatum	Cooke and Mass., Grev. XV. 97 (1887)	Stellate lycoperdon
1405	1300	„ 1621	L. substellatum	Berk. and Curt., in Mass. Mon. Lyc. Trans. R.M.S. 720 (1887)	Sub-stellate lycoperdon
1406	1305	„ 437	L. tephrum	Berk., in Mass. Mon. Lyc. Trans. R.M.S. 723 (1887)	Ash-coloured lycoperdon
1407	1286	„ 1607	L. violascens	Cooke and Mass., Mon. Lyc. Trans. R.M.S. 706 (1887)	Violet lycoperdon
					132. AREOLARIA.—Forq., Champ.
1408	1318	VII. 481	A. strobilina	Forq., Champ. Exot. 155 (1886)	Cone-like areolaria
					133. CASTOREUM.—Cooke and Mass.,
1409	1322	VII. 476	C. radicatum	Cooke and Mass., Grev. XV. 100 (1887)	Rooting castoreum
					134. XYLOPODIUM.—Mont.,
1410	1323	VII. 479	X. australe	Berk., Linn. Journ. XIII. 171 (1873)	Southern xylopodium
1411	1324	„ 478	X. ochroleucum	Cooke and Mass., Grev. XV. 95 (1887)	Whitish-ochre xylopodium
					135. FAVILLEA.—Fries,
1412	1325	VII. 487	F. argillacea	Fries, Fung. Nat. 32 (1848)	Clay-coloured favillea

OF AUSTRALIAN FUNGI—*continued.*

Number.	Habitat.						B.	Occurrence.	General Characters.
	W.A.	S.A.	T.	V.	N.S.W.	Q.			

Sp. Pl. II. 1183 (1753).—Bovista—*continued.*

1401	...	...	...	V.	...	...	...	Ground ...	Globose, sessile, passing abruptly into short tapering root. Peridium thick, minutely warted, becoming smooth.
1402	...	...	T.	V.	N.S.W.	Q.	B.	On stumps or soil, attached to branches, &c.	Densely tufted, pear shaped, membranous, covered with minute-pointed warts, brownish, rooting.
1403	...	...	...	V.		...	...	Ground ...	Globose, tapering downward, with slightly-raised reticulations, eventually disappearing and leaving polished surface.
1404	W.A.	...	...		...	...	Ground ...	Sessile, nearly globose. Peridium thin, covered at first with stout stellate spiny warts, falling away in patches and leaving smooth surface.
1405	..	...	...	...		Q.	...	Rotten wood, &c. ...	Globose, sessile, whitish, covered with delicate spines, which become smaller downwards.
1406	...	...	...	...		Q.	...	...	Globose, sessile, thick and rigid, brown, minutely velvety.
1407	...	...	...	V.		...	...	Ground ...	Globose, sessile, terminating in short slender root. Peridium papery, covered at first with minute warts, becoming smooth and shining.

Exot. 155 (1886).—Scleroderma, Phellorina.

| 1408 | ... | ... | ... | ... | | ... | Q. | ... | Ground ... | Globose and depressed, with stout angular scales above. Stem solid, rather woody. |

Grev. XV. 100 (1887).

| 1409 | ... | ... | T. | ... | | ... | ... | Ground ... | Tufted, nearly globose, confluent below in tough rooting stem. Outer peridium tawny, leathery; inner peridium at length horny. |

Ann. Sci. Nat. 3 Ser. IV. 364 (1845).

| 1410 | ... | S.A. | ... | V. | N.S.W. | ... | ... | Ground, trunk of Eucalyptus hemiphloia | Peridium volvate when young. Stem rooting, broken into scales. |
| 1411 | ... | ... | ... | ... | | Q. | ... | Ground ... | Stalked. Peridium globose, with large warts. Stem erect, thick, solid, with overlapping scales. |

Fung. Nat. 32 (1848).

| 1412 | ... | ... | ... | ... | | | | Ground | Peridium club shaped, simple, without special cortex, membranous above. |

Number.	Order's Number.	Saccardo's Number.	Scientific Name.	Authority for Name.	English Name.

ORDER XI.—SCLERODERMACEÆ,

136. SCLERODERMA.—Pers., Syn.

1413	1316	IX. 1142	S. australe	Mass., Grev. XVIII. 26 (1889) ...	... Southern scleroderma ...
1414	1311	VII. 446	S. Bovista	Fries, S.M. III. 40 (1829) ...	... Bovista scleroderma ...
1415	1319	„ 474	S corium...	Grav., in Duby. Bot. Gall. II. 892 (1830)	... Leathery scleroderma ...
1416	1310	„ 459	S. Genster	Fries, S.M. III. 46 (1849) ...	.. Earth-star scleroderma
1417	1321	„ 1629	S. olivaceum	De Ton., Sacc. Syll. VII. 489 (1888)	... Olive scleroderma ...
1418	1314	„ 454	S. Pandanaceum ...	F. v. M., Linn. Journ. XIII. 171 (1873) ...	... Pandanus scleroderma... ...
1419	1320	„ 461	S. phæotrichum ...	De Ton., Sacc. Syll. VII. 139 (1888) ...	... Dusky-haired scleroderma ...
1420	...	„ ...	S. pileolatum ...	Kalch., Linn. Soc. N.S.W. VII. 565 (1882)	... Capped scleroderma ... ...
1421	1317	IX. 1146	S. umbrinum ...	Cooke and Mass., Grev. XIX. 45 (1890) ...	... Umber scleroderma ...
1422	1313	VII. 447	S. verrucosum	Pers., Syn. 154 (1801) ...	... Warted scleroderma ... ...
1423	1312	„ 445	S. vulgare	Fries, S.M. III. 46 (1829)	... Common scleroderma ... ...

137. POLYSACCUM.—D. C., Fl. Fr. V. 103 (1815).—

1424	1334		P. album ...	... Cooke and Mass , Grev. XX. 36 (1891) ...	... White polysaccum
1425	1333	VII. 490	P. australe	... Lev., Ann. Sci. Nat. Ser. 3, IX. 136 (1848)	... Southern polysaccum ...
1426	1332	„ 1632	P. confusum	.. Cooke, Grev. XVI. 76 (1889)	... Confused polysaccum ...
1427	1328	„ 491	P. crassipes	... D. C., Fl. Fr. V. 103 (1815)	... Thick-stalked polysaccum
1428	1335	„ 500	P. degenerans ...	... Fries, Pl. Preiss. 139 (1846) ...	... Degenerating polysaccum
1429	1331	„ 499	P. marmoratum ...	... Berk., Linn. Journ. XIII. 171 (1873) ...	... Marbled polysaccum ...
1430	1327	„ 1633	P. microcarpum	... Cooke and Mass., Grev. XVI. 28 (1887) ...	... Small-fruited polysaccum ...
1431	1326	„ 494	P. pisocarpium ...	... Fries, S.M. III. 54 (1829) ...	... Pea-fruited polysaccum
1431A	„	„ „	P. pisocarpium, var. acaule ...	D. C., Fl. Fr. V. 103 (1815)	... Stemless polysaccum ...
1432	1330	„ 501	P. tuberosum ...	... Fries, Linn. V. 694 (1830)	... Tuberous polysaccum ...
1433	1329	„ 489	P. turgidum ...	... Fries, S.M. III. 53 (1829) ...	... Turgid polysaccum ...

138. ARACHNION.—Schwein.,

| 1434 | 1336 | VII. 507 | A. Drummondi | ... Berk., Linn. Journ. XVIII. 389 (1887) | ... Drummond's arachnion |

139. PAUROCOTYLIS.—Berk.,

| 1435 | 1338 | VII. 512 | P. echinosperma ... | ... Cooke, Grev. VIII. 59 (1879) | ... Spiny-spored paurocotylis ... |

OF AUSTRALIAN FUNGI—*continued*.

Number.	Habitat.						B.	Occurrence.	General Characters.
	W.A.	S.A.	T.	V.	N.S.W.	Q.			

FRIES, S.M. III. 5 (1829).

Fung. 150 (1801).—Lycoperdon, Mycenastrum.

1413	...	...	...	...	...	Q.	...	Soil ...	Nearly globose, sessile. Peridium thick, externally minutely felty, dirty ochre, with rooting base.
1414		...		V.	N.S.W.	Q	B.	Sandy ground	Obovoid to spherical, somewhat stalked or possibly with rooting base, warty, turning yellow.
1415	...	...	...	V.	...	Q.	...	Sandy ground	Nearly globose, whitish, then greyish brown, leathery, splitting above in stellate manner.
1416	W.A.	S.A.	T.	V.	N.S.W.	Q.	B.	Sandy ground	Sessile, nearly globose, opening at top in stellate fringes.
1417	...	...	...	...	N.S.W.	Q.	...	Ground ...	Nearly globose, sessile, lead colour above, whitish below ; sterile threads and mass of spores olive.
1418	...	...	...	...	...	Q.	...	Ground ...	Globose, sessile, yellow, tessellate above.
1419	W.A.	...	...	...	N.S.W.	Q.	...	Ground ...	Nearly globose ; sterile threads and mass of spores dark purple brown.
1420	...	...	...	...	N.S.W.	...	...	... ...	Peridium globose, opaque, umber beneath, concave, size of hazel-nut, with slender stem.
1421	...		...	...		Q.	...	Ground ...	Peridium globose, coarsely wrinkled below, dirty pale ochre. Stem coarsely furrowed, dark brown, passing into dense bulbous mass.
1422	...	...	...	...		Q.	B.	Sandy ground	Peridium rounded, at first hard, then fragile, covered with rather warted cortex, dingy yellowish.
1423	W.A.	S.A.	T.	V.	N.S.W.	Q.	B.	Ground under trees, &c.	Almost sessile, deformed. Peridium corky, citron then reddish brown, growing pale with age.

Lycoperdon, Lycoperdoides.

1424	...		...	...	...	Q.	...	Ground	Peridium globose, white, polished and shining, tapering below into very short stout irregular stem-like base.
1425	W.A.	...	...	...	...	Q.	...	Ground ...	Stem rooting, almost cylindrical, shining, blackish brown, dilated into like-coloured nearly globose peridium.
1426	...	S.A.	...	V.			...	Ground	Peridium nearly globose, slightly narrowed below into short thick stem, or pear shaped, olivaceous umber with polygonal cavities.
1427	W.A.	S.A.	...	V.	...		...	Immersed in sand ...	Peridium from spherical to clavate, pale ochre becoming darker ; stem-like base stout, minute cavities.
1428	W.A.	...	...	...			...	River banks	Club shaped with rooting stem. Peridium simple, ochrey-tan colour.
1429	W.A.	S.A.	...	...	N.S.W.		...	Ground ...	Peridium somewhat globose, tapering to stem-like base, dirty ochre, marbled with darker patches.
1430	...		...	V.		Q.	...	Ground	Peridium nearly globose, coarsely tubereled, ochrey brown, stout, with stem-like base, bright citrine.
1431	W.A.	...	V.		Q.		B.	Sandy ground	Peridium nearly globose, passing down into short stem-like base, reddish brown, tinged olive.
1431a	...	...	...	...	...	Q.	...	Sandy ground	Without distinct stem.
1432	...	...	...	...	N.S.W.	Q.	...	Ground ...	Peridium nearly globose or deformed, with short stem-like base, ochrey ; cavities large, angular.
1433	W.A.	...	...	...			...	Sandy soil	Peridium cylindric to club shaped, covered at first with fine spider-like web, dark umber, passing into stem-like base.

Syn. Car. 14 (1822).

1434	W.A.	...	...	...			...	Attached to *Lavellinia cycnopotamia*	Globose, depressed a little, pale.

Fl. N. Zeal. II. 188 (1855).

1435	...	...		V.				Trunks	Globose, depressed, tawny flesh colour, spores spiny.

SYSTEMATIC ARRANGEMENT

Number.	Cooke's Number.	Saccardo's Number.	Scientific Name.	Authority for Name.	English Name.

ORDER XII.—HYMENOGASTRACEÆ,

140. OCTAVIANIA.—Vitt., Mon.

1436	1340	IX. 1150	O. alveolata	...	Cooke and Mass., Grev. XVI. 2 (1887) ...	...	Alveolate octaviania ...	...
1437	1341	VII. 529	O. Archeri	Berk., Fl. Tasm. II. 269 (1860) ...	...	...	Archer's octaviania ...	...
1438	1339	„ 578	O. australiense	Berk., in Cooke's Handb. Austr. Fung. 246 (1892)	Australian octaviania ...	...

141. RHIZOPOGON.—Fries,

| 1439 | 1342 | VII. 534 | R. luteolus | Fries, Symb. Gast. 5 (1848) ... | Yellowish rhizopogon ... |

142. HYMENOGASTER.—Vitt., Mon. Tub. 80 (1831).—

1440	1343	VII. 560	H. Klotzschii	...	Tul., Fung. Hyp. 64 (1851) ...	...	...	Klotzsch's hymenogaster	...
1441	1344	563	H. lycoperdineus ...	Vitt. Mon. Tub. 22 (1831)	...	Puff-ball hymenogaster	...
1442	1345	„ 564 bis.	H. Moselei	De Ton., Sacc., Syll. VII. 172 (1888) ...	...	Moseley's hymenogaster	...

143. HYDNANGIUM.—Wallr., in Corda

| 1443 | 1346 | H. brisbanensis | Berk. and Br., in Cooke's Handb. Austr. Fung. 247 (1892) | Brisbane hydnangium | ... |
| 1444 | 1347 | H. tasmanicum ... | ... | Kalch., in Grev. XIX. 95 (1891) ... | ... | Tasmanian hydnangium | ... |

144. GAUTIERIA.—Vitt.,

| 1445 | 1348 | G. Drummondi ... | ... | Berk., in Cooke's Handb. Austr. Fung. 247 (1892) | Drummond's gautieria | ... |

OF AUSTRALIAN FUNGI—*continued.*

Number.	W.A.	S.A.	T.	V.	N.S.W.	Q.	R.	Occurrence.	General Characters.

VITT. MON. TUB. 11 (1831).

Tub. 15 (1831).—Hydnangium.

Number.	W.A.	S.A.	T.	V.	N.S.W.	Q.	R.	Occurrence.	General Characters.
1436	...	...	...	...	N.S.W.	...	...	In the ground	Somewhat globose or irregular, whitish, then ochrey. Spores globose, alveolate.
1437	...	..	T.	...		...	...	Ground ...	Inversely egg-shaped, small, with large sterile base, no fibrils.
1438	...	...	...	V.		Q.	...	Under tea tree	Somewhat globose, irregular. Peridium thin, wrinkled, ochrey.

Symb. Gast. 5 (1818).—Hysterangium, Tuber.

Number.	W.A.	S.A.	T.	V.	N.S.W.	Q.	R.	Occurrence.	General Characters.
1439	...	...	...	V.		...	B.	In sandy soil	Deformed, usually spherical, kidney or egg shaped; fibrils rooting becoming yellowish.

Rhizopogon, Splanchnomyces, Hymenangium.

Number.	W.A.	S.A.	T.	V.	N.S.W.	Q.	R.	Occurrence.	General Characters.
1440	W.A.	...	...	...	...	...	B.	In ground	Obovate, fibrillose at base. Peridium membranous, whitish, somewhat, downy.
1441	...	...	...	...	...	Q.	...	Ground ...	Gregarious, with strong odour of garlic. Peridium rounded, deformed, white, then brownish.
1442	...	...	...	...	N.S.W.	...	...	Soil	Almost globose, tapering at base, citrine yellow, smooth.

Ic. Fung. V. 28 (1842).—Octaviania.

Number.	W.A.	S.A.	T.	V.	N.S.W.	Q.	R.	Occurrence.	General Characters.
1443	...	...	...	V.		Q.	...	In soil	Almost globose. Peridium thick, dry, wrinkled, reddish brown.
1444	...	...	T.	...		...	...	Ground ...	Nearly globose. Peridium thick, dark brown, angular when dry.

Mon. Tub. 25 (1831).

Number.	W.A.	S.A.	T.	V.	N.S.W.	Q.	R.	Occurrence.	General Characters.
1445	W.A.	...	...	...	...	...	...	In soil ...	Nearly globose, small; cells wavy. Spores ellipsoid.

GENERAL CLASSIFICATION OF UREDINES.

GROUP III.—UREDINES, BRONGN.

ORDER XIII.—UREDINACE.E—Parasitic. Spores usually of more than one kind and not all unicellular.

ARRANGEMENT OF GENERA (9).

Section 1. Amerosporæ. Sacc. and De Toni—Teleutospores continuous, one-celled.

Genera (3)—
    145. Uromyces, Link. | 146. Melampsora, Cast. | 147. Cronartium, Fries.

Section 2. Didymosporæ, Sacc. and De Toni—Teleutospores bilocular.

Genus (1)—
    148. Puccinia, Pers.

Section 3. Phragmosporæ, Sacc. and De Toni—Teleutospores 3, or many-celled.

Genera (2)—
    149. Phragmidium, Link. | 150. Hamaspora, Korn.

Imperfect Forms—

Genera (3)—
    151. Æcidium, Pers. | 152. Roestelia, Reb. | 153. Uredo, Pers.

SYSTEMATIC ARRANGEMENT

Number.	Cooke's Number.	Saccardo's Number.	Scientific Name.	Authority for Name.	English Name.

GROUP III.—UREDINES.—
ORDER XIII.—UREDINACEÆ,
145 UROMYCES.—Link, Berl. Mag. VII. 28 (1816).—

1446	1732	VII. 1928	U. Betæ	Kuehn, Bot. Zeit. 540 (1869) ...	Beet uromyces ... ...
1447	1738 bis.	„ 2033	U. Bulbinis	Thuem. Fl. 410 (1877)	Bulbine uromyces ... ...
1448	1737	„ 1982	U. digitatus	Wint. Rev. Myc. 209 (1886) ...	Digitate uromyces ... ...
1449	1738	IX. 1212	U. Diploglottidis ...	Cooke and Mass., Grev. XVII. 55 (1889)	Diploglottis uromyces ...
1450	1735	VII. 1980	U. fusisporus	Cooke and Mass., Grev. XVI. 2 (1887) ...	Spindle-spored uromyces ...
1451	...	„ 1940	U. Junci	Tul., Ann Sci. Nat. 146 (1854)	Rush uromyces ...
1452	...		U. Kuehnii	Krueger, Bericht. Versuchs. Java I. (1891)	Kuehn's uromyces (Cane rust) ...
1453		VII. 1209	U. Limosellæ	Ludw., Dict. Hedw. 182 (1888) ..	Limosella uromyces ...
1454	1740	„ 2066	U. Microtidis	Cooke, Grev. XIV. 12 (1885) ...	Microtis uromyces ...
1455	1741	„ 2071	U. orchidearum ...	Cooke and Mass., Grev. XVI. 74 (1888) ...	Orchid uromyces ...
1456	1736	IX. 1203	U. phyllodii	Cooke and Mass., Grev. XVII. 70 (1889)...	Phyllode uromyces ...
1457	1742	VII 2100	U. puccinioides	Berk. and F. v. M., Linn. Journ. XIII. 173 (1873)	Puccinia-like uromyces ..
1458	1739	IX. 1204	U. Tepperianus	Sacc., Hedw. 126 (1889)	Tepper's uromyces
1459	1731	VII. 1925	U. Trifolii	Lev. in Winter's Die Pilze I. 159 (1884) ...	Clover uromyces
1460	1733	„ 1955	U. vesiculosa	Wint., Hedw. 2 (1885)...	Vesicular uromyces ...

146. MELAMPSORA.—Cast.,

1461	1743	VII. 2107	M. Lini	Tul., Ann. Sci. Nat. 93 (1854) ...	Flax melampsora ...
1462	1744	„ 2124	M. Nesodaphnes	Berk. and Br., Linn. Trans. II. 67 (1883)	Nesodaphnes melampsora
1463	1745	„ 2123	M. phyllodiorum ...	Berk. and Br., Linn. Trans. II. 67 (1883)	Phyllode melampsora ...

147. CRONARTIUM.— Fries.

| 1464 | 1746 | VII. 2137 | C. Asclepiadeum ... | Fries, Obs. Myc. I. 220 (1815) ... | Asclepiad Cronartium ... |

148. PUCCINIA.—Pers.,

1465	1760	VII. 2218	P. Acetosæ	Karn., Hedw. 184 (1876)	Sorrel puccinia
1466	1751	„ 2174	P. ægra	Grove, Journ. Bot. 274 (1883) ...	Sick puccinia...
1467	1768	„ 2494	P. Alyxiæ	Cooke and Mass., Grev. XVI. 2 (1887)	Alyxia puccinia
1468	1757	„ 2211	P. Apii ...	Corda, Icon. VI. 30 (1854)	Celery puccinia
1469	1765	„ 2337	P. aucta ...	Berk. and F. v. M., Linn. Journ. XIII. 173 (1873)	Abundant puccinia ...

OF AUSTRALIAN FUNGI—*continued*.

Number.	Habitat.						B.	Occurrence.	General Characters.
	W.A.	S.A.	T.	V.	N.S.W.	Q.			

BRONGN. DICT. V. 33 (1824).

BRONGN. DICT. V. 33 (1824).

Uredo, Trichobasis, Puccinia, Æcidium.

Number	W.A.	S.A.	T.	V.	N.S.W.	Q.	B.	Occurrence.	General Characters.
1446	...	S.A.	...	V.	...	...	B.	Leaves of *Beta* ...	Cluster cups on circular or oblong yellow spots. Uredospore pustules cinnamon. Teleutospore pustules dark brown.
1447	...	...	...	V.	N.S.W.	...	...	Leaves of *Bulbine bulbosa*	Pustules on both surfaces, small, densely crowded, covered by cuticle, brown.
1448	...	S.A.	...	...	...	...	...	Leaves of *Acacia notabilis*	Pustules in centre of circular spots, bounded by narrow brown line, black. Teleutospores with finger like processes.
1449	...	...	...	...	...	Q.	...	Fading leaves of *Diploglottis*	Pustules scattered, convex, minute, at length splitting, pale brown, on circular greenish spots.
1450	...	...	...	V.	...	...	...	Phyllodes of *Acacia salicina*	On both surfaces. Pustules disc-like, bursting through, black, surrounded by ruptured cuticle.
1451	...	...	...	V.	...	...	B.	*Juncus maritima* ...	Cluster cups cup shaped with whitish torn edges. Uredospore pustules on brownish spots. Teleutospore pustules round or elongated.
1452	...	...	...	...	N.S.W.	...	...	Leaves of Sugar cane	Mostly on under surface, spots finally orange. Pustules in streaks, elongated, narrow, brownish or blackish, bursting through, often, run together.
1453	...	S.A.	...	...	...	...	...	Leaves of *Limosella*	Cluster cups on both sides, scattered or gregarious. Teleutospore pustules mixed with cluster cups.
1454	...	...	...	...	N.S.W.	...	...	Leaves of *Microtis porrifolia*	On both surfaces. Pustules gregarious, dark brown, surrounded by epidermis.
1455	...	...	...	...	N.S.W.	...	...	Leaves of *Chiloglottis diphylla*	Pustules blistered, at length bursting, brown.
1456	...	...	...	...	...	Q.	...	Phyllodes of *Acacia*	Pustules minute, circular, compact, brown, crowded on blistered spots, at length naked.
1457	...	S.A.	T.	V.	...	Q.	...	Leaves and flower stalks of *Goodenia* and *Selliera*	Cluster cups on brown circular spots (*Æcidium Goodeniacearum*, Berk.). Pustules blistered.
1458	...	S.A.	...	V.	...	...	...	Living branches of *Acacia salicina, A. myrtifolia,* and *A. hakioides*	Long and broad, flattened, growing beneath cuticle and casting off bark, bright cinnamon.
1459	...	...	...	V.	...	...	B.	Clover ...	Cluster cups circularly arranged, pale orange. Uredospore clusters chestnut brown. Teleutospore cluster-smaller.
1460	...	S.A.	...	...	...	...	...	Leaves and stems of *Zygophyllum ammophilum*	Pustules scattered or gregarious, often run together, covered by ashy vesicular epidermis.

Obs. II. 18 (1843).—Uredo, Cæoma.

Number	W.A.	S.A.	T.	V.	N.S.W.	Q.	B.	Occurrence.	General Characters.
1461	...	S.A.	...	V.	N.S.W.	...	B.	Leaves of *Linum marginale* and *L. usitatissimum*	Uredospore pustules scattered, rounded, orange, minute. Teleutospore pustules at first red brown, then almost black.
1462	...	...	...	...	...	Q.	...	Fruit of *Nesodaphnes obtusifolia*	Spore masses powdery, shaggy, ochrey.
1463	...	...	...	...	...	Q.	...	Phyllodes of *Acacia*	Pustules in tubercles ; spores arising from delicate filaments, granulated.

Obs. Myc. I. 220 (1845).—Erineum, Uredo, Cæoma.

Number	W.A.	S.A.	T.	V.	N.S.W.	Q.	B.	Occurrence.	General Characters.
1464	...	...	...	...	...	Q.	...	Leaves of *Jacksonia scoparia*	Uredospore pustules on under surface, scattered or clustered, brown. Teleutospore pustules yellowish.

Tent. Disp. Meth. 38 (1797).—Uredo, Æcidium, Cæoma, Trichobasis, Uromyces.

Number	W.A.	S.A.	T.	V.	N.S.W.	Q.	B.	Occurrence.	General Characters.
1465	...	...	...	...	...	Q.	...	Leaves, stems, &c., of *Rumex*	Pustules of both forms scattered, minute on leaves and irregularly rounded, oblong on stems and leaf stalks.
1466	...	...	...	V.	...	...	B.	Leaves of *Viola hederacea*	Cluster cups scattered, white, spores orange yellow. Uredospore pustules on yellow spots. Teleutospore pustules similar.
1467	...	...	...	V.	...	...	...	Leaves of *Ayxia bixifolia*	On under surface. Pustules disc-like, compact, dark brown, girt by ruptured epidermis.
1468	...	...	...	V.	...	...	B.	Celery ...	Cluster cups causing long yellow swellings on stem. Uredospore pustules large, cinnamon brown. Teleutospore pustules blackish brown.
1469	...	S.A.	...	V.	N.S.W.	...	...	Leaves of *Lobelia anceps, L. pedunculata, L. platycalyx*	Cluster cups occupying entire surface of leaves or leaf stalks, ochrey. Teleutospore pustules blistered.

Number.	Cooke's Number.	Saccardo's Number.	Scientific Name.	Authority for Name.	English Name
					148. PUCCINIA.—Pers.,
1470	1769	VII. 2506	P. Berkeleyana	... De Toni, in Sacc. Syll. VII. 717 (1888)	... Berkeley's puccinia ... ...
1471			P. Burchardiæ	... Ludw., Zeit. f. Pflkrk. III. 137 (1893)	Burchardia puccinia ...
1472			P. Carissæ	... Cooke and Mass. XXII. 37 (1893)	... Carissa puccinia
1473	...	VII. 2162	P. Epilobii	... D. C., Fl. Fr. VI. 61 (1805)	... Epilobium puccinia
1474	...	„ 2409	P. Geranii	... Corda. Icon. IV. 12 (1854)	... Geranium puccinia
1475	1753	„ 2101	P. graminis	... Pers, Disp. Fung. 39 (1797)	... Grass puccinia (Rust in wheat) ...
1476	1747	„ 2150	P. Helianthi	... Schw., Syn. Car. 73 (1822)	... Sunflower puccinia ...
1477	1767	„ 2403	P. heterospora	... Berk. and Curt., Linn. Journ. X. 356 (1869)	... Heterosporons puccinia ...
1478	1749	„ 2210	P. Hieracii	... Mart., Fl. Mosq. 226 (1812) ...	... Hawk-weed puccinia ...
1479		...	P. Juncephila	... Cooke and Mass. XXII. 37 (1893)	... Rush-loving puccinia ...
1480	1761	VII. 2240	P. Kalchbrenneri ...	... De Toni, Sacc. Syll. VII. 645 (1888)	... Kalchbrenner's puccinia ...
1481		...	P. Kochiæ ...	... Mass., Grev. XXII. 17 (1893) ...	... Kochia puccinia ...
1482	1752	VII. 2169	P. Lagenophoræ ...	... Cooke, Grev. XIII. 6 (1894)	... Lagenophora puccinia ...
1483	1759	...	P. Ludwigii	Tepper, Bot. Centr. Blatt. 6 (1890)	... Ludwig's puccinia ...
1484		VII. 2205	P. Magnusiana	Korn., Hedw. 179 (1876) ...	Magnus' puccinia ...
1485	1766	„ 2368	P. Malvacearum ...	Mont., in Gay's Hist. Chil. VIII. 43 (1845)	Mallow puccinia (Hollyhock fungus)
1486	...		P. munita	... Ludw., Zeit. f. Pflkrk. II. 133 (1892) ...	Protected puccinia ...
1487	1787	VII. 2202	P. obscura	... Schrot., Nuor. Giorn. Bot. Ital. IX. 256 (1875)	Obscure puccinia ...
1488	1756	„ 2204	P. Phragmitis	Korn., Hedw. 179 (1876)	Phragmitis puccinia ...
1489	1755	„ 2195	P. Poarum	Niels., Bot. Tids. II. 26 (1877) ...	... Poa puccinia ...
1490	1748	„ 2157	P. Prenanthis	... Fckl., Symb. Myc. 45 (1875)	Prenanthes puccinia ...
1491	1734	„ 2252	P. Pruni ...	... Pers. Syn. Fung. 226 (1808)	... Plum puccinia (Peach and Plum leaf rust)
1492	1770		P. rimosa	Link, Winter, Hedw. 28 (1880) ...	... Cracking puccinia ...
1493	1754	VII. 2194	P. Rubigo-vera	Wint., Die Pilze. I. 217 (1884) ...	True-rust puccinia ...
1494	1758	„ 2214	P. Rumicis-scutati	Wint., Die Pilze. I. 187 (1884) ...	Rumex puccinia ...
1494b	„	IX. 1236	P. Rumicis-scutati, Muehlenbeckia	var. Cooke, Grev. XIX. 47 (1890)	Muehlenbeckia puccinia
1495	1764	„ 1280	P. Saccardoi	Ludw., Hedw. 362 (1889) ...	... Saccardo's puccinia ...
1496	1762	VII. 2289	P. Sorghi	... Schw., N. Am. Fung. 295 (1831)	Sorghum puccinia ...
1497	...		P. Tepperi	Ludw., Zeit. f. Pflkrk. II. 130 (1892)	... Tepper's puccinia ...
1498	1750	VII. 2163	P. Violæ ...	Winter, Die Pilze. I. 215 (1881)	Violet puccinia ...
1499	1763	„ 2301	P. Wurmbseæ	Cooke and Mass., Grev. XVI. 71 (1888)...	Wurmbsea puccinia ...

OF AUSTRALIAN FUNGI—continued.

Tent. Disp. Meth. 38 (1797).—Uredo, Æcidium, Cæoma, Trichobasis, Uromyces—continued.

Number	W.A.	S.A.	T.	V.	N.S.W.	Q.	B.	Occurrence	General Characters.
1470	...	...		V.	...	...	...	Leaves of Dichondra repens	Pustules minute on lower, rarely on upper surface, scattered, resembling a cluster cup.
1471	...	S.A.	...	V.		...	...	Leaves of Burchardia umbellata	Pustules bursting through, circular or elliptical, black. Uredospores and teleutospores.
1472	...	...	...	...		Q.	...	Living leaves of Carissa ovata	On under surface. Only teleutospores seen. Pustules small, gregarious on circular spots, dark brown.
1473	...	...		V.			B.	Leaves of Epilobium glabellum	Pustules small, roundish, rather crowded, soon naked, surrounded by torn epidermis, dark brown.
1474	...	...		V.		...	...	Leaves of Pelargonium australe	Pustules small, brown, powdery, on under surface. Teleutospore pustules black.
1475	...	S.A.	T	V.	N.S.W.	Q.	B.	Avena and Triticum	Cluster cups forming circular, reddish-yellow spots. Uredospore pustules linear (Uredo linearis). Teleutospore pustules elongated, black.
1476	...	...		V.	N.S.W.	Q.	...	Sunflower leaves ...	Cluster cups crowded or circular, spores orange red. Uredospore pustules minute, chestnut brown. Teleutospore pustules dark brown or black.
1477	...	...			N.S.W.	Q.	...	Leaves of Abutilon crispum and A. avicennæ	Spots purplish or yellow. Pustules on under surface, minute, soon naked, brown.
1478	...	...	...			Q.	B.	Hypochæris glabra	Pustules oblong, mostly solitary, reddish brown, girt by ruptured epidermis.
1479	...	...		V.		...	...	Juncus ...	Pustules rusty, powdery, elliptical or confluent. Uredospores and teleutospores intermixed.
1480	...	...		V.			...	Leaves of Helichrysum	Pustules on both surfaces, scattered or gregarious, covered at first, then free, ochrey.
1481	...	...		V.			...	Leaves of Kochia sedifolia	On both surfaces of leaf. Pustules discoid, blackish brown, girt by torn epidermis.
1482	...	...		V.			...	Living leaves of Lagenophora Billardieri	Cluster cups on upper surface. Uredospore pustules small, brown. Teleutospore pustules dark brown.
1483	...	...		V.	...		...	Leaves of Rumex Brownii	Uredospores pale yellow brown. Teleutospore pustules minute, circular girt by ruptured epidermis.
1484	...	S.A.	...	...		...	...	Arundo phragmites	Cluster cups mostly on under surface of leaves. Uredospore pustules orange brown. Teleutospore pustules black.
1485	...	S.A.	...	V.	N.S.W.	Q.	B.	Leaves and stems of Althæa rosea and Malva rotundifolia	Pustules greyish brown, compact, round, elongated on stems, scattered, pale reddish brown.
1486	...	S.A.	...	...		...	...	Leaves of Hydrocotyle hirta	On under surface. Pustules cinnamon brown. Teleutospores protected by white pseudo-peridium.
1487	...	...		V.		...	B.	Bellis perennis ...	Cluster cups on large roundish spots, mostly on upper surface (Æcidium Bellidis).
1488	...	...		V.		...	B.	Arundo phragmites. &c.	Cluster cups on circular red spots. Uredospore pustules large, dark brown. Teleutospore pustules sooty black.
1489	...	...		V.	N.S.W.	...	B.	Poa species ...	Cluster-cup spots yellow. Uredospore pustules small, orange. Teleutospore pustules black.
1490	...	S.A.	...	V.	N.S.W.		B.	Lactuca, &c. ...	Cluster cups in circular or elongated patches. Uredospore pustules reddish brown. Teleutospore pustules blackish.
1491	...	S.A.	T.	V.	N.S.W.	Q.	D.	Peach, Plum, and Almond leaves, also on fruit of Peach	Uredospore pustules light brown, small, round, crowded. Teleutospore pustules almost black.
1492	...	...	...	V.		...	...	Isolepis nodosa ...	Producing narrow cracks, often encircling stem and causing brown spots.
1493	...	S.A.	T.	V.	N.S.W.	Q.	B.	Cereals, Poa annua, &c.	Cluster cups on large circular spots. Uredospore pustules rust colour (Uredo rubigo-vera). Teleutospore pustules black, covered a long time.
1494	...	...	...	...		Q.	...	Leaves, leaf stalks, and stems of Rumex	Pustules scattered or arranged in circle, rounded or elongated, girt by torn epidermis, brown.
1404A	...	...	...	V.			...	Leaves of Muehlenbeckia adpressa	Pustules on upper surface, scattered, blistered at first.
1495	...	S.A.	...	V.	...		...	Leaves of Goodenia geniculata	Cluster cups in groups, on brownish spots. Teleutospore pustules rounded or elongated.
1496	...	...			N.S.W.	Q.	...	Maize ...	Uredospore pustules on both surfaces, reddish brown (Uredo Maydis). Teleutospores not noted.
1497	...	S.A.		...		...	...	Arundo phragmites	Generally resembling P. phragmitis, but the teleutospores differ in size and shape, and are yellowish to pale yellowish brown.
1498	...	...		V.			B.	Different species of Viola	Cluster cups on all green parts (Æcidium Violæ). Uredospore pustules brown. Teleutospore pustules black.
1499	...	...			...	Q.	...	Leaves of Wurmbsea dioica	Pustules elongated, blistered, dark brown. Uredospores brown. Teleutospores darker.

Number.	Cooke's Number.	Saccardo's Number.	Scientific Name.	Authority for Name.	English Name.

149. Phragmidium.—Link. Sp. II. 84 (1824).—

1500	1771	VII. 2621	P. Barnardi	Plow. and Wint., Rev. Myc. 208 (1886) ...	... Barnard's phragmidium ...
1501	1773	„ 2616	P. Potentillæ	... *Karst.*, Fung. Fenn. 94 (1887) ...	... Potentilla phragmidium ...
1502	1772	„ 2622	P. subcorticium	... *Winter*, Die Pilze. 1. 228 (1884)	... Sub-cortical phragmidium ...

150. Hamaspora.—Körn.

| 1503 | 1774 | VII. 2630 | H. longissima | ... Körn., Hedw. XVI. 23 (1877) ... | ... Very long hamaspora ... |

151. Æcidium.—Pers., in Gmel.

1504	1780	VII. 2857	A. Apocyni	... Schwein, Syn. Car. 68 (1822)	... Apocynum æcidium ...
1505	1776	„ 2719	A. Barbareæ	... D. C., Fl. Fr. II. 244 (1815)	... Barbarea æcidium ...
1506	1779	„ 2815	A. Compositarum ...	... Mart., Erl. 314 (1817) ...	... Composite æcidium ...
1507	1785	„ 2970	A. cystoseiroides ...	Berk., Fl. Tasm. 270 (1860) ...	... Cystoseira-like æcidium ...
1508	1786	„ 2100	A. Goodeniacearum	Berk., Linn. Journ. XIII 173 (1873)	... Goodenia æcidium ...
1509	1781	„ 2864	A. Nymphoidis	D. C., Fl. Fr. II. 597 (1815)	... Nymphoides æcidium ...
1510	1782	„ 2879	A. Plantaginis	... Ces., Erb. Critt. Ital 247 (1878)...	... Plantain æcidium ...
1511	1775	„ 2707	A. Ranunculacearum	... D. C., Fl. Fr. V. 97 (1805) ...	... Buttercup æcidium ...
1512	1790	„ 2313	A. Senecionis	Desm., Ann. Sci. Nat. 243 (1806)	... Groundsel æcidium ...
1513	1778	„ 2779	A. soleniiforme	Berk., Fl. Tasm. II. 270 (1860) ...	... Solenia-like æcidium ...
1514	1789	„ 2496	A. Urticæ	... Schum., Fl. Sæll. II. 223 (1801)	... Nettle æcidium ...
1515	1783	„ 2487	A. Veronicæ	... Berk., Grev. XI. 97 (1883)	... Speedwell æcidium ...

152. Roestelia.—Robert.,

| 1516 | 1791 | VII. 2974 | R. polita ... | Berk., Linn. Journ. XIII. 174 (1873) | ... Polished roestelia ... |

153. Uredo.—Pers.,

1517	1792	VII. 2999	U. angiosperma	... Thuem., Myc. Austr. IV. 95 (1880)	Angiospermous uredo ... ...
1518	1801	...	U. armillata ...	... Ludw., Bot. Centr. 6 (1890)	... Collared uredo ...
1519	1799	VII. 2210	U. Cichoracearum ...	... D. C., Fl. Fr. II. 229 (1815)	... Chicory uredo... ...
1520	1800	„ 3139	U. Clematidis	... Berk, Hook., Journ. VI. 205 (1854)	... Clematis uredo
1521	1798		U. leguminum	... Desm., Ann. Sci. Nat X. 310 (1838) ...	... Legume uredo
1522	1793		U. notabilis	... Ludw., Bot. Centr. 5 (1890)	... Notable uredo
1523			U. pallidula	... Cooke and Mass., Grev. XXII. 37 (1893)	... Pallid uredo...
1524	1795	VII. 3101	U. Rhagodiæ	Cooke and Mass., Grev. XV. 99 (1887) ...	... Rhagodia uredo
1525	1796	„ 3111	U. Spyridii	... Cooke and Mass., Grev. XV. 99 (1887) ...	.... Spyridium uredo
1526	1794	„ 3000	U. Wurmb-eæ = U. Anguillariæ	Cooke, Grev. XIV. 11 (1885)	Wurmbea uredo

OF AUSTRALIAN FUNGI—*continued.*

Number.	Habitat.						B.	Occurrence.	General Characters.
	W.A.	S.A.	T.	V.	N.S.W.	Q.			

Uredo, Cæoma, Puccinia, Lycoperdon, Hamaspora.

1500	...	S.A.	...	V.	...	...	...	Leaves of *Rubus parvifolius*	Uredospore pustules scattered on under surface. Teleutospore pustules scattered or gregarious.
1501	...	...	...	V.	...	...	B.	Leaves of *Acæna Sanguisorbæ*	Spots roundish, orange yellow. Uredospores pustules orange red. Teleutospore pustules black.
1502	...	...	...	V.	...	...	B.	Rose leaves	Uredospore pustules on under surface, scattered or crowded, yellow. Teleutospore pustules black.

Hedw. 22 (1877).—Phragmidium, Uredo.

| 1503 | ... | ... | ... | ... | ... | Q. | ... | Leaves of *Rubus Moluccanus* | Uredospores pustules on under surface, scattered or gregarious, bright orange. Teleutospore pustules gregarious, pale ochre. |

Syst. 1472 (1791).—Cæoma, Uredo, Puccinia, Trichobasis.

1504	...	...	...	...	...	Q.	...	Leaves of *Tabernæmontana orientalis*	Spots thin, circular, large, orange, pale beneath.
1505	...	...	...	...	N.S.W.	...	B.	Crucifers...	Cluster cups on both surfaces on ruddy spots in irregular clusters, large, distorting leaf.
1506	...	...	...	V.	N.S.W.	Q.	B.	Leaves of *Senecio Vellegioides* and other *Compositæ*	Spots purplish, nearly round and run together. Cluster cups crowded on the spots in circular patches. Probably stage of *Puccinia Hieracii* and others.
1507	...	...	T.	...	...	...	...	*Opercularia*	Pustulate, deforming the leaves. Cluster cups immersed.
1508	...	S.A.	...	V.	N.S.W.	Q	...	Leaves of *Selliera, Gaudenia,* and *Scævola*	Spots circular, brown beneath. Cluster cups scattered. May belong to *Puccinia paucinioides.*
1509	...	...	...	...	...	Q.	...	Leaves of *Limnanthemum indicum*	Cluster cups gregarious, disposed without order on rounded spots or in concentric zones.
1510	...	...	...	V.	N.S.W.	...	...	Leaves of *Plantago*...	Spots mostly small, sometimes broad. Cluster cups loosely scattered on both surfaces.
1511	...	...	T.	V.	...	...	B.	Leaves of *Ranunculus rionlaris* and *R. inundatus*	Cluster cups on under surface, in circular or elongated clusters, cup shaped.
1512	...	...	...	V.	N.S.W.	...	...	*Senecio* ...	Cluster cups on brown spots often bordered with black, arranged in clusters without definite order.
1513	...	...	T.	...	...	...	...	*Goodia latifolia*	Spots circular, brown. Spores orange.
1514	...	...	...	V.	...	...	B.	Nettles, &c.	Referred to *Puccinia curicis.* Cluster cups arranged in single or double series on yellow or red spots, on leaves, or stems.
1515	...	...	...	V.	...	...	...	*Veronica* ...	Scattered half-immersed cluster cups with margins fringed.

Fl. Ncom. 330 (1804).

| 1516 | ... | ... | ... | V. | ... | Q. | ... | Branches of *Muchlenbeckia Cunninghamii* and *Jacksonia scoparia* | Ochrey, cylindrical, polished, delicately downy. |

in Usteri. n. Ann. IX. 16 (1795).

1517	W.A.	...	...	...	...	...	...	Leaves of *Hakea* ...	Pustules on both surfaces, large, commonly disposed about a circle, powdery, brown.
1518	...	...	...	V.	...	...	...	*Juncus pallidus* ...	Pustules red brown, run together, surrounded by ruptured epidermis.
1519	...	...	...	V.	...	Q.	B.	*Bidens pilosa* ...	Referred to *Puccinia Hieracii.* Spots very minute. Pustules on both surfaces, scattered, small, circular.
1520	...	...	...	V.	...	Q.	...	Leaves of *Clematis aristata* and *C. microphylla*	Pustules on under surface, solitary or gregarious, pale yellow, more or less rounded, flattened.
1521	...	...	...	...	...	Q.	...	Pods of *Acacia* ...	Pustules rounded, solitary, rather large, girt by ruptured epidermis.
1522	...	S.A.	...	V.	...	...	...	Phyllodes of *Acacia notabilis*	Pustules large, red brown, seated on distorted inflated tubercle.
1523	...	...	...	...	...	Q.	...	Leaves, twigs, and legumes of *Cassia*	Pustules on both surfaces, pallid, convex, gregarious, splitting irregularly, and then girt by ruptured epidermis.
1524	...	...	...	V.	...	...	...	Leaves of *Rhagodia Billardieri*	Pustules on under surface, scattered, covered a long time, at length torn, brown.
1525	...	...	...	V.	...	...	...	Leaves of *Spyridium parvifolium*	Pustules on under surface, scattered, yellowish, powdery.
1526	...	...	...	...	N.S.W.	...	...	Leaves of *Urambsea dioica*	Pustules on both surfaces, gregarious, blistered, long covered by epidermis.

GENERAL CLASSIFICATION OF PYRENOMYCETES.

GROUP IV.—PYRENOMYCETES, FRIES.

ARRANGEMENT OF ORDERS (18).

14. HYPOCREACEÆ—Simple or compound. Receptacles rather fleshy or waxy, bright coloured, never carbonaceous.
15. XYLARIACEÆ—Stroma erect, compound. Receptacles carbonaceous.
16. DOTHIDEACEÆ—Composite, leathery or carbonaceous, blackish.
17. MELOGRAMMACEÆ—Receptacles formed from the stroma, or confluent with it.
18. DIATRYPACEÆ—Receptacles immersed in a heterogeneous stroma.
19. VALSACEÆ—Receptacles distinct, circinating, or in a single row.
20. EUTYPACEÆ—Receptacles immersed in stroma, densely gregarious for the most part.
21. CUCURBITARIACEÆ—Receptacles tufted or gregarious; erumpent, carbonaceous.
22. SUPERFICIALES—Receptacles distinct from each other, superficial or nearly so.
23. PERTUSACEÆ—Receptacles emergent, smooth, flattened at base.
24. LOPHIOSTOMACEÆ—Receptacles nearly superficial, opening compressed.
25. CERATOSTOMACEÆ—Receptacles for the most part immersed or sometimes nearly superficial.
26. OBTECTACEÆ—Receptacles innate in bark, and covered by cuticle.
27. CAULICOLACEÆ—Immersed, innate, observed mostly on dead stems of herbaceous plants.
28. FOLIICOLACEÆ—Receptacles minute, membranous, innate, growing mostly on leaves.
29. MICROTHYRIACEÆ—Simple, receptacles nearly superficial, membranous or carbonaceous.
30. PERISPORIACEÆ—Receptacles membranous, leathery or somewhat carbonaceous, wholly closed.
31. HYSTERIACEÆ—Receptacles more or less elongated, leathery or somewhat carbonaceous.

ORDER XIV.—HYPOCREACEÆ, DE NOT.

ARRANGEMENT OF GENERA (15).

Sub-order 1. Hypocreoideæ, Cooke—Composite forms.

Genera (6)—

154. Claviceps, Tul.	156. Epichloe, Fries.	158. Hypocrella, Sacc.
155. Cordyceps, Fries.	157. Hypocrea, Fries.	159. Polystigma, Pers.

Sub-order 2. Nectrieæ, Cooke—Simple or tufted forms.

Genera (9)—

160. Sphærostilbe, Tul.	163. Hypomyces, Fries.	166. Gibberella, Sacc.
161. Nectria, Fries.	164. Dialonectria, Sacc.	167. Lisiella, Cooke.
162. Calonectria, De Not.	165. Ophionectria, Sacc.	168. Melanospora, Corda.

ORDER XV.—XYLARIACEÆ, COOKE.

Genera (7)—

169. Xylaria, Hill.	172. Ustulina, Tul.	174. Daldinia, De Not. and Ces.
170. Porouia, Willd.	173. Nummularia, Tul.	175. Hypoxylon, Bull.
171. Kretzschmaria, Fries.		

ORDER XVI.—DOTHIDEACEÆ, NITS. AND FCKL.

ARRANGEMENT OF GENERA (7).

Sub-order 1. Dothideoideæ.

Genera (5)—

176. Phyllachora, Nits.	178. Montaguella, Speg.	180. Darwiniella, Speg.
177. Dothidella, Speg.	179. Bagnisiella, Speg.	

Sub-order 2. Rhytismoideæ.

Genus (1)—
181. Rhytisma, Fries.

Sub-order 3. Stigmatoideæ.

Genus (1)—
182. Trabutia, Sacc. and Roum.

ORDER XVII.—MELOGRAMMACEÆ, NITS.

Genera (1)—

183. Sarcoxylon, Cooke.	185. Botryosphæria, De Not.	186. Melogramma, Tul.
184. Gibellia, Sacc.		

ORDER XVIII.—DIATRYPACEÆ, FRIES.

Genera (2)—
187. Diatrype. Fries.

| 188. Cœlosphæria, Sacc.

ORDER XIX.—VALSACEÆ, FRIES.

Genera (2)—
189. Valsa, Fries.

| 190. Eutypella, Nits.

ORDER XX.—EUTYPACEÆ, COOKE.

Genera (3)—
191. Cryptovalsa, Ces. and De Not. | 192. Cryptosphærella, Sacc. | 193. Eutypa, Tul.

ORDER XXI.—CUCURBITARIACEÆ, COOKE.

Genus (1)—
194. Gibberidea, Fckl.

ORDER XXII.—SUPERFICIALES, FRIES.

Genera (7)—
195. Byssosphæria, Cooke.
196. Lasiosphæria, Ces. and De Not.

197. Pleosphæria, Speg.
198. Venturia, De Not. and Ces.
199. Chætomium, Kunze.

200. Rosellinia, De Not.
201. Trematosphæria, Fckl.

ORDER XXIII.—PERTUSACEÆ, FRIES.

Genus (1)—
202. Conisphæria, Cooke.

ORDER XXIV.—LOPHIOSTOMACEÆ, SACC.

Genus (1)—
203. Lophiostoma, Ces. and De Not.

ORDER XXV.—CERATOSTOMACEÆ, FRIES.

Genus (1)—
204. Rhamphoria, Niessl.

ORDER XXVI.—ORTECTACEÆ, FRIES.

Genera (2)—
205. Massariella, Speg.

| 206. Didymosphæria, Fckl.

ORDER XXVII.—CAULICOLACEÆ, FRIES.

Genera (4)—
207. Physalospora, Niessl.
208. Didymella, Sacc.

| 209. Anthostomella, Sacc. | 210. Pleospora, Rabh.

ORDER XXVIII.—FOLIICOLACEÆ, FRIES.

Genera (3)—
211. Læstadia, Auersw.

| 212. Sphærella, Ces. and De Not. | 213. Sphærulina, Sacc.

ORDER XXIX.—MICROTHYRIACEÆ, SACC.

Genera (2)—
214. Microthyrium, Desm.

| 215. Micropeltis, Mont.

ORDER XXX.—PERISPORIACEÆ, FRIES.

Genera (13)—
216. Podosphæra, Kunze.
217. Sphærotheca, Lev.
218. Erysiphe, Hedw.
219. Eurotium, Link.
220. Asterina, Lev.

221. Asterella, Sacc.
222. Dimerosporium, Fckl.
223. Parodiella, Speg.
224. Meliola, Fries.

225. Zukalia, Sacc.
226. Asteridium, Sacc.
227. Capnodium, Mont.
228. Antennaria, Link.

ORDER XXXI.—HYSTERIACEÆ, CORDA.

Genera (8)—
229. Aulographum, Lib.
230. Glonium, Muhl.
231. Lembosia, Lev.

232. Hysterium, Tode.
233. Tryblidiella, Sacc.
234. Rhytidhysterium, Speg.

235. Platychcilus, Cooke.
236. Hysterographium, Corda.

Number.	Cooke's Number.	Saccardo's Number.	Scientific Name.	Authority for Name.	English Name.

| Number. | Cooke's Number. | Saccardo's Number. | Scientific Name. | Authority for Name. | English Name. |

GROUP IV.—PYRENOMYCETES.—

ORDER XIV.—HYPOCREACEÆ,

154. CLAVICEPS.—Tul.,

| 1527 | | II. 5005 | C. purpurea | *Tul.*, Ann. Sci. Nat. XX. (1853) | Purple claviceps ... (Ergot) |

155. CORDYCEPS.—Fries,

1528	1488	II. 5012	C. entomorrhiza	*Fries*, S.V.S. 381 (1849)	... Insect-root cordyceps ...
1528a	1488		C. entomorrhiza, var. Menesteridis	*Cooke*, Handb. Aust. Fung. 277 (1892)	... Menesteridis cordyceps ...
1529	1486	II. 5030	C. Gunnii	*Berk.*, Hook., Lond. Journ. VII. 577 (1848)	Gunn's cordyceps ...
1530	1487	IX. 4013	C. Hawkesii ...	... *Gray*, Not. Insects (1858)	Hawkes' cordyceps ...
1531		II. 5038	C. ophioglossoides ...	*Link*, Hand b. III. 347 (1833)	... Ophioglossum-like cordyceps ...
1532	...	„ 5041	C. Taylori	... *Sacc.*, Mich. I. 320 (1878)	... Taylor's cordyceps ...

156. EPICHLŒ.—Fries,

| 1533 | 1489 | II. 5050 | E. cinerea | *Berk. and Br.*, Linn. Journ. XIV. 111 (1875) | Ashy epichlœ ... ... |

157. HYPOCREA.—Fries,

1534	1490	II. 4897	H. cerebriformis	*Berk.*, Linn. Journ. XIII. 179 (1873)	... Brain-like hypocrea ...
1535	1492	„ 4875	H. citrina	*Fries*, S.V.S. 185 (1849) ...	... Lemon-yellow hypocrea ..
1536	1493	„ 4834	H. rufa	*Fries*, S.V.S. 383 (1849)	Reddish hypocrea
1537	1491	„ 4828	H. semiorbis	*Berk.*, Fl. Tasm. II. 278 (1860) ...	... Semicircular hypocrea

158. HYPOCRELLA.—Sacc.,

| 1538 | 1495 | | H. axillaris | *Cooke*, Grev. XX. 4 (1891) | Axillary hypocrella |
| 1539 | 1494 | II. 5001 | H. discoidea | *Sacc.*, Mich. I. 322 (1878) | Discoid hypocrella ... |

159. POLYSTIGMA.—D. C.,

| 1540 | 1496 | IX. 3892 | P. australicus | *Sacc.*, Bull., Soc. Myc. Fr. V. 119 (1889) ... | Australian polystigma ... |

160. SPHÆROSTILBE.—Tul.,

1541	1497	II. 4817	S. cinnabarina		*Tul.*, Carp. III. 103 (1865)	Cinnabar sphærostilbe ...
1542	1500	„ 4825	S. dubia ...	... *Berk.*, Linn. Journ. XVIII. 389 (1881) ...	Doubtful sphærostilbe ...
1543	1498	„ 4820	S. hypocreoides	... *Kalch. and Cooke*, Grev. IX. 26 (1880) ...	Hypocrea-like sphærostilbe
1544	1499	IX. 3912	S. micro-spora	... *Cooke and Mass.*, Grev. XVI. 4 (1887)	Small-spored sphærostilbe

OF AUSTRALIAN FUNGI—*continued.*

Number.	Habitat.						B.	Occurrence.	General Characters.
	W.A.	S.A.	T.	V.	N.S.W.	Q.			

FRIES, S.M. I. 51 (1821).

DE NOT. OSS. PIR. GIORN. BOT. I. (1844).

Ann. Sci. Nat. Ser. III., XX. 43 (1853).—Sphæria.

| 1527 | ... | ... | | V. | | | B. | On *Lolium perenne,* L. temulentum, Triticum sativum, &c. | Stroma stalked, erect, arising from a sclerotium, club headed. |

S.M. II. 324 (1821).—Sphæria.

1528		...		V.	...	...	B.	Insect larvæ (*Tinea,* &c.)	Fleshy. Club somewhat globose, brown. Stem thin, very long.
1528A	...			V.	...	...	...	Larvæ of *Menestenidis,* &c.	Club elliptic, reddish, at first powdery. Stem thickened upwards.
1529	...	...	T.	V.	N.S.W.	...	...	Larvæ of some *Cossus* or *Hepialus*	Fleshy. Club cylindrical, yellow, blackening above. Stem elongated, white.
1530	...	...	T.			...	...	Larvæ of insects ...	Cylindrical, tapering and truncate at apex. Stem flexuous, sometimes forked with three or four clubs.
1531	...	...	...		...	...	B.	... ...	Solitary, rarely tufted, simple, rarely branched, fleshy, yellow within. Club oblong. Stem olive, then blackening.
1532	...	...	...	V.	N.S.W.	...		Larvæ of insects ...	Stems tufted, running together in net-like manner. Clubs reddish yellow, delicately velvety.

S.V.S. 381 (1849).

| 1533 | ... | ... | ... | ... | | Q. | ... | Various grasses, preferably *Sporobolus* | Encircling stems of grasses, dark ashy, and dotted with the darker openings of immersed receptacles. |

S.V.S. 383 (1849).

1534		S.A.		...			...	...	Trunks	Cushion shaped, wrinkled, lobed, fawn, substance thick.
1535	...		T.	...			...	B.	Soil, mosses, rotting leaves, &c.	Fleshy, spread out, lemon yellow, prominent openings of receptacles brownish.
1536	...	...	T.	...			...	B.	Wood or bark ...	Gregarious, hemispherical when moist, collapsing when dry, wrinkled, reddish, soft and fleshy.
1537	...	...	T.	...			...	...	Bark and wood ...	Hemispherical, rather fleshy, ochrey, darker than the bark. Receptacles immersed, openings minute.

Mich. I. 322 (1878).—Hypocrea.

| 1538 | ... | ... | | ... | ... | Q. | | Grasses (*Eragrostis stricta*) | Inversely club shaped, seated in the upper axils, black, opaque, minutely granular with the openings of the receptacles. |
| 1539 | ... | | ... | ... | | Q. | | Leaves ... ... | Circular, separating from matrix, scarlet. Receptacles rather prominent. |

Fl. Fr. V. 161 (1815).

| 1540 | ... | • | ... | ... | V. | | | Leaves, rarely stems, of *Leguminosæ* | Immersed, rather swollen, covering half or entire leaf, dull rosy, rather fleshy. |

Carp. III. 103 (1865).

1541	...	...	...	...	N.S.W	Q.	...	Bark	Receptacles at base of conidia-bearing layers, small, sessile, globose, smooth, orange red.
1542	...	...	...	...		Q.	...	Bark of *Ægiceras* ...	Only Stilbum form known, bearing conidia.
1543	...	...	...	...		Q.	...	Bark	Pale rose, convex. Receptacles as in Hypocrea, associated with club-shaped conidia bearers.
1544	...	...		...	V.		...	Bark	Receptacles associated with conidia bearers, minute, scattered, ovate, orange.

Number.	Cooke's Number.	Saccardo's Number.	Scientific Name.	Authority for Name.	English Name.
					161. NECTRIA.—Fries,
1545	1502	II. 4670	N. coccinea	Fries, S.V.S. 368 (1849)	... Scarlet nectria ... ...
1546	..	IX. 3858	N. ferruginea	... Cooke, Grev. XIII. 8 (1884) ...	... Rusty nectria... ... ...
1547	1501	II. 4561	N. fusarioides	Berk., Fl. Tasm. II. 279 (1860) ...	... Fusarium-like nectria ...
1548	1503	„ 4703	N. tasmanica	... Berk., Fl. Tasm. II. 279 (1860) ..	... Tasmanian nectria
1549	1504	„ 4678	N. zealandica	... Cooke, Grev. VIII. 65 (1879)	... New Zealand nectria ... ...
					162. CALONECTRIA.—De Not.,
1550		II. 6174	C. otagensis	Sacc. Syll. II. LXVIII. (1883) ...	... Otago calonectria ... ...
					163. HYPOMYCES.—Fries, Pl. Homon. 105 (1825).—
1551	1508	II. 4622	H. aurantius	Fckl., Symb. Myc. 183 (1875)	Golden hypomyces ... ...
1552	1505	„ 4614	H. chrysospermus ...	... Tul., Sel. Fung. Carp. III. 51 (1865)	Golden-seeded hypomyces ...
1553	1509	...	H. membranaceus ...	... Cooke, Handb. Austr. Fung. 281 (1892) ...	Membranous hypomyces ...
1554	1506	II. 4617	H. rosellus	... Tul., Sel. Fung. Carp. III. 45 (1865)	Rosy-red hypomyces ... ...
1555	1507	„ 4648	H. tomentosus	... Fries, in Grev. IV. 15 (1875) ...	Downy hypomyces ... ...
					164. DIALONECTRIA.—Sacc.
1556	1511	II. 4733	D. quisquilaris	Cooke, Handb. Austr. Fung. 282 (1882)	... Rubbish-loving dialonectria ...
1557	1510	„ 4721	D. sanguinea	... Fries, S.V.S. 388 (1849) ...	... Blood-red dialonectria...
1558	1512	„ 4742	D. tephrothele	... Berk., Fl. Tasm. II. 278 (1860)...	... Dark-nippled dialonectria ...
					165. OPHIONECTRIA.—Sacc.,
1559	1513	II. 5001	O. agaricicola	... Sacc. Syll. II. 563 (1883)	Agaricus-growing ophionectria ...
					166. GIBBERELLA.—Sacc., Mich. I. 43 (1878).—
1560	1514	II. 4977	G. Saubinetii	... Sacc., Mich. I. 513 (1878)	... Saubinet's gibberella ... ...
					167. LISIELLA.—Cooke,
1561	1515	IX. 3804	L. passiflorae	... Cooke, Grev. XVI. 5 (1887)	... Passion-flower lisiella :. ...
					168. MELANOSPORA.—Corda,
1562	1516	II. 4599	M. caprina	Sacc. Syll. II. 462 (1883) ...	... Shaggy melanospora ...

OF AUSTRALIAN FUNGI—*continued.*

Number	W.A.	S.A.	T.	V.	N.S.W.	Q.	B.	Occurrence.	General Characters.

S.V.S. 387 (1849).—Sphæria.

Number	W.A.	S.A.	T.	V.	N.S.W.	Q.	B.	Occurrence.	General Characters.
1545	...	...	T.	V.	N.S.W.	Q.	B.	Bark and dead branches	Receptacles in tufts, on a convex yellowish layer, ovoid, bright red, sometimes ochrey red.
1546	...	...	...	V.		...	...	Living leaves, bracts, &c. of *Styphelia*	Bursting through, tufted. Receptacles waxy, almost globose, dark brown, at length naked.
1547	...	...	T.	...		...	...	Dead bark	Pale crimson. Receptacles ovate to pap-like, with a bloom half immersed in umber-coloured layer.
1548	...	...	T.	...	...	...		Dead bark	Tufted, red. Receptacles ovate, with pap-like openings, often arising from circular disc.
1549	...	...		V.	...	...	...	Bark	Tufted, brick red, bursting through, pustules convex. Receptacles almost globose, pap-like, soon concave.

Comm. Critt. 11. 477 (1867).—Nectria.

1550	...	...	...	...	...	Q.	...	Twigs of *Capparis Mitchelli*	Receptacles densely tufted, pale-apricot colour, with openings a little deeper in colour, papillate.

Sphæria, Hypocrea, Nectria.

1551	...	...	...	...	...	Q.	B.	On *Polyporus*, &c.	Conidia-bearing mycelium white, then orange. Receptacles seated on fluffy ochrey base, often white at margin, golden yellow or orange.
1552	W.A.	...	...	V.		Q.	B.	On *Boleti* chiefly and *Polyporus*	Conidia-bearing mycelium penetrating matrix. Receptacles closely packed in a rough layer, pale yellow brown.
1553	...	...	...			Q.	...	On *Polyporus*	Forming at first fine filaments, gradually becoming membranous, tan coloured.
1554	W.A.	...	...			...	B.	On *Polyporus*, &c.	Receptacles gregarious, at first white, then bright-red mycelium.
1555	...	...	T.			...	...	On Agarics	Stratum white, delicate downy.

Syll. II. 490 (1883).—Nectria, Sphæria.

1556	...	...	...	V.	...	...	...	Bark, chips, &c.	Scattered, umber coloured or somewhat orange. Receptacles crowded here and there.
1557	W.A.	...	...			...	B.	Wood and bark	Receptacles scattered, egg shaped, openings pap-like, blood red.
1558	...	...	T.	...		...	...	*Hypoxylon*	Receptacles scattered, crimson, egg shaped, openings darker and pap-like.

Mich. I. 323 (1878).—Nectria.

1559	...	...	T.	V.	...	...	...	Putrid Agarics	Vermilion. Receptacles egg shaped, with fibrous swollen texture.

Gibbera, Botryosphæria.

1560	...	...	T.	V.	...	...	B.	Herbaceous stems	Receptacles gregarious, growing together in tufts, somewhat membranous, warted, folded, blue.

Grev. XVI. 5 (1887).—Gibberella.

1561	...	...	...	...	...	Q.	...	Stems of *Passiflora*	Receptacles bursting through, collected in small clusters, globose, substance bright blue.

Icon. Fung. I. 24 (1837).—Sphæria.

1562	...	...	T.	...		...	B.	Wood and chips	Receptacles globose, shaggy, white, openings turning blackish.

111

SYSTEMATIC ARRANGEMENT

Number.	Code No. Number.	Saccardo's Number.	Scientific Name.	Authority for Name.	English Name.

ORDER XV.—XYLARIACEÆ,

169. XYLARIA.—Hill. Hist. Plant. 62 (1773).—

1563	1543	IX. 2175	X. agariciformis	Cooke and Mass , Grev. XVII. 81 (1889)	Agaricoid xylaria ... ...
1564	1532	I. 1178	X. allantoidea	Berk., Linn. Journ. X. 380 (1869)	Allantoid xylaria
1565	1533	., 1219	X. anis pleura	Mont., Syll. 688 (1856)	Unequal-ribbed xylaria
1566	1545	„ 1211	X. aphrodisiaca	Welw. and Curr., Fung. Angol. 280 (1867)	Aphrodisiac xylaria
1567	1517	IX. 2137	X. australis ...	Cooke, Grev. XI, 84 (1883)	Southern xylaria
1568	1531	I. 1216	X. castorea ...	Berk., Fl. N. Zeal. II. 204 (1855) ...	Beaver xylaria
1569	1528	IX. 2157	X. cerebriformis ...	Cooke, Grev. XI. 86 (1883) ...	Brain-like xylaria
1570	1541	„ 2173	X. cinnabarina ..	Cooke and Mass., XV. 101 (1887) ...	Vermilion xylaria
1571	1514	I. 1239	X. corniformis	Fries, S.V S. 381 (1849)	Horn-shaped xylaria ... ...
1572	1535	II. 5963	X. cretacea	Berk. and Br., Linn. Trans. 405 (1879) ...	Chalky xylaria ... ...
1573	1529	IX. 2165	X. cynoglossa	Cooke, Grev. XII. 1 (1883)	Dog-tongue xylaria ... ...
1574	1534	I. 1223	X. dealbata	Berk. and Curt., Exot. Fung. 254 (1853)	Bleached xylaria ... ...
1575	1546	., 1283	X. digitata	Grev., Fl. Ed. 356 (1824)	Digitate xylaria ...
1576	1540	IX. 2172	X. elastica	Cooke, Grev. XVI. 4 (1887)	Elastic xylaria ... ...
1577	1536	„ 2169	X. ellipsospora ...	Cooke and Mass., Grev. XVI. 33 (1887)	Elliptical-spored xylaria ...
1578	1533	I 1155	X. fistulosa ...	Fries, Nov. Symb. 126 (1851) ...	Hollow xylaria ...
1579	1539	...	X. gigas ...	Cooke, Handb. Austr. Fung. 287 (1892)...	Gigantic xylaria ...
1580	1526	I. 1158	X. gracilis ...	Sacc. Syll. I. 317 (1882)	Graceful xylaria ...
1581	1519	„ 1189	X. grammica ...	Mont., Syll. 689 (1856)	Lined xylaria ..
1582	...	„ 1303	X. hippotrichoides ...	Sacc. Syll. I. 344 (1882)	Horsehair like xylaria ...
1583	1547	„ 1260	X. Hypoxylon ...	Grev., Fl. Ed. 353 (1824)	Hypoxylon xylaria ... : (Candle-snuff fungus)
1584	1518	„ 1228	X. involuta	Klotsch, Linn. VII. (1832)	Involute xylaria ...
1585	1530	IX. 2171	X. lobata	Cooke, Grev. XI. 86 (1883)	Lobed xylaria ...
1586	1525	„ 2154	X. ovispora ...	Cooke and Mass., Grev. XV. 101 (1887)...	Egg-spored xylaria ... ...
1587	1537	I 1167	X. phosphorea	Berk., Linn. Journ. XIII. 177 (1873)	Phosphorous xylaria ...
1588	1527	„ 1150	X. polymorpha	Grev., Fl. Ed. 35 (1824) ...	Polymorphous xylaria ...
1588A	...	...	X. polymorpha, var. pachy-stroma	Sacc., Myc. Austr. 13 (1890) ...	Thick stroma xylaria ...
1589	...		X. Renderi ...	F. v. M , Grev. XXII. 17 (1893)	Reader's xylaria ...
1590			X. rhizophila	Cooke and Mass., XXII. 37 (1893)	Root-loving xylaria ... ...
1591	1520	I. 1234	X. rhopaloides ...	Mont., Ann. Sci. Nat. III. 99 (1855) ...	Club-like xylaria ...
1592	1522	., 1230	X. rhytidophlœa ...	Mont., Syll. 687 (1856)	Wrinkled xylaria ...
1593	1521	„ 1222	X. Schweinitzii ...	Berk. and Curt., Exot. Fung. 284 (1853)	Schweinitz's xylaria ...
1594	1524	„ 1288	X. scopiformis	Mont., Ann. Sci. Nat. XIII. 319 (1840) ...	Stalk-like xylaria ...
1595	1542	„ 1925	X. tuberiformis	Berk , Fl. N. Zeal. II. 204 (1855)	Tuber-like xylaria ... ...
1596	1523	1209	X. zealandica	Cooke, Grev. VIII. 66 (1879) ...	New Zealand xylaria ...

OF AUSTRALIAN FUNGI—*continued.*

COOKE, GREV. XIII. 9 (1884).

Sphæria, Clavaria, Hypoxylon, Rhizomorpha.

Number.	W A	S.A.	T	V.	N.S.W.	Q	R.	Occurrence.	General Characters.
1563	...	S.A.	...	...	...	...	...	Stumps ...	Half globose, glaucous, dotted with black openings. Stem equal, with barren black ring round it.
1564	W.A.		...	V.		Q	...	Stumps ...	Club shaped, encrusted with black, leathery. Stem very short. Receptacles minute, globose.
1565	...	...	...	...		Q.	...	Dead wood	Solitary, woody, very hard. Club inversely egg shaped as well as stem, scaly.
1566	...	...	...	...		Q	...	Rotten trunks	Tufted, smooth. Clubs somewhat cylindrical, brown to ashy, substance white.
1567	...	...	...	...		Q.	...	Wood ...	Club shaped, thickened upwards, brown. Stem elongated, blackening.
1568	...	...	T.	...		Q.	...	Rotten wood	Stem short, spongy and velvety at first, at length naked, black. Clubs obtuse, ovate, compressed.
1569			...	...		Q.	...	Wood ...	Large, corky, stalked, sooty brown. Stem woody, smooth, furrowed. Clubs somewhat elliptic, brain-like.
1570	...		...	...		Q.	...	Wood ...	Corky, somewhat globose, brown, vermilion under thin skin. Stem obsolete.
1571	...	...	T.	...			R.	Prostrate trunks ...	Cylindrical, thick, brown, then black, horn shaped.
1572	...		...	...		Q.	...	Trunks ...	Nearly globose, stalked, white, rather wrinkled, netted with thin brown lines.
1573	...	...	...	...	...	Q.	...	Wood ...	Umber, tongue shaped, shortly stalked, flesh white.
1574	...		...	...	...	Q.	...	Rotten trunks	Club elliptical, tapering downwards into short cylindrical stem, covered with whitened crust.
1575	...	...	...	V.	N.S.W.	...	B	Rotten wood	Erect, thick, brown, velvety, tapering towards apex, sometimes divided into forked branches; short stem.
1576	...	...	...	V.		Q.	...	Rotten wood	Corky to elastic, nearly globose, or hemispherical, sessile, becoming black.
1577	...	...	T.	...		...	...	Rotten wood	Club shaped, obtuse, black. Stem shortened, smooth. Receptacles immersed, not prominent.
1578	...	...	...	...		Q.	..	Trunks ...	Corky, simple, club shaped, with varnished crust, black, hollow, confluent with short stem.
1579	...	...	...	...	N.S.W.	Q.	...	Stumps, &c.	Large, ochrey, then brown. Clubs oval or irregular, wrinkled. Stem stout and irregular.
1580	...	...	...	...	...	Q.	...	Wood of *Acacia harpophylla*	Leathery. Stem smooth, forked, tufted. Clubs cylindrical, narrow, wrinkled, black.
1581	...	...	...	V.		Q.	...	Trunks ...	Large, corky, club shaped, rigid, sooty black, becoming whitish. Clubs with longitudinal lines running into one another.
1582	...	...	...	...	...	...	R.		Somewhat tufted, thread-like, ascending, black, branched.
1583	...	...	...	...	N.S.W.	Q.	B.	Stumps ...	Erect, compressed, and dilated, black, shaggy about base. Stem usually shorter than club.
1584	...	...	...	...	N.S.W.	Q.	...	Woods ...	Leathery, club shaped, ochrey to brownish yellow or fawn, tapering below into long slender stem.
1585	...	...	...	...		Q.	...	Wood	Large, corky, shell shaped, sessile, circumference lobed, brown, lobes rounded.
1586	...	...	...	...		Q.	...	Stumps ...	Leathery, black, stalked, erect, forked above, tapering downwards into smooth stem.
1587	...	...	...	V.		...	...	Trunks ...	Reddish brown, small. Stem short, cylindrical, streaked, expanding upwards into short club.
1588	...	...	...	...	N.S.W.	Q.	B.	Wood ...	Clubs in clusters, rarely solitary erect, thick, smooth, brown, then black, variously shaped.
1588A	...	...	...	...		Q.	...	Trunks ...	Stroma or receptacle-bearing layer very hard and thick above middle branching into finger-like processes.
1589	...	...	...	V.		...	...	Sandy desert	Black, globose or broadly elliptical, crowned by short spine, mealy, with white conidia. Stem erect.
1590	...	...	...	...		Q.	...	Roots of herbs and grasses	Stroma club shaped, divided nearly to base into two to six clubs. Clubs spoon shaped, flattened, bright brown.
1591	...	...	...	...	...	Q.	...	Putrid wood	Clubs cylindrical, obtuse, tapering downwards into short smooth stem.
1592	...	...	...	V.	...	Q.	...	Wood ...	Horny, compressed, obtuse or horn shaped, black. Stem very short and wrinkled in a netted manner.
1593	...	...	...	V.	N.S.W.	...	...	Rotten wood	Club elliptic, obtuse, corky, compact. Stem elongated, smooth, slightly cracked.
1594	...	...	...	...		Q.	...	Decaying fruit of *Flindersia australis*	Simple, slender. Clubs cylindrical, acute at apex, black. Stem as long as club, compressed.
1595	..	...	...	...		Q.	...	Rotten wood	Corky, almost globose, wrinkled, cap like, thick. Stem short or obsolete.
1596	...		...	...		Q.	...	Rotten wood	Simple, slender, stalked, black. Clubs cylindrical, wrinkled. Stem smooth, channelled.

Number.	Cooke's Number.	Saccardo's Number.	Scientific Name.	Authority for Name.	English Name.
					170. PORONIA.—Willd.,
1597	1549	I. 1322	P. edipus	Mont., Syll. 209 (1856)	Swollen-stalked poronia
1598	1530	„ 1323	P. pileiformis	... Fries, Nov. Symb. 129 (1851) ...	... Cap-shaped poronia ...
1599	154*	„ 1321	P. punctata	... Fries, S.V.S. 382 (1849) ...	... Punctate poronia ...
					171. KRETZSCHMARIA.—Fries,
1600	1552	I. 1519	K. angolensis	... Sacc. Syll. IX. 565 (1891)	... Angola kretzschmaria ...
1601	1551	„ 14*9	K. cetrarioides	Sacc. Syll. IX. 567 (1891)	... Cetrarium-like kretzschmaria ...
					172. USTULINA.—Tul., Sel. Fung.
1602	1553	I. 1328	U. vulgaris	Tul., Sel. Fung. Carp. II. 23 (1863)	... Common ustulina ...
					173. NUMMULARIA.—Tul., Sel. Fung.
1603	1558	IX. 2295	N. australis	... Cooke, Grev. XII. 6 (1883)	... Southern nummularia ..
1604	1554	„ 2292	N. Baileyi	... Cooke, Grev. XII. 6 (1883) ...	... Bailey's nummularia ...
1605	1556	I. 1524	N. Bulliardi	Tul. Sel. Fung. Carp. II. 43 (1863) ...	Bulliard's nummularia
1606	1557	., 1105	N. exutans	... Cooke, Handb. Austr. Fung. 291 (1892)...	Shedding nummularia
1607	1555	„ 1528	N. lutea ...	Nits. Pyr. Germ. 59 (1867) ...	Yellow nummularia ...
1608	1559	„ 1112	N. microplaca	... Cooke, Grev. XIII. 13 (1884) ...	... Cake-like nummularia
1609	1560	IX. 2300	N. pusilla	Sacc., Hedw. (1889) ...	... Small nummularia
					174. DALDINIA.—De Not. and Ces.,
1610	1561	I. 1515	D. concentrica	... Ces. and De Not., Schema Sfer. Ital. 198 (1870) ...	Concentric daldinia ...
1611	1562	„ 1516	D. vernicosa	Ces. and De Not., Schema Sfer. Ital. 198 (1870) ...	Varnished daldinia ...
					175. HYPOXYLON.—Bull.,
1612	1573	I. 1384	H. annulatum	... Mont., Syll. 213 (1856) ...	... Annulate hypoxylon ...
1613	1580	„ 1449	H. Archeri	... Berk., Fl. Tasm. II. 280 (1860)	... Archer's hypoxylon ...
1614	1567	„ 1337	H. argillaceum	Berk., Outl. 387 (1860)	... Clay-coloured hypoxylon ...
1615	...	„ 1433	H. atro-purpureum	Fries, N.V.S. 384 (1849)	... Dark-purple hypoxylon ...
1616	1576	„ 1113	H. capnodes	... Cooke, Grev. XI. 147 (1883)	... Capnodium-like hypoxylon ...
1617	1566	„ 1333	H coccineum	Bull., Champ. I. 174 (1798)	... Brick-red hypoxylon ...
1618	1571	„ 1370	H. cohærens	... Fries, S.V.S. 384 (1849)	... Cohering hypoxylon ...
1619	1581	IX. 2264	H. ellipticum	Cooke and Mass., Grev. XVII. 70 (1889)...	... Elliptic hypoxylon ...
1620	...	„ 2230	H. flavo-fuscum ...	Berk. and Br., Linn. Trans. II. 222 (1887)	... Yellow-brown hypoxylon ...
1621	1569	I. 1368	H. fuscum	Fries, S.V.S. 384 (1849)	... Brown hypoxylon ...
1622	1575	IX. 2254	H. hæmatites	Lev., Grev. XI. 133 (1883)	... Orange-red hypoxylon ...
1623	1572	I. 1435	H. hæmato-stroma	... Mont., Syll. 737 (1856)	... Blood-red stroma hypoxylon ...

OF AUSTRALIAN FUNGI—*continued.*

Number.	W.A.	S.A.	T.	V.	N.S.W.	Q.	B.	Occurrence.	General Characters.

Fl. Berol. 400 (1787).—Sphæria, Peziza.

Number.	W.A.	S.A.	T.	V.	N.S.W.	Q.	B.	Occurrence.	General Characters.
1597	...	...	...	V.	N.S.W.	Q.	...	Dung ...	Erect, externally blackish, simple or branched. Stem long, clubbed at base, expanded into cup at apex.
1598	...	...	...	...		Q.	...	Rotten wood	Branched, nearly even, branches ending in pileiform cups, blackish.
1599	W.A.	S.A.	T.	V.		...	B.	Dung ...	Erect, simple, at first club shaped, soon cup shaped, tapering into long black downy stem, disc white, punctate with black openings.

S.V.S. 409 (1849).—Rhopalopsis, Daldinia, Hypoxylon.

Number.	W.A.	S.A.	T.	V.	N.S.W.	Q.	B.	Occurrence.	General Characters.
1600	...	...	...	...		Q.	...	Rotting bark ...	Club shaped, black, bright, and shining. Receptacles thickly crowded, openings minute.
1601	...			...		Q.	...	Trunks ...	Resembling in habit and mode of growth the Lichen *Cetraria tristis*, fringed, fringes passing into receptacles at apex.

Carp. II. 23 (1861).—Sphæria, Hypoxylon.

Number.	W.A.	S.A.	T.	V.	N.S.W.	Q.	B.	Occurrence.	General Characters.
1602	...		...	...		Q.	B.	Branches and logs	Spread out, large, thick, wavy, at length quite black, carbonaceous, hollow within.

Carp. II. 42 (1863).—Hypoxylon, Sphæria, Authostoma, Diatrype.

Number.	W.A.	S.A.	T.	V.	N.S.W.	Q.	B.	Occurrence.	General Characters.
1603	...	...	...	...	N.S.W.		...	Branches	Developed within bark, then bursting through, unpolished, black. Receptacles nearly globose, small.
1604	...	...	...	...		Q.	...	Wood ...	Bursting through, circular, cup shaped, disc rough with prominent openings of receptacles, which are immersed in centre.
1605	...	...	T.				B.	Branches	Bursting through, circular or oval, broadly expanded, black without and within.
1606	...	...	...	V.			...	Branches	Broadly expanded, marginate, black, bursting through, and throwing off epidermis.
1607	...	...	...	...		Q.	B.	Wood	Superficial on decorticated wood, circular, thick, cup shaped, surface brown, then blackish.
1608	...	...	...	...		Q.	...	Bark ...	Thin, circular, sparingly dotted with minute pap-like openings of receptacles, black.
1609	...	S.A.	...	...		...	...	Branches of *Bursaria spinosa*	Small, flattened, becoming black, shining. Receptacles crowded, oblong, openings point-like.

Schema Sfer. Ital. I. 197 (1870).—Hypoxylon, Sphæria.

Number.	W.A.	S.A.	T.	V.	N.S.W.	Q.	B.	Occurrence.	General Characters.
1610	W.A.	...	T.	V.	N.S.W.	Q.	B.	Trunks ...	Spherical or hemispherical, zoned internally with concentric layers, black or brown.
1611	...	...	...	V.		...	...	Trunks ...	Large, tapering below into thick stem, surface black, varnished. Receptacles in many rows, black.

Champ. I. 168 (1798).—Sphæria, Diatrype, Authostoma, Nummularia.

Number.	W.A.	S.A.	T.	V.	N.S.W.	Q.	B.	Occurrence.	General Characters.
1612	...	...	T.	...	...	...	...	Bark and wood ...	Hemispherical, confluent, blackening. Receptacles nearly globose, opening in centre of disc, with rather prominent annulate margin.
1613	...	...	T.	...		...	...	Rotten wood	Quite black, spread out. Receptacles nearly globose, wrinkled, opening in centre of depressed disc.
1614	...		T.	...		...	B.	Trunks ...	Bursting through, somewhat globose, clay coloured, turning black within. Receptacles small, ovate, crowded.
1615	...		...	...		...	B.	Wood	Stroma in wood turning black and widely spread out, circumference variously and often interrupted.
1616	W.A.	...	...	...		Q.	...	Branches	Spread out, greyish black, dotted with the prominent openings of the receptacles.
1617	...	...	T.	V.		...	B.	Branches	Bursting through, nearly globose, violet brown or fawn, then brick red. Receptacles minute, ovate, crowded.
1618	...	...	T.	...		...	B.	Branches	Bursting through, nearly globose or flattened, thick, dirty brown, then black. Receptacles large, globose.
1619	...	...	...	...		Q.	...	Decorticated wood	Parallel, elliptic, black, openings of receptacles minute, crowded, dotted.
1620	...	...	...	...		Q.	...	Roots and stumps of grasses	Convex, irregularly lobed, yellow brown, mealy; mouths black, prominently punctate.
1621	...	...	...	...		Q.	B.	Bark ...	Bursting through, spot-like, hemispherical, purple brown then black. Receptacles globose, crowded.
1622	...	...	...	...		Q.	...	Wood ...	Expanded, crustaceous, wrinkled, bright orange red, at length rusty. Receptacles thickly crowded.
1623	...	...	...	...		Q.	...	Dark	Irregularly spread out, confluent, purplish black. Receptacles immersed rather prominent, layer blood red.

Number.	Cooke's Number.	Saccardo's Number.	Scientific Name.	Authority for Name.	English Name.
					175. HYPOXYLON.—Bull.,
1623A	...	IX. 2232	H. hæmato-stroma, var. hæmatozonnm	Sacc., Pug. Austr. 13 (1890) ...	... Blood-red zoned hypoxylon ...
1624	...	„ 2227	H. hians	Berk. and Cooke, Grev. XI. 129 (1883) ...	... Gaping hypoxylon
1625	1574	I. 1414	H. marginatum ...	Berk., Linn. Journ. X. 385 (1869)	... Marginate hypoxylon
1626	1570	„ 1376	H. multiforme	Fries, S.V.S. 384 (1849) ...	... Multiform hypoxylon ..
1627	1577	„ 1456	H. oodes ...	Berk. and Br., Linn. Journ. 122 (1875) ...	... Egg-like hypoxylon ...
1628	1564	„ 1535	H. placentæforme	Berk. and Curt., Linn. Journ. 383 (1869)	... Placenta-shaped hypoxylon ...
1629	1578	„ 1534	H. punctulatum ...	Berk. and Rav., Grev. IV. 94 (1875)	... Punctulate hypoxylou ...
1630	1568	„ 1344	H. rutilum ...	Tul., Sel. Fung. Carp. II. 38 (1863) ...	... Reddish hypoxylon
1631	1563	„ 1341	H. sclerophæum ...	Berk. and Curt., Linn. Journ. XIII. 177 (1873) ...	Hard-dusky hypoxylon ...
1632	1579	„ 1448	H. serpens	Fries, S.V.S. 384 (1849)	Creeping hypoxylon ...
1633	1565	IX. 2212	H. stratosum	... Sacc., Pug. Austr. 13 (1890) ...	... Stratose hypoxylon ...

ORDER XVI.—DOTHIDEACEÆ,

					176. PHYLLACHORA.—Nits,
1634	1585	IX. 4083	P. Alpiniæ	... Sacc. and Berl., Misc. Myc. II. 5 (1884) ...	Alpinia phyllachora .. ...
1635	1586	„ 4091	P. anceps	Sacc., Hedw. 156 (1890)	... Two-sided phyllachora ...
1636	1588	„ 4089	P. Fimbristylis ...	Sacc. Syll. IX. 1025 (1891) ...	... Fimbristylis phyllachora
1637	1582	II. 5132	P. graminis ...	Fckl., Symb. Myc. 216 (1869) ...	... Grass phyllachora ...
1638	1589	„ 5144	P. junci ...	Fckl., Symb. Myc. 216 (1869) ...	... Rush phyllachora ...
1639	1587	... ...	P. maculata	... Cooke, Grev. XX. 4 (1891)	... Spotted phyllachora ...
1640		IX. 4088	P. nervisequia	... Winter, Hedw. 9 (1885)	... Nerve-following phyllachora ...
1641	1583	II. 5093	P. rhytismoides ...	Sacc. Syll. II. 594 (1883)	Rhytisma-like phyllachora
1642	1584	„ 5184	P. Trifolii	Fckl., Symb. Myc. 218 (1869) ...	Clover phyllachora
					177. DOTHIDELLA.—Speg.,
1643	1590	IX. 4124	D. apiculata	Sacc. and Berl., F. Austr. 4 (1885)	Apiculate-spored dothidella
1644	1592	... ...	D. inæqualis	... Cooke, Grev. XX. 5 (1891)	Unequal dothidella ...
1645	1591	II. 5260	D. tephrosia	Sacc. Syll. II. 630 (1883)	Ash-coloured dothidella
					178. MONTAGNELLA.—Speg.,
1646	1593	IX. 4165	M. Eucalypti	Cooke and Mass., Grev. XVI. 5 (1887) ...	Eucalypt montagnella
1647	1594		M. rugulosa	Cooke, Grev. XX. 5 (1891) ...	Wrinkled montagnella

OF AUSTRALIAN FUNGI—*continued.*

Number.	Habitat.						B.	Occurrence.	General Characters.
	W.A.	S.A.	T.	V.	N.S.W.	Q.			

Champ. I. 168 (1798).—Sphæria, Diatrype, Anthostoma, Nummularia—*continued.*

Number.	W.A.	S.A.	T.	V.	N.S.W.	Q.	B.	Occurrence.	General Characters.
1623	...	...	...	...		Q.		Wood	Rather thick, externally, minutely and densely papillate, clay to red internally.
1624	...		T.	...		...	...	Wood ...	Hemispherical or nearly globose, superficial, black, shining dusky black within. Cup-shaped depression around opening of receptacle.
1625	...	...	...	...		Q.	B.	Bark and wood ...	Hemispherical, confluent, finally black, openings of receptacles singly in distinct marginate disc.
1626	W.A.	...	T.	V.		...	B.	Bark or wood	Bursting through, hemispherical, thick, variously shaped, often deformed. Receptacles rather large, globose.
1627	...	...	...	...		Q.	...	Rotten wood	Broadly expanded, coffee coloured, blackening. Receptacles globose, sparingly confluent.
1628	...	...	...	...		Q.	...	Old trunks	Large. margin inflexed, substance black, surface rusty to black. Receptacles oblong, immersed.
1629	...	...	...	...		Q.	...	Rotten branches ...	Very broadly expanded, black, girt by ruptured epidermis. Receptacles small, ovoid, crowded.
1630	...	...	...	V.		Q.	...	Bark and wood ...	Bursting through, cushion shaped, irregular form when young, clay colour to bright red, old red brown or dark red.
1631	...	S.A.	...	...		...	...	Trunks ...	Expanded, cushion shaped, thick, surface rust coloured, substance black. Receptacles oblong.
1632	...	...	...	...		Q.	B.	Rotten wood	On wood, rarely on bark, spread out in narrow thin crust, dark brown or black. Receptacles large, thickly crowded.
1633	...	...	...	...		Q.		Bark of dead trees	Hemispherical, large, corky to woody, sooty brown. Receptacles in outer layer black, oblong.

NITS AND FCKL., SYMB. 214 (1869).

Fckl., Smyb. Myc. 216 (1869).—Dothidea, Sphæria.

Number.	W.A.	S.A.	T.	V.	N.S.W.	Q.	B.	Occurrence.	General Characters.
1634	...	...	...	...		Q.	...	Fading leaves of *Alpinia carulea*	Spots brown, then pitch black, elongated, running together here and there.
1635	...	S A.	...	...			...	Stems of *Scirpus nodosus*	Elongated, immersed, making matrix brownish. Receptacles parallel, globose, sporidia unequal-sided.
1636	...	...	...	...		Q.	...	*Fimbristylis*	Black, covered by epidermis, openings granulate.
1637	...	...	...	V.			B.	Leaves of grass, dying or dead	Distinct or run together in parenchyma of leaf, covered by shining blackened epidermis.
1638	...			V.			B.	*Juncus* ...	Internal, brown, epidermis ultimately brownish and blackish, cracked lengthwise.
1639	...			V.			...	Leaves of Eucalypts	Gregarious, on blistered tawny spots of living leaves, black, half immersed.
1640	...	...	...	...		Q.	...	Living and dry leaves of *Cordyline terminalis*, var. *Cannæfolia*	Elongated, lance shaped, running parallel with veins, shining black, in brown spots.
1641	...	...	...	...		Q.		Phyllodes of *Acacia penninervis* and leaves of Figs	Immersed, black, shining, pustules black, minute, warty, warts with openings.
1642	...			V.			B.	Leaves of clover ...	Internal, forming brownish spots, at first producing conidia.

Fung. Arg. Pug. IV. 186 (1882).—Phyllachora, Sphæria.

Number.	W.A.	S.A.	T.	V.	N.S.W.	Q.	B.	Occurrence.	General Characters.
1643	...	...	...	...		Q.	...	Fading leaves of *Labea dealbata*	On ochrey-brown spots, loosely gregarious, at first covered with epidermis, shining black.
1644	...	...	...	V.			...	Dead leaves of Eucalypts	Bursting through on both surfaces, nearly circular, black, shining.
1645	W.A.	...	...				...	Leaves ...	Internal, circular, plane then convex, ash coloured, openings point-like, black.

Fung. Arg. Pug. IV. 188 (1882).

Number.	W.A.	S.A.	T.	V.	N.S.W.	Q.	B.	Occurrence.	General Characters.
1646	...			V.				Dead leaves of Eucalypts	Circular, convex, shining, black.
1647				V				Leaves of Eucalypts	On upper or under service, thin, somewhat circular, black, wrinkled.

Number.	Cooke's Number.	Saccardo's Number.	Scientific Name.	Authority for Name.	English Name.
					179. BAGNISIELLA.—Speg., F. Arg.
1648	1597	...	B. catervaria	...	*Cooke*, Handb. Austr. Fung. 299 (1892) ...	... : Crowded bagnisiella ... ...
1649	1596	IX. 4026	B. eudopyria	...	Sacc., Hedw. 155 (1890)	Fiery bagnisiella ...
1650, 1595	., 4025		B. rugulosa	...	Cooke, Grev. XIX. 45 (1890) ...	... Wrinkled bagnisiella ... ...
					180. DARWINIELLA.—Speg.,
1651	159~	II. 4167	D. globulosa	*Sacc.* Syll. IX. 1049 (1891)	... Globose darwiniella ... ...
					181. RHYTISMA.—Fries,
1652	1600	VIII.3027	R. filicinum ...	...	Berk. and Br., Linn. Journ. XIV. 130 (1875)	... Fern rhytisma ... ...
1653	1599	X. 4655	R. hypoxanthum ...	...	Berk. and Br., Fung. Brisb. II. 71 (1883)	... Buff-coloured rhytisma ...
					182. TRABUTIA.—Sacc. and
1654	1603	IX. 2439	T. Eucalypti	...	Cooke and Mass., Grev. XVII. 43 (1888)	... Eucalypt trabutia ... ...
1655	1601	...	T. parvicapsa	...	Cooke, Grev. XX. 5 (1891)	... Small-receptacled trabutia ...
1656	1602		T. phyllodiæ		Cooke and Mass., Grev. XIX. 60 (1891)...	... Phyllode trabutia ...
					ORDER XVII.—
					183. SARCOXYLON.—Cooke,
1657	1605	I. 1231	S. compunctum ...	...	*Cooke*, Grev. XIII. 107 (1885) ...	... Punctate sarcoxylon ... ... :
					184. GIBELLIA.—Sacc.,
1658	1606	IX. 2470	G. dothideoides ...	...	Sacc. and Berl., Misc. Myc. II. 23 (1885)	Dothidea-like gibellia... ... :
					185. BOTRYOSPHÆRIA.—Ces. and De Not.,
1659	...		B. hypoxyloidea ...	...	Cooke, Grev. XIII. 102 (1885)... ...	... Hypoxylon-like botryosphæria...
					186. MELOGRAMMA.—Tul., Carp. II. 81 (1863).—
1660	1607	I. 2814	M. rubricosa	...	*Cooke*, Handb. Austr. Fung. 301 (1892) ...	... Red melogramma ... ...
					ORDER XVIII.—
					187. DIATRYPE.—Fries, p.p. Nits
1661	1610	I. 715	D. chlorosarca		Berk. and Br., Linn. Journ. 123 (1875) ...	... : Green-fleshed diatrype ...
1662	1608	„ 740	D. glomeraria	...	Berk., Fl. N. Zeal. II. 205 (1855) ...	... : Ball-like diatrype ...
1663	1609	„ 705	D. stigma	...	*Fries*, S.V.S. 385 (1849)	... Stigma diatrype ...
					188. CÆLOSPHÆRIA.—Sacc.,
1664		II. 3890 IX. 1856	C. leptosporoides ...	...	Wint., Hedw. 2 (1883)...	Leptospora-like cælosphæria ...

OF AUSTRALIAN FUNGI—*continued.*

Number.	Habitat.						B.	Occurrence.	General Characters.
	W.A.	S.A.	T.	V.	N.S.W.	Q.			

Pag. III. 22 (1882).—Phyllachora, Dothidea.

1648	...	...	...	...	...	Q.	...	Leaves of *Ficus* ...	Pustules minute, crowded in orbicular or irregular spots, openings pap-like.
1649	...	...	...	V.	...	...	...	Leaves of *Myoporum platycarpum*	Minute, disc shaped, black, crowded here and there, surface wrinkled, leathery, substance bright fiery orange.
1650	...	...	...	V.	...	...	...	Leaves of Eucalypts	On upper or under surface, gregarious, globose, black, wrinkled.

F. Fueg. 279 (1887).—Dothidea.

| 1651 | ... | ... | T. | ... | ... | ... | ... | Leaves of *Tasmania aromatica* | On both surfaces, globose, wrinkled, black, opaque. |

S.M. II. 569 (1822).—Marchalia.

| 1652 | ... | ... | ... | ... | ... | Q. | ... | Fronds of *Alsophila* | Spots rather circular, thin. Cells elongated, wavy, thin. |
| 1653 | ... | ... | ... | ... | ... | Q. | ... | Leaves of *Cudrania javanensis* | Spots irregular, thickened, buff coloured, layer shining black, margin distinct. |

Ronm., Rev. Myc. 27 (1881).

1654	...	...	T.	V.	...	...	...	Leaves of Eucalypts	Leathery, somewhat circular, convex, wrinkled, black, shining.
1655	...	...	...	V.	...	...	...	Phyllodes of *Acacia*	Internal. Receptacles on brown elliptical spots, crowded, small, black, shining.
1656	...	...	...	V.	...	...	...	Phyllodes of *Acacia longifolia*	Receptacles convex, brown, four to ten, seated on circular spots, with pore at apex.

MELOGRAMMACEÆ, NITS.

Grev. XIII. 107 (1885).

| 1657 | ... | ... | ... | ... | ... | Q. | ... | Prostrate trunks ... | Globose, deformed, constricted at base, smooth, pale tan, punctate with black openings. |

Misc. Myc. II. 12 (1885).

| 1658 | ... | ... | ... | ... | ... | Q. | ... | Bark ... | Cushion shaped, loosely gregarious, black, paler within, openings of receptacles point-like. |

Schem. Sfer. Comm. I. 211 (1861).

| 1659 | ... | ... | ... | ... | ... | ... | ... | Branches ... | Soon superficial, with habit and appearance of *Hypoxylon*, but not carbonaceous. Receptacles very small. |

Venturia, Hypoxylon, Sphæria, Valsaria.

| 1660 | W.A. | ... | ... | ... | ... | Q. | B. | Bark, dead wood ... | Deformed, tubercled, wrinkled and cracked, reddish. Receptacles immersed, black, shining. |

DIATRYPACÆ, FRIES.

Pyr. Germ. 64 (1867).—Sphæria.

1661	...	...	...	...	...	Q.	...	Branches and bark of trees	Small, pustulate, circular, green inside.
1662	...	...	...	V.	...	...	...	Branches of *Rhipigonum parviflorum*	Bursting through, angular or run together, and elongated. Receptacles ovate.
1663	...	...	...	...	...	Q.	B.	Branches ...	Receptacle-bearing layer, bursting through, long and broad, at length black. Receptacles ovoid.

Myc. Ven. Spec. 115 (1873)

| 1664 | ... | ... | ... | ... | ... | ... | ... | Branches ... | Receptacles gregarious, at first moist and globose, at length dry and cup shaped, dark brown. |

Number.	Cooke's Number.	Saccardo's Number.	Scientific Name.	Authority for Name.	English Name.
					ORDER XIX.—
					189. VALSA.—Fries,
1665	...	I. 476	V. decorticans	... Fries, S.V.S. 412 (1849)	... Decorticating valsa ... ...
1666	1612	,, 498	V. echidna	... Cooke, Grev. IX. 4 (1880)	... Hedgehog valsa ... ...
					190. EUTYPELLA.—Nits,
1667	1611	I. 571	E. stellulata	... Sacc. Syll. I. 149 (1882)	... Stellate eutypella ... ...
					ORDER XX.—EUTYPACEÆ,
					191. CRYPTOVALSA.—Ces. and De Not.,
1668	1613	I. 702	C. elevata	... Sacc. Syll. I. 191 (1882)	... ... Elevated cryptovalsa ... ...
					192. CRYPTOSPHÆRELLA.—Sacc.,
1669	...	IX. 1940	C. Macrozamia	... Sacc. Syll. IX. 471 (1891)	... Macrozamia cryptosphærella ...
					193. EUTYPA.—Tul., Sel. Fung.
1670	1614	I. 637	E. lata ...	... Tul., Sel. Fung. Carp. II. 56 (1861)	... Broad eutypa ... ...
1671	1616	,, 632	E. ludibunda	... Sacc., Mich. I. 13 (1878) ...	... Sportive eutypa ...
1672	1615	IX. 1926	E. polyscia	... Berl. and Vogl., in Sacc. Syll. IX. 467 (1891)	... Shadowy eutypa ...
					ORDER XXI.—CUCURBITARIACEÆ,
					191. GIBBERIDEA.—Fckl.,
1673	1617	II. 3637	G. Archeri	... Cooke, Handb. Aust. Fung. 304 (1892)	... Archer's gibberidea ...
1674	1618	IX. 3340	G. plagia ...	... Sacc. Syll. IX. 820 (1891) ...	... Striped gibberidea ...
					ORDER XXII.—
					195. BYSSOSPHÆRIA.—Cooke,
1675	1620	IX. 2450	B. acanthostroma ...	... Cooke. Handb. Aust. Fung. 305 (1892)	Thorny byssosphæria ... ...
1676	1619	I. 916	B. aquila ...	Cooke, Grev. XV. 122 (1887)	... Eagle byssosphæria ... ...
					196. LASIOSPHÆRIA.—Ces. and De Not.,
1677	1622	IX. 5469	L. larvæspora	Cooke and Mass., Grev. XIX. 83 (1891) ...	Larva-spored lasiosphæria ...
1678	1621	II. 3568	L. ovina ...	Ces. and De Not., Schema Sfer. 229 (1870)	... Woolly lasiosphæria ...
					197. PLEOSPHÆRIA.—Speg.,
1679	1623	II. 3927	P. pulvinula	Sacc. Syll. II. (1883)	... Cushion-shaped pleosphæria ...

OF AUSTRALIAN FUNGI—*continued*.

Number.	Habitat.						B.	Occurrence.	General Characters.
	W.A.	S.A.	T.	V.	N.S.W.	Q.			

VALSACEÆ, FRIES.

S.V.S., 410 (1849).—Sphæria.

| 1665 | ... | ... | ... | ... | ... | Q. | ... | Dry branches | ... Stroma somewhat circular or often oval, abruptly tapering into a disc of openings, becoming dusky, at length black. |
| 1666 | ... | ... | ... | ... | N.S.W. | Q. | ... | Dark | ... Bursting through. Receptacles nestling in a white powdery layer. |

Pyr. Germ. 163 (1867).—Valsa, Sphæria.

| 1667 | ... | ... | ... | V. | ... | ... | B. | Branches... | ... Somewhat round, immersed; receptacle-bearing layer, white or dirty white; openings radiately stellate. |

COOKE, GREV. XIV. 93 (1886).

Schema Sfer. It. 29 (1870).—Diatrype, Sphæria, Eutypa.

| 1668 | W.A. | ... | T. | ... | ... | ... | B. | Dead branches | ... Elongated, emergent, black or grey. Receptacles gregarious, globose, immersed in wood. |

Syll. I. 186 (1882).—Sphæria.

| 1669 | ... | ... | ... | ... | ... | Q. | | Fruit of *Macrozamia Hopei* | Receptacles under epidermis, at first scattered, then crowded, black, springing from dark-brown filaments. |

Carp. II. 52 (1861).—Sphæria, Valsa.

1670	...	...	T.	...	...	...	B.	Wood and bark	... Receptacle-bearing layer long and broad, innate in wood or bark, brown or ashy. Receptacles immersed, sphæroid.
1671	...			V.		...	...	Branches...	... Receptacle-bearing layer spreading. Receptacles globose, externally black, at first mealy white.
1672	...			...	...	Q.	...	Epicarp of *Cucurbita lagenaria*	Receptacle-bearing layer black, point-like. Receptacles immersed.

COOKE, GREV. XV. 83 (1887).

Symb. Myc. 168 (1869).—Zignoella, Sphæria, Melanomma.

| 1673 | ... | ... | T. | ... | ... | ... | ... | Rotten wood | Crowded. Receptacles wrinkled, at length collapsing and cup shaped. |
| 1674 | ... | ... | ... | V. | ... | ... | ... | Living twigs of *Cassinia aculeata* | Receptacles densely crowded, at length run together in large patches, globose, black, shining, smooth. |

SUPERFICIALES, FRIES.

Grev. VII. 84 (1879).—Scortechinia, Rosellinia, Sphæria.

| 1675 | ... | ... | ... | ... | ... | Q. | ... | Wood and bark | ... Finely filamentous. Receptacles very small, globose, crowded, black, not pap-like. |
| 1676 | ... | | T. | | | | ... | B. | Wood and bark | ... Receptacles gregarious or densely crowded, globose, dark brown, pap-like. |

Schema Sfer. 55 (1870).—Sphæria, Leptospora.

| 1677 | ... | ... | ... | V. | ... | ... | Bark | ... Receptacles loosely gregarious, globose, covered with mealy fluffy sulphur-coloured coat. Sporidia long, spindle shaped, with fifteen to nineteen partitions. |
| 1678 | ... | ... | ... | V. | ... | B. | Rotten wood | ... Receptacles gregarious, almost globose, covered with mealy fluffy lemon-coloured coat. |

Fung. Arg. IV. 65 (1882).—Sphæria, Coniochæta, Lasiosphæria.

| 1679 | W.A. | ... | ... | ... | | ... | ... | Rotten wood | ... Scattered, somewhat globose, collapsed and depressed at length, hairy, black. |

Number.	Cooke's Number.	Saccardo's Number.	Scientific Name.	Authority for Name.	English Name.
					198. VENTURIA.—De Not. and Ces.,
1680	1624	I. 2311	V. circinans	... Sacc., Mich. I. 499 (1868)	... Circinate venturia ... ...
					199. CHÆTOMIUM.—Kunze,
1681	1626	IX. 1997	C. cymatotrichum	Cooke, Grev. XII. 21 (1883) ...	Wavy-haired chætomium ...
1682	1625	I. 793	C. comatum	Fries, S.M. III. 253 (1832)	.. Hairy chætomium ... ...
					200. ROSELLINIA.—De Not.,
1683	1627	I. 970	R. inspersa	Sacc. Syll. I. 265 (1882)	... Scattered rosellinia ... ...
1684	1628	IX. 2055	R. tremellicola	Cooke and Mass., Grev. XVIII. 6 (1889)	... Tremella rosellinia ... ...
					201. TREMATOSPHÆRIA.—Fckl.,
1685	1629	IX. 3307	T. congesta	... Berl. and Vogl., in Sacc. Syll. IX. 811 (1891)	... Congested trematosphæria ...

ORDER XXIII.—

202. CONISPHÆRIA.—Cooke,

Number.	Cooke's Number.	Saccardo's Number.	Scientific Name.	Authority for Name.	English Name.
1686	1632	IX. 3539	C. australica	... Cooke and Mass., Handb. Austr. Fung. 307 (1892)	Australian conisphæria ...
1687	1631	...	C. erumpens	... Cooke, Handb. Austr. Fung. 307 (1892)	... Erumpent conisphæria ...
1688	1630	IX. 3527	C. subcorticalis ...	... Cooke, Handb. Austr. Fung. 307 (1892)	... Underbark conisphæria ...

ORDER XXIV.—LOPHIOSTOMACEÆ,

203. LOPHIOSTOMA.—Ces. and De Not.,

Number.	Cooke's Number.	Saccardo's Number.	Scientific Name.	Authority for Name.	English Name.
1689	1633	II. 5405	L. Schomburgkii ...	... Cooke, Handb. Austr. Fung. 307 (1892)	... Schomburgk's lophiostoma ...

ORDER XX .—

204. RHAMPHORIA.—Niessl.,

Number.	Cooke's Number.	Saccardo's Number.	Scientific Name.	Authority for Name.	English Name.
1690	1634	IX. 3654	R. tenella	... Sacc., Hedw. 155 (1890)	... Delicate rhamphoria ... ...

ORDER XXVI.—

205. MASSARIELLA.—Speg.,

Number.	Cooke's Number.	Saccardo's Number.	Scientific Name.	Authority for Name.	English Name.
1691	1635	I. 2707	M. australis	Sacc. Syll. I. 716 (1882)	... Southern massariella ... ...
					206. DIDYMOSPHÆRIA.—Fckl.,
1692	1638	IX. 3000	D. Banksiæ	Cooke, Grev. XIX. 90 (1891)	... Banksia didymosphæria ...
1693	1637	„ 2979	D. conoidella	Sacc. and Berl., Misc. Myc. II. 26 (1885)	... Conical didymosphæria ...

OF AUSTRALIAN FUNGI—*continued*.

Number.	W.A.	S.A.	T.	V.	N.S.W.	Q.	B.	Occurrence.	General Characters.
				Habitat.					

Schema Sf. I. 225 (1870).—Perisporium, Stigmatea.

| 1680 | ... | ... | ... | V. | ... | ... | B. | Leaves of *Geranium* | Receptacles clustered in patches or spots, hardly circinating, hairs thickened at base. |

Myc. Heft. I. 15 (1817).—Sphæria.

| 1681 | ... | ... | ... | ... | | Q. | ... | Leaves of *Solanum Dallachyi* | Gregarious. Receptacles depressed globose, woolly, sooty olive. |
| 1682 | ... | S.A. | ... | V. | | | B. | Rotting grass, &c. ... | Receptacles gregarious, nearly ovoid, thin black hairs radiating from base. |

Giorn. Bot. Ital. II. 334 (1847).—Sphæria.

| 1683 | W.A. | ... | ... | ... | ... | ... | ... | Rotten wood ... | Crowded or scattered, black. Receptacles nearly globose, wrinkled |
| 1684 | ... | ... | ... | ... | ... | Q. | ... | *Tremella fuciformis* | Receptacles scattered, globose, superficial, black, pap-like, smooth. |

Symb. Myc. 161 (1869) emended.—Psilosphæria, Melanomma.

| 1685 | ... | ... | ... | ... | N.S.W. | ... | ... | Bark ... | Gregarious, crowded, black. Receptacles convex, smooth, pierced with a pore. |

PERTUSACEÆ, FRIES.

Grev. XVI. 87 (1888).—Zignoëlla.

1686	...	...	...	V.	...	...	...	Naked branches ...	Receptacles scattered, half immersed, rather conical, base buried in wood.
1687	...	...	...	V.	...	...	...	Twigs ...	Scattered or collected together, erumpent. Receptacles globose, smooth, black.
1688	...	...	...	...	...	...	...	Inside dead bark of trees	Scattered. Receptacles half immersed, pierced, black.

SACC., MICH. I. 333 (1879).

Schem. Sfer. 45 (1870).—Schizostoma, Sphæria.

| 1689 | ... | ... | ... | ... | ... | Q. | ... | Wood ... | Receptacles large, free, black, openings linear. |

CERATOSTOMACEÆ, FRIES.

Notiz. 44 (1876).

| 1690 | ... | S.A. | ... | ... | ... | ... | ... | Rotten wood of *Eucalyptus viminalis* | Receptacles almost superficial, or base buried in wood, small, globose, black, thinly carbonaceous. |

OBTECTACEÆ, FRIES., S.V.S. (1849).

Fung. Arg. Pug. 2 (1880).—Massaria.

| 1691 | ... | ... | ... | V. | ... | ... | ... | Bark ... | Scattered, covered, inconspicuous. Receptacles depressed. |

Symb. Myc. 140 (1869) emended.

| 1692 | ... | ... | ... | V. | ... | ... | ... | Living leaves of *Banksia* | Spots circular, pale with indistinct brown margin. Receptacles few, black, bursting through. |
| 1693 | ... | ... | ... | ... | ... | Q. | ... | Dead branches of *Capparis* | Receptacles loosely gregarious, globose, then conical, black. |

Number.	Cooke's Number.	Saccardo's Number.	Scientific Name.	Authority for Name.	English Name.
			ORDER XXVII.—CAULICOLACEÆ,		
				207. PHYSALOSPORA.—Niessl., Notiz.	
1694	1639	I. 1660	P. gregaria	Sacc. Syll. I. (1882) ...	Gregarious physalospora
1695		., 1723	P. labecula	Sacc. Syll. I. 447 (1882)	Stain physalospora ...
1696	...	IX. 2427	P. Sacchari	Sacc. Syll. IX 599 (1891)	Sugar-cane physalospora
				208. DIDYMELLA.—Sacc., Mich. I. 377 (1878).—	
1697	1640	I. 2171	D. Bryoniæ	Rehm., Ascom. III. 99 (1881) ...	Bryonia didymella
1698	1636	., 2126	D. cladophila	Sacc. Syll. I. 545 (1882)	Branch-loving didymella
				209. ANTHOSTOMELLA.—Sacc., Consp.	
1699	1641		A. Lepidospermæ...	Cooke, Grev. XX. 5 (1891)	Lepidosperma anthostomella ...
				210. PLEOSPORA.—Rabh., Herb. Myc.	
1700	1643	II. 3776	P. Aucubæ	Lamb., Fl. Myc. Belg. II. 268 (1889)	Aucuba pleospora
1701	1642	., 3730	P. herbarum	Rabh., Herb. Myc. Ed. II. 547 (1858) ...	Herb pleospora
			ORDER XXVIII.—FOLIICOLACEÆ,		
				211. LÆSTADIA.—Auersw., in Hedw. 177	
1702	..	IX. 2568	L. Dammaræ	Sacc. Syll. IX. 586 (1891)	Dammara læstadia ...
1703	1644	., 2578	L. destructiva	Berl. and Vogl., in Sacc. Syll. IX. 588 (1891)	Destructive læstadia ...
1704	" bis.	...	L. Litseæ	Cooke, Grev. XX. 65 (1892)	Litsea læstadia
1705	...	IX. 2316	L. Melaleucæ	Sacc. Syll. IX. 581 (1891)	Melaleuca læstadia
1706	1645	., 2400	L. phyllodiæ	Cooke, Handb. Austr. Fung. 510 (1892) ...	Phyllode læstadia
				212. SPHÆRELLA.—Ces. and De Not.,	
1707	1651	IX. 2556	S. Alyxiæ	Cooke and Mass., Grev. XVI. 5 (1887) ...	Alyxia sphærella
1708	...	I. 1921	S. atra ...	Sacc. Syll. I. 498 (1882)	Black sphærella
1709	1650	IX. 2578	S. Banksiæ	Cooke and Mass., Grev. XVI. 114 (1888)	Banksia sphærella
1710	1647	...	S. cryptica	Cooke, Grev. XX. 8 (1891)	Hidden sphærella
1711	1649	I. 1906	S. Euonymi	Auersw., Myc. Eur. 10	Euonymus sphærella ...
1712	...	., 1951	S. Fragariæ	Sacc. Syll. I. 505 (1882)	Strawberry sphærella...
1713	...		S. Goodiæfolia	Cooke, Grev. XXI. 38 (1892) ...	Goodiæ-leaf sphærella

OF AUSTRALIAN FUNGI—*continued*.

Number.	Habitat.						B.	Occurrence.	General Characters.
	W.A.	S.A.	T.	V.	N.S.W.	Q.			

FRIES, S.M. II. 503 (1823).

Kr. Pyr. 10 (1876).—Sphæria.

1694	...	...	...	...	...	Q.	...	Stems of *Ricinus communis*	Receptacles densely gregarious, covered by epidermis, globose, black, white within.
1695	...	...	...	...			...	Leaves of *Acacia verticillata*	Innate, upper or under surface spot-like, black, shining. Receptacles nearly globose, black within.
1696	...			...		Q.	...	Leaves of Sugar cane	Receptacles scattered or gregarious, minute, black, bursting through, filaments dark brown.

Sphæria, Sphærella, Didymosphæria.

| 1697 | ... | ... | ... | ... | | Q. : B. | Twigs of cucurbitaceous plant | Receptacles beneath epidermis, at length almost free, gregarious, very minute, black. |
| 1698 | ... | S.A. | ... | ... | | ... | Branches of Grape vine | Receptacles loosely gregarious, covered with bleached cuticle, hemispherical, black, leathery. |

Gen. Pyr. Ital. 8 (1875).

| 1699 | ... | ... | ... | V. | ... | | ... | *Lepidosperma* | Receptacles on bleached elongated spots, with dark-brown border, globular. |

Ed. II. 347 (1858).—Sphæria, Phoma.

| 1700 | ... | ... | ... | V. | ... | ... | ... | Leaves of *Aucuba japonica* | Receptacles spherical, immersed, black, scattered upon somewhat circular brown spots. |
| 1701 | ... | ... | ... | V. | ... | ... | B. | Herbaceous stems | Receptacles somewhat gregarious, at first covered, then almost naked, spherical, depressed. |

FRIES. S.M. II. 513 (1823).

(1869).—Sphærella, Physalospora.

1702	...		...	...	...	Q.	Leaves of *Dammara robusta = Agathis*	Spots pale brown or yellow, margin amber, then dark brown. Receptacles black, internal.
1703	...	...	...	V.	N.S.W.	Q.	Leaves of Lucerne	Receptacles minute, seated on brown spots; raised above general surface.
1704	...	...	...	...		Q.	Leaves of *Litsea* ...	Spots irregular or somewhat circular on upper surface, becoming pale, with broad brown border.
1705	...			...	N.S.W.	... ...	Leaves of *Melaleuca*	Spots circular, minute, brown, convex. Receptacles black, in crowded spots.
1706	...		...	V.		... ...	Phyllodes of *Acacia suaveolens*	Receptacles scattered, very thin, immersed, black, covered by blackened cuticle.

Schema Sfer. Ital. 62 (1870).—Sphæria.

1707	...	...	...	V.	...	... ...	Dead leaves of *Alyxia buxifolia*	On both sides. Receptacles gregarious, arising from within, nearly globose, covered by blackened cuticle.
1708	...	...	...	...		...	Leaves of *Grevillea*	Gregarious or confluent. Receptacles globose, white within, covered by a black expanded layer.
1709	...	...	...	V.	...	...	Fading leaves of *Banksia integrifolia*	On upper surface, spots none. Receptacles gregarious, black, pierced with a pore.
1710	...		...	V.	...	...	Fading leaves of Eucalypts	On both sides, spots reddish brown, large. Receptacles immersed in substance of leaf and hidden.
1711	...		...	V.	...	...	Dead leaves of *Euonymus*	On under surface. Receptacles black, beneath epidermis, globose, on greyish spots.
1712	...	S.A.	...	V.	N.S.W.	... ...	Leaves of Strawberry	Spots becoming purple, then pale towards the centre. Receptacles very minute, globular, black.
1713	...	...	...	V.		... ...	Leaves of *Goodia latifolia*	Spots circular, brown, surrounded by darker line. Receptacles gregarious, minute.

Number.	Cooke's Number.	Saccardo's Number.	Scientific Name.	Authority for Name.	English Name.
				212. SPHÆRELLA.—Ces. and De Not.,	
1714	1653	I. 2048	S. graminicola	Fckl, Symb. Myc. 101 (1875) ...	Grass-growing sphærella
1715	1646	...	S. nubilosa	Cooke, Grev. XIX. 61 (1891) ..	Cloudy sphærella
1716	1648	IX. 2598	S. rubiginosa	Cooke, Grev. XIV. 90 (1886) ...	Rubiginous sphærella
1717	1652	I. 2028	S. smilacicola	Cooke, Grev. VI. 146 (1878)	Smilax sphærella
				213. SPHÆRULINA.—Sacc.,	
1718	1654	IX. 3189	S. Camelliæ	Cooke, Handb. Austr. Fung, 312 (1892) ...	Camellia sphærulina

ORDER XXIX.—MICROTHYRIACEÆ,

Number.	Cooke's Number.	Saccardo's Number.	Scientific Name.	Authority for Name.	English Name.
				214. MICROTHYRIUM.—Desm., Ann.	
1719	1655	IX. 4196	M. amygdalinum ...	Cooke and Mass., Grev. XIX. 90 (1891)...	Amygdalina microthyrium
				215. MICROPELTIS.—Mont., Ann.	
1720	1656	II. 5390	M. applanata	Mont., Ann. Sci. Nat. XVII. 122 (1842) ...	Flattened micropeltis
1720A	...	IX. 4236	M. applanata, var. depauperata	Sacc. and Berl., Rev. Myc. (1885)	Impoverishing micropeltis

ORDER XXX.—PERISPORIACEÆ,

Number.	Cooke's Number.	Saccardo's Number.	Scientific Name.	Authority for Name.	English Name.
				216. PODOSPHÆRA.—Kunze,	
1721	...	I. 3	P. tridactyla	De Bary, Syst. Erys, in Hedw. 68 (1871)	Three-fingered podosphæra (Powdery mildew of apple)
				217. SPHÆROTHECA.—Lev., Ann. Sci.	
1722	1658	I. 6	S. pannosa	Lev., Ann. Sci. Nat. XV. 138 (1851)	Cloth-like sphærotheca (Rose blight)
				218. ERYSIPHE.—Hedw., Lev.	
1723	...	I. 70	E. communis	Fries, S.V.S. 406 (1849)	Common erysiphe
1724	...	„ 74	E. graminis	D. C., Fl. Fr. VI. 106 (1815)	Grass erysiphe (Grass mildew)...
1725	1657	IX. 1571	E. vitigera	Cooke and Mass., Grev. XV. 98 (1887)	Vine-growing erysiphe
				219. EUROTIUM.—Link, Berl.	
1726	1659	I. 101	E. herbariorum	Link, Sp. Pl, I. 79 (1824)	Herbarium eurotium
1727	1660	„ 104	E. lateritium	Mont., Syll. 918 (1856)	Brick-red eurotium

OF AUSTRALIAN FUNGI—*continued.*

Number.	W.A.	S.A.	T.	V.	N.S.W.	Q.	B.	Occurrence.	General Characters.

Schema Sfer. Ital. 62 (1870).—Sphæria—*continued.*

1714				V.				Leaves of grass	On upper surface. Receptacles growing from within, prominent, small, black, occupying entire surface.
1715				V.				Living leaves of Eucalypts	On under surface, spots circular or irregular, greyish brown, soon falling away. Receptacles very minute.
1716						Q.		Leaves of *Pittosporum rubiginosum*	On upper surface. Receptacles scattered, minute, point-like, prominent, black.
1717						Q.		Leaves of *Dioscorea* and *Smilax*	Spots reddish brown with black margin. Receptacles point-like, conical, black.

Mich. 1. 399 (1878).

| 1718 | | | | V. | | | | Living leaves of Camellia | Spots on upper surface, turning brownish, indeterminate. Receptacles scattered. |

SACC. SYLL. II. 658 (1883).

Sci. Nat. XV. 137 (1841).

| 1719 | | S.A. | | | | | | Living leaves of Eucalyptus amygdalina | Receptacles gregarious or scattered, on both surfaces, membranous, very dark brown. |

Sci. Nat. XVII. 122 (1842).

| 1720 | | | | | | | | | |
| 1720a | | | | | | Q. | | Languid leaves of Eucalyptus tereticornis | Receptacles on both surfaces, convex to flattened, black, finally opening in centre. |

FRIES, S.V.S. 375 (1849).

Myc. Heft. II. 111 (1817).—Alphitomorpha.

| 1721 | | | | V. | N.S.W. | | | Young leaves and shoots of Apple | On both surfaces of leaf, forming a white felt, and the spores so numerous as to make it powdery; conidia barrel shaped. |

Nat. XV. 138 (1851).—Alphitomorpha.

| 1722 | | | | V. | N.S.W. | Q. | B. | Rose leaves | Mycelium woolly, then cloth-like, white, persistent. *Oidium leucoconium* is the conidial stage. |

Ann. Sci. Nat. XV. 161 (1851).

1723				V.			B.	Leaves of various plants	Mycelium spreading, arachnoid, evanescent or persistent. Receptacles scattered or gregarious, minute.
1724				V.	N.S.W.		B.	Leaves and stems of various grasses	Mycelium spreading, fluffy to woolly, ochrey white, jointed, persistent. Early or conidial stage is *Oidium monilioides*, Link.
1725				V.				Vine leaves	On both sides of leaf. Mycelium fluffy, persistent. Receptacles gregarious, very minute, spherical.

Mag. III. 31 (1809).—Mucor.

| 1726 | | | | | | Q | B. | Plants in herbaria, decaying organic matter, &c. | Mycelium creeping, fluffy, branched, uncoloured. Conidial stage is *Aspergillus glaucus*, Link, with glaucous conidia. |
| 1727 | W.A. | | | | | Q. | | Leaves of *Piperomia* | Mycelium dense, woolly, orange yellow. Receptacles membranous, yellow, then ochre, immersed. |

Number	Cooke's Number	Sac___'s Number.	Scientific Name	Authority for Name.	English Name.
					220. ASTERINA.—Lev., Ann.
1728	1661	IX. 1609	A. Baileyi	Berk. and Br., Linn. Trans II. 71 (1883)	Bailey's asterina
1729	1662	„ 1622	A. correicola	Cooke and Mass., Grev. XVI. 5 (1887) ...	Correa asterina ...
1730	...		A. hoveafolia ...	Cooke and Mass., Grev. XXII. 36 (1893)	Hovea-leaved asterina
1731	1665	IX. 1607	A. microthyrioides	Winter, Hedw. 3 (1885) ...	... Microthyrium-like asterina
1732	1663	I. 200	A. pelliculosa	Berk., Antarc. Crypt. II. 155 (1847)	Pellicle asterina ...
1733	1664	IX. 1621	A. platystoma	Cooke and Mass., Grev. XVIII. 6 (1889)	Broad-mouthed asterina
1734	1668	I. 198	A. reptans	Berk. and Curt., Linn. Trans. X. 373 (1869)	... Creeping asterina
					221. ASTERELLA.—Sacc.
1735	1666	IX. 1690	A. Alsophilæ	Cooke and Mass., Grev. XVIII. 81 (1890)	Alsophila asterella ... ...
1736	1667	„ 1679	A. subcuticulosa ...	... Cooke, Grev. XVII. 81 (1889) ...	Subcuticular asterella ...
					222. DIMEROSPORIUM.—Fckl.
1737	1670	IX. 1710	D. Ludwigianum ...	Sacc., Hedw. 127 (1889)	... Ludwig's dimerosporium
1738	1671		D. parvulum	Cooke, Grev. XX. 5 (1891)	Small dimerosporium ...
1739			D. secedens	... Sacc., Hedw. 57 (1893)	Seceding dimerosporium
1740			D. strigosum	Sacc., Grev. XXI. 68 (1893)	Hispid dimerosporium
					223. PARODIELLA.—Speg.,
1741	...	IX. 1723	P. Banksiæ	Sacc. and Bizz., Syll. IX. 410 (1891)	Banksia parodiella ... ...
1742	1604	I. 2711	P. Perisporioides	Speg., Fung. Arg. Pug. I. 178 (1880)	Perisporia-like parodiella
					224. MELIOLA.—Fries,
1743	1673	I. 287	M. amphitricha ...	Fries, Elench. II. 109 (1828) ...	Amphitrichous meliola ...
1744		„ 281	M. cia lotricha	Lev., Ann. Sci. Nat. V. 266 (1846)	Brai ch-haired meliola
1745	1672	279	M. corallina	Mont., Syll. 910 (1856)	Coralline meliola
1746	1676	IX. 1758	M. densa	Cooke, Grev. XII. 85 (1884)	... Dense meliola
1747	1679	I. 237	M. mollis	Berk. and Br., Linn. Trans. XIV. 136 (1875)	Soft meliola ...
1748	1674	„ 291	M. Musæ	Mont., Syll. 905 (1856)...	... Musa meliola
1749	1678	IX. 1752	M. octospora	Cooke, Grev. XI. 38 (1882)	Eight-spored meliola ...
1750	1675	I. 294	M. orbicularis	Berk. and Curt., Linn. Trans. X. 392 (1869)	Orbicular meliola
1751		„ 301	M. polytricha	Kalch. and Cooke, Grev. VIII. 72 (1879)	Many-haired meliola
1752	1680	„ 310	M. Tetraceræ	Thuem., Symb. Myc. Aust. II. 92 (1878)	Tetracera meliola

OF AUSTRALIAN FUNGI—*continued.*

Number.	Habitat.						B.	Occurrence.	General Characters.
	W.A.	S.A.	T.	V.	N.S.W.	Q.			

Sci. Nat. III. 59 (1845).

1728	...	...	...	...	...	Q.		Leaves of *Hakea* *baccu* and other shrubs	Mycelial threads brown, knotted, branching, forming reddish-brown patches. Receptacles minute, wrinkled, brown.
1729	...			V.				Living leaves of *Correa Lawrenciana*	Circular black spots on leaves. Receptacles convex to flat, black, crowded on spots.
1730	...					Q.		Leaves of *Howea longifolia*	Spots black, or with brown centre, nearly circular. Receptacles usually arranged in a ring, black.
1731			...	V.				Leaves of *Eucalyptus pilularis*	On under surface. Receptacles scattered or loosely gregarious, very minute, black, wrinkled, margin brown.
1732		...	...			Q.	...	Leaves of *Trema aspera*	Mycelium forming a pellicle, spot-like, black. Receptacles globose, depressed, black.
1733	...					Q.	...	Living leaves of *Castanospermum australe*	Mycelium thin, more or less circular, tree-like, black. Receptacles convex to flattened, black.
1734	...		...	...		Q.	...	Leaves of *Eugenia*	Mycelium thin, rather netted. Receptacles minute, formed from the radiating cells.

Syll. I. 42 (1882).

| 1735 | ... | ... | ... | ... | ... | Q. | ... | Fronds of *Alsophila Rebeccæ* | Receptacles membranous, disc-like, nearly circular, mostly running together in oblong or irregular pitchy-black patches. |
| 1736 | ... | ... | ... | V. | N.S.W. | | ... | Fading and dead leaves of *Aster argophyllus* | Receptacles thin, flattened, without mycelium, black or brown. |

Symb. Myc. 89 (1869).

1737	...	S.A.	...	V.			...	Fading leaves of *Lagenophora Bill-ardieri*	Mycelium forming pale sooty spots on both surfaces. Receptacles crowded here and there, globular, superficial, dark sooty brown.
1738		...				Q.	...	Living leaves of *Trema aspera*	Mycelium brown, sparse, radiating, on irregular black spots. Receptacles minute, nearly globose.
1739	...					Q.	...	Leaves of living plants	Broadly expanded, pitch black, easily separating, capnodium-like. Receptacles thickly clustered, globose, shining black.
1740	...		...	...	...	...	...	...	(Belongs to genus *Dinemasporium.*)

Fung. Arg. Png. I. 178 (1880).—Dothidea.

| 1741 | ... | ... | ... | ... | N.S.W. | ... | ... | Lower surface of languid leaves of *Banksia marginata* | Receptacles globose, black, superficial, like dots, densely clustered. Mycelium almost none. |
| 1742 | ... | | | V. | N.S.W. | | ... | Living leaves of *Leguminosæ* | Receptacles globose, black, superficial, thickly clustered, often covering entire surface, sooty olive. |

Elench. II. 100 (1828).—Sphæria, Dothidea.

1743	...	...		V.	...	Q.	...	Leaves of *Cupania Eucalyptus, Flindersia, Acacia, &c.*	Lower surface. Mycelium spot-like, radiating from centre, continuous, black. Receptacles globose, surrounded by black rigid erect appendages.
1744	...			V.				Leaves of *Eugenia*	Mycelium spread out, with radiating branched circumference. Receptacles globose, with erect bifid appendages.
1745	...	...			...		...	Leaves ...	Both surfaces. Mycelium spot-like, spots circular, black. Receptacles large, globose, surrounded by rigid shiny black appendages.
1746	...	...				Q.		Leaves of *Eucalyptus*	Under rarely upper surface, forming circular very black velvety spots. Receptacles globose, black, surrounded by crowded erect appendages.
1747	...	...		...	...	Q.	...	Leaves ...	Mycelium of soft black threads. Receptacles globose, appendages erect, brown.
1748	...	...		...		Q.	...	*Musæ* ...	Spot like, large, black tufts. Receptacles very minute and inconspicuous; appendages erect, simple.
1749	...	...		...		Q.	...	Leaves of *Tristania conferta*	Spots circular, minute, velvety. Receptacles medium sized; appendages erect.
1750	...	...		...		Q.	...	Branches and leaves	Spots thick, orbicular. Receptacles globose ; appendages flexuous, curved.
1751	...			...		Q.	...	Leaves of *Callistemon*	On under or both surfaces, black, spot-like. Mycelium spread out, radiating, conidia bearing. Receptacles globose, with erect acute wavy appendages.
1752	...			...		Q.	...	Leaves of *Tretavera Wathiana*	Both surfaces, spots more or less circular, black, fading. Receptacles carbonaceous, globose.

132

SYSTEMATIC ARRANGEMENT

Number.	Cooke's Number.	Saccardo's Number.	Scientific Name.	Authority for Name.	English Name.
					225. ZUKALIA.—Sacc.
1753	1677	IX. 1792	Z. loganiensis	*Sacc. and Berl.*, Syll. IX. 431 (1891)	Logan zukalia
					226. ASTERIDIUM.—Sacc.
1754	1669	IX. 1808	A. Eucalypti	... Cooke and Mass., Grev. XVI. 74 (1888)	Eucalyptus asteridium ...
					227. CAPNODIUM.—Mont., Ann.
1755	1683	I. 324	C. australe	... Mont., Syll. 916 (1856)	Southern capnodium
1756	1681	„ 346	C. citri ...	Berk. and Desm., Journ. Hort. Lond. IV. 11 (1849)	Citrus capnodium ...
1757	1682	„ 329	C. elongatum	Berk. and Desm., Journ. Hort. Lond. IV. 251 (1849)	Elongated capnodium
1758	1684	„ 323	C. salicinum	Mont., Syll. 915 (1856)	Willow capnodium ...
1759	...	...	C. Walteri	... Sacc., Hedw. 58 (1893)	Walter's capnodium ... ...
					228. ANTENNARIA.—Link,
1760	1687	I. 362	A. Robinsoni	... Berk. and Mont., in Mont. Syll. 1066 (1856) ...	Robinson's antennaria
1761	1686	„ 364	A. scoriadea	Berk., Hook., Journ. 70 (1845)...	Drossy antennaria ...
1762	1688	„ 366	A. semiovata	... Berk. and Br., Ann. Nat. Hist. XIII., 2 Ser. 468 (1854)	Semi-ovate antennaria

ORDER XXXI.—HYSTERIACEÆ,

					229. AULOGRAPHUM.—Lib., Crypt.
1763	1475	IX. 4344	A. Eucalypti	Cooke and Mass., Grev. XVIII. 6 (1889)	... Eucalypt aulographum ...
1764	1474	„ 4343	A. melioloides	... Cooke and Mass., Grev. XVIII. 6 (1889)	Meliola-like aulographum ...
					230. GLONIUM.—Muhl., Cat.
1765	1476	II. 5586	G. stellatum	Muhl., Cat. Am. 101 (1813)	... Stellate glonium ...
1766	1477	„ 5607	G. tardum	... Sacc. Syll. II. 737 (1883)	... Slow-opening glonium ...
					231. LEMBOSIA.—Lev., Ann.
1767	1478	IX. 4353	L. graphdioides	Sacc. and Berl., Misc. Myc. II. 6 (1885)	Graphium-like lembosia ...
1768	...	„ 4354	L. orbicularis	Winter, Hedw. 29 (1885) ...	... Orbicular lembosia ...
					232. HYSTERIUM.—Tode, Fung.
1769	1479	II. 5634	H. pulicare	Pers., Syn. 98 (1801) ...	... Flea-like hysterium ... ...
					233. TRYBLIDIELLA.—Sacc.
1770	1489	II. 5694	T. rufula	... Sacc. Syll. II. 737 (1883)	... Reddish tryblidiella ...
					234. RHYTIDHYSTERIUM.—Speg.,
1771	1481	IX. 4378	R. Scortechinii	... Sacc. and Berl., Misc. Myc. II. 7 (1885)	... Scortechini's rhytidhysterium ...
					235. PLATYCHEILUS.—Cooke,
1772	1482		P. cæspitosus	Cooke and Mass., Handb. Austr. Fung. 409 (1892)	Tufted platycheilus ... ...
					236. HYSTEROGRAPHIUM.—Corda,
1773	1483	II. 5759	H. elongatum	*Corda*, Icon. V. 77 (1842)	.. Elongated hysterographium ...
1774	1484	„ 5771	H. hiascens	Rehm., Ascom. 314 (1881)	... Gaping hysterographium ...
1775	1485	„ 5768	H. Roussellii	*Sacc.* Syll. II. 779 (1883)	Roussel's hysterographium ...

OF AUSTRALIAN FUNGI—*continued*.

Number.	Habitat.						n.	Occurrence.	General Characters.
	W.A.	S.A.	T.	V.	N.S.W.	Q.			

Syll. IX. 431 (1891).—Meliola.

| 1753 | ... | ... | ... | ... | ... | Q. | ... | Leaves of *Smilax* ... | Upper surface. Mycelium thin, widely spreading. Receptacles globose, black, sparingly beset with sooty-brown bristles. |

Syll. I. 25 (1882).

| 1754 | ... | ... | ... | V. | | | ... | Dead leaves of *Euca-lyptus amygdalina* | Very thickly gregarious. Receptacles minute, disc-like, flattened, black. |

Sci. Nat. 3 XI. 233 (1848).—Fumago.

1755	W.A.	...	...	...	...	Q.	...	Branches of Conifers, *Cycas*, &c.	Involved, velvety. Mycelium of branched necklace-like fibres. Receptacles somewhat forked, obtuse.
1756	...	...	...	V.			...	Leaves of *Citrus* ...	Scattered, bristle-like. Mycelium branched, necklace-like and netted. Receptacles elongated.
1757	S.A.	...	...	...	...	Q.	...	Leaves, &c.	Bristle-like. Receptacles elongated, pointed, and apex fringed.
1758	...	...	...	...	...	Q.	B.	*Xanthoxylon* ...	Mass of dark-brown branched threads, bearing receptacles, fleshy, club to horn shaped.
1759	...	...	...	V.			...	Branches and living leaves of *Bursaria spinosa*	Black, broadly expanded, separating as a pellicle; threads creeping, branched, sooty brown.

Schrad. Journ. III. 16 (1810).

1760	...	...	...	V.	...	Q.		Ferns, &c.	Mycelium expanded, cloth-like, fibres very thin, elongated. Receptacles oblong.
1761	W.A.	...	...	V.	N.S.W.	...		Branches, &c.	Spongy, black, woolly tufts, filaments necklace-like or even. Receptacles elliptic.
1762	...	...	...	...		Q.	B.	Ferns, &c.	Mycelium dense, black, forming cloth-like coating on leaves. Receptacles curved.

CORDA ANL. 142 (1842).

Arl. 272 (1837).—Schizothyrium.

| 1763 | ... | ... | ... | V. | ... | ... | ... | Dead leaves of *Eucalyptus* | On both surfaces, spots circular, reddish brown. Receptacles gregarious, minute, linear or run together, black. |
| 1764 | ... | | | | | Q | ... | Leathery leaves, living and languid | Spots black, circular or run together. Receptacles gregarious, elongated, linear, flexuous, black. |

Am. 101 (1813).—Solenarium, Hysterium.

| 1765 | ... | ... | T. | ... | ... | | | Rotten wood | Receptacle-bearing layer spread out irregularly, dark brown. Receptacles diverging from centre in a radiate manner for an inch or two. |
| 1766 | ... | | T. | | | | | Leaves of *Cyathodes straminea* | Receptacles elliptical, obtuse, opening slowly. |

Sci. Nat. III. 58 (1845).

| 1767 | ... | | ... | ... | | Q. | | Leaves of *Olea paniculata* | Receptacles gregarious, linear oblong or forked, black, receptacle-bearing layer obsolete. |
| 1768 | ... | | | | | | ... | Leaves of *Eucalyptus pilularis* | Spots somewhat circular, black, distinctly defined, and for the most part on upper surface. Receptacles elongated, shining black. |

Meckl. II. 4 (1790).—Hysterographium.

| 1769 | ... | ... | ... | ... | N.S.W. | ... | B. | Bark | Receptacles scattered or gregarious, superficial, variable in form, mostly oblong, striate lengthwise, black. |

Syll. II. 757 (1883).—Tryblidium, Hysterium.

| 1770 | ... | | | | ... | Q. | ... | Bark of trees | Bursting through, wavy or triangular, black, swollen lips, disc turning red. |

Fung. Arg. Pug. IV. 191 (1882).

| 1771 | ... | | | | ... | Q. | ... | Bark of trees | Receptacles scattered, superficial, oblong to elongated, somewhat leathery, disc reddish to brown. |

Handb. Austr. Fung. 309 (1892).—Triblidiopsis, Tryblidium.

| 1772 | ... | ... | ... | V. | ... | ... | ... | Bark | Tufts scattered, bursting through, black, hemispherical, leathery. |

Icon. V. 31 (1842).—Hysterium.

1773	W.A.	...	...	...	...	B.		Decorticated wood	Receptacles superficial, on black spot-like crust, oblong, straight, black.
1774	...	...	...	V.	...	Q.	...	Rotten wood	Receptacles elongated, linear, straight, parallel, black, with narrow fissure, lips stout.
1776	...	...	...	V.			B.	Wood	Receptacles bursting through, at length superficial, oblong to linear, in parallel lines, black.

GENERAL CLASSIFICATION OF DISCOMYCETES.

GROUP V.—DISCOMYCETES, FRIES.

ARRANGEMENT OF ORDERS (10).

32. CYTTARIACEÆ—Receptacle sub-globose.
33. HELVELLACEÆ—Receptacle vertical, stalked; cap club or mitre shaped.
34. PEZIZACEÆ—Receptacle cup shaped or disc shaped, sessile or stalked, fleshy or waxy.
35. ASCOBOLACEÆ—Receptacle plane or convex, sessile or sub-sessile, fleshy.
36. DERMATACEÆ—Receptacle concave or plane, sessile or somewhat stalked, corky, leathery, or horny.
37. BULGARIACEÆ—Receptacle top, cup, or disc shaped, gelatinous, becoming cartilaginous or horny.
38. STICTACEÆ—Receptacle immersed, usually bright coloured, waxy.
39. PHACIDIACEÆ—Receptacle immersed, usually blackish, waxy.
40. PATELLARIACEÆ—Receptacle superficial, often blackish, leathery to horny.
41. GYMNOASCEÆ—No proper receptacle.

Genus (1)—
  237. Cyttaria, Berk.

ORDER XXXII.—CYTTARIACEÆ, LEV.

ORDER XXXIII.—HELVELLACEÆ, SCHW.

Genera (5)—
  238. Morchella, *Linn.*
  239. Helvella, Linn.

| 240. Mitrula, Fries. | 242. Geoglossum, Pers. |
| 241. Leotia, *Fries*. | |

ORDER XXXIV.—PEZIZACEÆ, FRIES.

Genera (23)—
  243. Rhizina, Fries.
  244. Geopyxis, Pers.
  245. Peziza, *Linn.*
  246. Otidea, Pers.
  247. Discina, Fries.
  248. Pyronema, Carus.
  249. Humaria, Fries.
  250. Phillipsia, Berk.

251. Sarcoscypha, Fries.	259. Mollisia, Fries.
252. Trichoscypha, Cooke.	260. Tapesia, Pers.
253. Lachnea, Fries.	261. Trichopeziza, Fckl.
254. Ciboria, Fckl.	262. Dasyscypha, Fries.
255. Helotium, Fries.	263. Phaeopeziza, Sacc.
256. Phialea, Fries.	264. Belonidium, Mont. and Dur.
257. Pseudohelotium, Fckl.	265. Erinella, Sacc.
258. Chlorosplenium, Fries.	

Genus (1)—
  266. Ascobolus, Pers.

ORDER XXXV.—ASCOBOLACEÆ, BOND.

ORDER XXXVI.—DERMATACEÆ, FRIES.

Genera (3)—
  267. Urnula, Fries.

| 268. Cenangium, Fries. | 269. Tympanis, Tode. |

ORDER XXXVII.—BULGARIACEÆ, FRIES.

Genera (3)—
  270. Ombrophila, Fries.

| 271. Orbilia, Fries. | 272. Coryne, Tul. |

Genus (1)—
  273. Stictis, Pers.

ORDER XXXVIII.—STICTACEÆ, FRIES.

ORDER XXXIX.—PHACIDIACEÆ, FRIES.

Genera (3)—
  274. Pseudopeziza, Fckl.

| 275. Fabraea, Sacc. | 276. Coccomyces. De Not. |

ORDER XL.—PATELLARIACEÆ, FRIES.

Genera (2)—
  277. Patinella, Sacc.

| 278. Karschia, Kœrb. |

Genus (1)—
  279. Exoascus, Fckl.

ORDER XLI.—GYMNOASCEÆ, Bar.

SYSTEMATIC ARRANGEMENT

Number.	Cooke's Number.	Saccardo's Number.	Scientific Name.	Authority for Name.	English Name.

GROUP V.–DISCOMYCETES.–FRIES,

ORDER XXXII.–CYTTARIACEÆ,

237. CYTTARIA.—Berk.,

| 1776 | 1352 | VIII. 1 | C. Gunnii | ... Berk., Hook., Lond. Journ. 576 (1848) ... | Gunn's cyttaria ... ... |

ORDER XXXIII.–HELVELLACEÆ,

238. MORCHELLA.—Link,

1777	1354	VIII. 10	M. conica...	... Pers., Champ. Com. 257 (1818) ...	... Conical morel ...
1778	1355	„ 13	M. deliciosa	... Fries, S.M. II. 8 (1821)	... Delicious morel
1779	1353	„ 8	M. esculenta	... Pers., Syn. 618 (1801)...	... Esculent morel
1780	1356	„ 23	M. semilibera	D. C., Fl. Fr. II. 212 (1815)	... Half-free morel

239. HELVELLA.—Linn.

| 1781 | 1357 | VIII. 62 | H. monachella | ... Fries, S.M II. 18 (1821) | ... Monkish helvel ... ... |

240. MITRULA.—Fries,

| 1782 | 1358 | VIII. 118 | M. vinosa | ... Berk., Fl. Tasm. II. 273 (1860) | Wine-coloured mitrula ... |

241. LEOTIA.—Fries,

| 1783 | 1359 | VIII.2510 | L. lubrica | ... Pers., Syn. 613 (1801) | ... Slimy leotia |

242. GEOGLOSSUM.—Pers.,

1784	1364	VIII. 141	G. australe	... Berk., in Cooke Myco. 6 (1879)	... Southern geoglossum ... ...
1785	1363	„ 141	G. glabrum	... Pers., Syn. 608 (1801) ...	Smooth geoglossum ... ...
1786	1360	„ 150	G. hirsutum	... Pers, Syn. 608 (1801) ...	... Hairy geoglossum ... ...
1787	1362	138	G. Muelleri	... Cooke, Myco. 4 (1879) ...	Mueller's geoglossum ...
1788	1365	„ 145	G. nigritum	... Cooke, Myco. 205 (1879)	... Black geoglossum ...
1789	1366	„ 147	G. Peckianum	... Cooke, Myco. 5 (1879)	.. Peck's geoglossum ... ...
1790	1361	„ 149	G. Walteri	... Berk., in Cooke Myco. 4 (1879) ...	... Walter's geoglossum ...

ORDER XXXIV.–PEZIZACEÆ, FRIES,

243. RHIZINA.—Fries,

| 1791 | 1367 | VIII 182 | R. ferruginea | Phil., Grev. XVI. 74 (1888) | Rusty rhizina .. |

244. GEOPYXIS.—Pers.,

| 1792 | 1368 | VIII. 210 | G. aluticolor | Berk., Linn. Journ XIII. 176 (1873) | Tan-coloured geopyxis |

OF AUSTRALIAN FUNGI—*continued.*

Number.	W.A.	S.A.	T.	V.	N.S.W.	Q.	n.	Occurrence.	General Characters.

EPICR. I. (1836).

LEV. CONSID. MYC. 117 (1840).

Linn. Trans. XIX. 37 (1841).

| 1776 | ... | | T. | V. | | | ... | Living branches of *Fagus Cunninghamii* | Gregarious, globose or pear shaped, at length hollow, tapering below, without distinct stem. |

SCHWARTZ, SUMM. VEG. SCAND. (1814).

Berl. Mag. III. 41 (1809).—Phallus, Helvella.

1777	...	S.A.	T.	V.	N.S.W.	...	B.	Shady places	... Oblong conic, attached at base, brown, bay to black. Stem whitish, cylindrical. *Edible.*
1778			...	V.		...	B.	Grassy places	... Rather cylindrical, acute, livid yellow. Stem short. *Edible.*
1779	...	...			...		B.	Moist places	... Round or ovate, attached at base to stem, dingy yellow. Stem white, inflated, faint odour. *Edible.*
1780	...	...		V.	N.S.W.		B.	Grassy places	... Conical, free to middle, yellowish to dirty tawny when dry. Stem whitish. *Edible.*

Sp. Pl., 1649 (1763).—Phallus, Boletus.

| 1781 | | | T. | ... | | | B. | Woods, on sandy ground | Bent downwards, lobed, attached at base, somewhat bay brown. Stem hollow, white. |

S.M. I. 491 (1822).

| 1782 | | | T. | | | | ... | Rotten wood | Vinous purple, slender, linear to club shaped. Stem thread-like, straight. |

S.M. II. 29 (1821).—Elvela.

| 1783 | ... | ... | T. | V. | ... | ... | B. | Moist ground | ... Gregarious, jelly-like, swollen, greenish yellow. Stem hollow, yellow. |

Obs. I. 11 (1795).—Clavaria.

1784	...	...	T.	V.	...			Among moss	Smooth, dry, brown to black, clubs compressed, almost distinct from scaly stem.
1785	...	...	V.		Q.	B.		Grassy places	Somewhat gregarious, smooth, dry, blackish. Stem slender, crooked, scaly.
1786		...	V.			B.		Among grass	Hairy, black. Club often elongated, compressed. Stem erect, cylindrical.
1787	...		V.				...	Grassy places	Smooth, rather viscid, blackening. Club compressed, equal in length to stem.
1788			V.					Grassy places	Tufted, fragile, black, hollow. Clubs rather compressed, equal in length to slender stem.
1789	...	...				...		Moist ground	Smooth, somewhat viscid, blackening.
1790	...	V.				...		Stems of *Dicksonia*	Hairy, dark brown, blackening. Clubs spoon shaped, compressed. Stem slender.

S.M. II. 38 (1821).

Number.	Cooke's Number.	Saccardo's Number.	Scientific Name.	Authority for Name.	English Name.

245. PEZIZA.—*Linn.*, Sp. Pl. (1753).—Ciboria, Dasyscypha, Discina, Geopyxis, Humaria,

1793		VIII. 343	P. applanata	... *Fries*, S.M. II. 64 (1821)	... Flattened peziza
1794	1369	„ 253	P. aurantia	.. Pers., Obs. II. 76 (1796)	... Orange-coloured peziza
1795	1371	„ 293	P. badia ...	... Pers., Obs. II. 78 (1796) ...	... Chestnut-brown peziza
1796	1375	„ 341	P. brunneo-atra ...	Desm., Des. Esp. Nouv. 9 (1836)	Dark-brown peziza
1797	1372	„ 307	P. cochleata	... Linn., Sp. Pl. 1625 (1753) ...	... Cochleate peziza
1798	1370	„ 279	P. Drummondii	... Berk., Hook., Journ. 71 (1845) ...	... Drummond's peziza
1799	...	„ 306	P. funerata	Cooke, Myco. (1879)	... Funereal peziza
1800	1374	„ 335	P. Saccardiana	Cooke, Myco. 174 (1879) ...	... Saccardo's peziza
1801	1373	„ 297	P. vesiculosa	Bull, Champ. 457 (1812) ...	... Swollen peziza

246. OTIDEA.—Pers.,

1802	...	VIII. 354	O. apophysata	Cooke and Phil., Grev. V. 60 (1876)	... Branching otidea ...
1803	1377	X. 4473	O. darjeelensis	Berk., Hook., Journ. 202 (1851)	Darjeeling otidea
1804	1378	VIII. 358	O. hirneoloides	*Sacc.* Syll. VIII. 96 (1889) ...	Hirneola-like otidea
1805	1379	„ 362	O. phlebophora	*Sacc.* Syll. VIII. 97 (1889) ...	Veined otidea

247. DISCINA.—Fries,

1806	1383		D. australica	Cooke, Handb. Austr. Fung. 255 (1892)	Australian discina
1807	1381	VIII. 377	D. lumbricalis	*Sacc.* Syll. VIII. 101 (1889)	Worm-like discina
1808	1380	„ 373	D. repanda	*Sacc.* Syll. VIII. 100 (1889)	... Repand discina
1809	1382	„ 391	D. venosa	*Sacc.* Syll. VIII. 104 (1889)	... Veined discina

248. PYRONEMA.—Carus,

1810	1384	VIII. 401	P. melaloma	*Fckl.*, Symb. Myc. 319 (1869)	... Black-bordered pyronema
1811	1385	„ 400	P. omphalodes	*Fckl.*, Symb. Myc. 319 (1869)	Navel-like pyronema ... ...

249. HUMARIA.—Fries,

1812	1393	VIII. 506	H. carbonigena ...	... *Sacc.* Syll. VIII. 130 (1889)	... Charcoal-growing humaria ...
1813	1395	„ 520	H. fusispora	... *Sacc.* Syll. VIII. 133 (1889)	... Spindle-spored humaria ...
1814	1388	„ 431	H. globifera	Cooke, Myco. (1879) ...	Globose-spored humaria ...
1814 bis	...	„ 503	H. granulata	Bull, Champ. 25s (1791)	Granulated humaria ...
1815	1392	„ 481	H. Hartmanni	... Phil., Grev. XVI. 5 (1887)	... Hartmann's humaria ...
1816	1387	„ 424	H. miltina	... *Cooke*, Handb. Austr. Fung. 256 (1892)	Crimson humaria
1817	1386	„ 416	H. miniata	... *Cooke*, Handb. Austr. Fung. 256 (1892)	Scarlet humaria
1818	1391	„ 455	H. Muelleri	... Berk., Linn. Journ. XIII. 176 (1873)	Mueller's humaria
1819	1390	„ 443	H. recurva	... *Cooke*, Handb. Austr. Fung. 257 (1892) ...	... Recurved humaria
1820	1394	„ 518	H. rutilans	... *Sacc.* Syll. VIII. 133 (1889)	Reddish humaria
1821	...	„ 588	H. scatigena	... *Sacc.* Syll. VIII. 147 (1889)	Springing humaria
1822	1396	„ 577	H. tenacella	... *Sacc.* Syll. VIII. 145 (1889)	Toughish humaria
1823	1347	„ 569	H. Thozetii	... ... *Sacc.* Syll. VIII. 144 (1889) ...	... Thozet's humaria ...

OF AUSTRALIAN FUNGI—*continued.*

Lachnea, Mollisia, Otidea, Patinella, Phialea, Pseudohelotium, Pyronema, Sarcoscypha, Urnula.

Number.	W.A.	S.A.	T.	V.	N.S.W.	Q.	B.	Occurrence.	General Characters.
1793	...	...	...	V.	...	...		Moist places	Sessile, depressed, reddish ; disc at length somewhat wrinkled, delicately frosted, fleshy.
1794	...	...	T.	V.	N.S.W.	...	B.	Ground ...	Gregarious, almost sessile, irregular, oblique, orange ; base prolonged into short stem.
1795	W.A.	S.A.	...		...		B.	Moist ground	Nearly sessile, entire, dark brown, fleshy base often passing into short stem.
1796	...		...	V.		...	B.	Ground ...	Sessile, solitary, largish, entire, fleshy, fragile, smooth, brownish black or dark brown.
1797	...	S.A.	T.	V.		Q.	B.	Ground ...	Often densely tufted and much twisted, sessile, large, fleshy, umber.
1798	W.A.	...	...	V.		...		Ground ...	Cup shaped, sessile, medium sized, tay brown.
1799	...	S A.	...	...		...		Immersed in sand	Bell shaped, brown, margin reflexed, somewhat lobed, thin, fragile.
1800	...	...	...	V.		...		Moist ground ...	Sessile, fleshy, fragile, concave, flesh red, margin often torn.
1801	...	S. A.	...	V.		...	B.	Ground, manure heaps, &c.	Large, entire, sessile, at first globose or top shaped, then bell shaped ; base of cup very fleshy.

M.E. I. 229 (1822).—Peziza.

Number.	W.A.	S.A.	T.	V.	N.S.W.	Q.	B.	Occurrence.	General Characters.
1802	...	...	...	...	...	Q.	B.	Moist places	Tufted or gregarious, sessile, lobed, margin indented, umber brown ; paraphyses peculiarly branched.
1803	...	...	...	V.		...		Ground ...	Expanded, somewhat cochleate, usually elongated on one side, umber.
1804	...	...	...	V.		Q.		Rotten wood	Sessile or very shortly stalked, red, white beneath.
1805	...	...	...	V.		...	B.	Ground ...	Cup shaped, oblique, ochrey yellow, finely powdery, with short stemlike base veined and ribbed.

S.V. 348 (1849).—Peziza.

Number.	W.A.	S.A.	T.	V.	N.S.W.	Q.	B.	Occurrence.	General Characters.
1806	...	...	...	V.		...		Ground ...	Large, cup shaped, then expanded, smooth, ochrey, tapering downwards into short thick rooting stem.
1807	...	...	...	V.		...		Ground ...	Large, cup shaped, at length expanded, internally pale brown, externally nearly smooth or mealy.
1808	...	...	...	V.	N.S.W.	Q.	B.	Rotten trunks and ground	Solitary or tufted, large, incised, repand or bent backwards, internally brown, externally whitish.
1809	...	...	...	V.		...	B.	Ground ...	Sessile or somewhat stalked, umber brown, externally whitish, rough with ribbed veins, strong nitrous odour.

Nov. Act. Cur. XVII. 370 (1835).

Number.	W.A.	S.A.	T.	V.	N.S.W.	Q.	B.	Occurrence.	General Characters.
1810	W.A.	...	...	V.			B.	Burnt ground	Sessile, crowded, dingy orange, margin with delicate black hairs or prominent cells.
1811	...		...	V.			B.	Burnt ground and cinder heaps	Sessile, crowded, often running together, minute, orange red or orange yellow.

S.M. II. 42 (1822).—Peziza.

Number.	W.A.	S.A.	T.	V.	N.S.W.	Q.	B.	Occurrence.	General Characters.
1812	...	...	T.	V.		...	B.	Charred ground ...	Gregarious, orange yellow, sessile, flexuous, slightly granular, margin wavy.
1813	...	...	T.			...	B.	Charred and heathy ground	Gregarious, rather crowded, sessile, hemispherical, yellow, downy ; spores spindle shaped.
1814	...	...	...	V.		...			Yellow, saucer shaped, margin turned in, sometimes lobed.
1814 bis	...	...	...			Q.	B.	Cow dung ...	Sessile, scattered or crowded, margin thick externally, tawny brown, and coarsely granular.
1815	...	...	...			Q.	...	Decayed branches ...	Gregarious, sessile or somewhat stalked, concave, margin splitting, disc pale crimson.
1816	...	...	...	V.		...	...	Sandy ground	Sessile, scattered, crimson, flattened, margin paler beneath, free.
1817	...	...	...	V.		...	B.	Among moss	Fleshy, firm, pitcher to plate like, scarlet.
1818	...	...	T.	V.		...	...	Ground ...	Scattered, sessile, cups irregular, marginate delicately downy, externally, disc crimson.
1819	...	...	T.	...		...	...	Ground ...	Nearly sessile, wavy, convex, recurved, bay brown, smooth.
1820	W.A	S.A.	...	...		...	B.	Ground among moss	Gregarious, nearly sessile, bell or beaker shaped ; disc orange yellow, externally paler, and slightly downy.
1821	...	...	...			Q.	...	Dung	Hemispherical, dark-wine colour, somewhat green when fresh, externally mealy white.
1822	...	...	...	V.		...	...	Ground ...	Sessile, slightly concave, smooth, umber brown, margin entire, flesh firm.
1823	...	...	...	...	N.S.W.	Q.	...	On *Nepenthes*	Dish shaped, fleshy, brown.

Number.	Cooke's Number.	Saccardo's Number.	Scientific Name.	Authority for Name.	English Name.
					250. PHILLIPSIA.—Berk.,
1824	1399	VIII. 608	P. polyporoides ...	Berk., Linn. Jonrn. XVIII. 388 (1881) ...	Polyporous-like phillipsia ...
1825	1398	„ 607	P. subpurpurea ...	Berk. and Br., Linn. Soc. N.S.W. 88 (1880)	Purplish phillipsia ... ...
					251. SARCOSYPHA.—Fries,
1826	1403	VIII. 638	S. bulbosa	Cooke, Handb. Austr. Fung. 259 (1892) ...	Bulbous sarcosypha ... ...
1827	1400	„ 618	S. coccinea	Sacc. Syll. VIII. 151 (1889)	Scarlet sarcosypha
1828	1401	„ 620	S. lepida ...	Sacc. Syll. VIII. 154 (1889) ...	Handsome sarcosypha ...
1829	1404	„ 657	S. melastoma	Cooke, Handb. Austr. Fung. 259 (1892)	Black-mouthed sarcosypha
1830	1402	„ 630	S. rhenana	Sacc. Syll. VIII. 157 (1889)	Woolly sarcosypha
					252. TRICHOSCYPHA.—Cooke,
1831	1405	VIII. 652	T. Hindsii	Sacc. Syll. VIII. 161 (1889)	Hind's trichoscypha ... ...
1832	1406	„ 647	T. tricholoma	Sacc. Syll. VIII. 160 (1889)	Hairy-edged trichoscypha
					253. LACHNEA.—Fries,
1833	1415	VIII. 738	L. alpina ...	Sacc. Syll. VIII. 180 (1889)	Alpine lachnea ...
1834	1410	„ 699	L. badio-herbis	Sacc. Syll. VIII. 173 (1889)	Bay-bearded lachnea ... ..
1835	1408	„ 772	L. confusa	Cooke, Handb. Austr. Fung. 260 (1892) ...	Confused lachnea ... ...
1836	1419	„ 735	L. coprogena ...	Sacc. Syll. VIII. 181 (1892) ...	Dung-borne lachnea ...
1837	1414	„ 730	L. dalmeniensis ...	Phil., Disc. 227 (1887) ...	Dalmeny lachnea ... ...
1838	1420	„ 741	L. Erinaceus	Sacc. Syll. VIII. 182 (1889)	Hedgehog lachnea
1839	1413	„ 705	L. hirta ...	Gill., Champ. 75 (1879)	Hairy lachnea
1840	1416	„ 722	L. lusatiæ	Sacc. Syll. VIII. 178 (1889)	Lusatian lachnea
1841	1411	„ 700	L. margaritacea	Sacc. Syll. VIII. 173 (1889)	Pearly lachnea
1842	1417	„ 725	L. scubalouta	Sacc. Syll. VIII. 179 (1889)	Refuse lachnea
1843	1409	„ 698	L. scutellata	Gill., Champ. 75 (1879)	Saucer-shaped lachnea
1844	1421	„ 711	L. stercorea	Gill., Champ. 76 (1879)	Dung lachnea ...
1845	1418	„ 728	L. theleboloides	Gill., Champ. 74 (1879)	Thelebolus-like lachnea
1846	1412	„ 701	L. umbrata ...	Phil., Disc. 222 (1887)...	Shaded lachnea ...
1846a	„	„	L. umbrata, var. pallida	Rehm., Asco. No. 456 (1873)	Pale lachnea ...
1847	1407	„ 687	L. vinoso-brunnea ...	Sacc. Syll. VIII. 171 (1889)	Vinous-brown lachnea
					254. CIBORIA.—Fckl.,
1848	1423	VIII. 829	C. firma ...	Fckl., Sym. Myc. 312 (1869) ...	Firm ciboria ... ...
					255. HELOTIUM.—Fries,
1849	1425	VIII. 910	H. citrinum	Fries, S.V. 355 (1849) ...	Lemon-yellow helotium
1850	1426	„ 914	H. claro-flavum	Berk., Outl. 372 (1860)	Light-yellow helotium
1851	1429	„ 925	H. epiphyllum	Fries, S.V. 356 (1849)	Leaf-growing helotium ...
1852	1427	„ 918	H. gratum	Cooke, Austr. Fung. 51 (1883) ...	Agreeable helotium ... ...
1853	1424	„ 876	H. nigripes	Fries, S.V. 356 (1849) ...	Black-stalked helotium
1854	1428	„ 1028	H. pateræforme	Cooke, Austr. Fung. 51 (1883) ...	Saucer-shaped helotium

OF AUSTRALIAN FUNGI—continued.

Number	Habitat W.A.	S.A.	T.	V.	N.S.W	Q.	B.	Occurrence.	General Characters.

Linn. Journ. XVIII. 388 (1881).—Peziza.

| 1824 | ... | ... | | | | Q. | | Dead stems of *Vitis* | Expanded, attached, thick, flesh colour. |
| 1825 | ... | | | | | Q. | | Wood ... | Plane, margin lobed, fixed at centre; disc purplish, brown when dry. |

S.M. II. 78 (1822).—Peziza, Lachnea, Macropodia.

1826	...	...	...	V.		...	B.	Ground in sandy soil	Hemispherical, turning ashy, minutely scaly, disc brown. Stem firm, tuberous at base.
1827		...	T.			...	B.	Rotten branches ...	Funnel shaped, externally whitish, downy, as well as stem, disc carmine. most handsome.
1828	S.A.					..		Ground ...	Funnel shaped, medium sized, with gradually tapering stem, disc crimson.
1829						Q.	B.	Old branches, &c. ...	Fleshy, almost globose, externally brick red, woolly, disc black. Stem short, with rooting black hairs.
1830	W.A.	...						Bare ground	Tufted, united in thick stem, whitish downy, nearly globose, margin turned in, disc orange.

Myc. 252 (1879).

| 1831 | ... | ... | ... | ... | | Q. | ... | Rotten wood | Bright red, cup shaped, externally with delicate bloom, tawny yellow. Stem tapering downwards. |
| 1832 | | ... | ... | ... | | Q. | ... | Rotten woo l | Stalked, fleshy, hemispherical, top shaped, yellow, hairy edged. Stem smooth. |

S.M. II. 77 (1822).—Peziza, Humaria, Sphærospora.

1833	...	...	V.		...	...		Cow dung	Gregarious, closel at first, then flattened, circular, margin orange yellow with jointed hairs.
1834	...	S.A.	...	N.S.W.		...	...	Rotten wood	Concave, disc vermilion, margin clad with very long hairs.
1835	...	...	...	...		Q.	...	Charcoal ...	Gregarious or crowded, sessile, almost spherical, brown, clad with short hairs.
1836	...	...	...	...		Q.	...	Dung ...	Nearly orange, invested with pale-bay obtuse hairs.
1837	...	...	...	V.			B.	Ground in shady woods	Sessile, fleshy, hemispherical, becoming expanded, bright yellow, fringed with long erect yellow hairs.
1838	...		...	...		Q.	..	Rotten wood	Gregarious, circular, depressed, ochrey white, externally beset with long bay-brown hairs.
1839	...	...	...	V.			B.	Ground among moss	Sessile, scattered, somewhat hemispherical, externally clad with brown hairs, disc scarlet.
1840	...	...	...	V.			...	Rotten wood	Gregarious, sessile, cup shaped, flattened at length, orange red, with erect brown hairs externally.
1841	...		...	V.		...	...	Rotten w ood	Sessile, hemispherical, at length expanded, vermilion, rough externally, with short brown hairs.
1842	...	...	...	V.		...	...	Dung ...	Scattered, sessile, fleshy, hemispherical, thickly clad externally, with septate brown hairs.
1843	W.A.	...	T.	V.	N.S.W.	Q.	B.	Rotten wood	Gregarious, sessile, flattened, vermilion red, rough towards margin, with long straight black hairs.
1844	...		T.	V.			B.	Dung	Gregarious, sessile, concave, dingy red or tawny, beset with brown septate hairs.
1845		...	...	V.			B.	Earth, &c.	Gregarious, spherical, then open, externally whitish, clad with pale hairs, disc pale yellow.
1846	...	...	...				B.		
1846a	...	...	...	V.			B.	Charred wood, &c....	Brownish flesh colour, with scattered obtuse hairs.
1847	...	...	...	...		Q.	...	Burnt ground	Sessile, hemispherical, flattened, vinous brown, rough, with short obtuse brown scattered hairs.

Sym. Myc. 311 (1869).—Peziza, Hymenoscypha.

| 1848 | ... | ... | T. | ... | | ... | B. | Rotting branches ... | Funnel shaped, then expanded, firm, pale brown. Stem long, tapering downwards, becoming blackish. |

S.V. 354 (1849).—Peziza, Phialea.

1849	...	...	T.	...		Q.	B.	Dead stumps and branches	Gregarious or crowded, shortly stalked or sessile, flattened, concave, lemon yellow, waxy.
1850	...		...	V.			B.	Decayed wood and fallen branches	Very minute, shortly stalked or sessile, clear yellow, smooth, margin somewhat lobed.
1851	...	...	...	...		Q.	B.	Dead leaves ...	Almost sessile, convex to plane, smooth, marginate, firm, pale ochrey.
1852	...	...	T.	...		..	...	Dead wood ...	Plane, transparent, marginate, shortly stalked, nearly orange. Stem paler, cylindrical.
1853	...	...	T.			...	...	Trunks and rotting leaves	Flattened, concave, pale, smooth, marginate. Stem longish, blackening.
1854	...	...	T.	...		...	...	Rotten wood ...	Ochrey, sessile, somewhat lobed, concave, somewhat wrinkled and delicately downy beneath.

142

SYSTEMATIC ARRANGEMENT

Number.	Cooke's Number.	Saccardo's Number.	Scientific Name.	Authority for Name.	English Name.

256. Phialea.—Fries,

1855	1432	VIII. 1018	P. Berggrenii	Sacc. Syll. VIII. 254 (1889)	... Berggren's phialea ...
1856	1434	„ 1104	P. byssogena	Sacc. Syll. VIII. 267 (1889)	... Thread-borne phialea ...
1857	1433	„ 1102	P. ceratina	... Sacc. Syll. VIII. 267 (1889)	... Horny phialea ...

257. Pseudohelotium.—Fckl., Symb. Myc. 298 (1869).—

| 1858 | 1435 | VIII.1215 | P. hyalinum | Fckl., Symb. Myc. 298 (1869) ... | ... Hyaline pseudohelotium ... |
| 1859 | 1436 | „ 1267 | P. ilicincolum ... | Sacc. Syll. VIII. 304 (1889) | ... Holly-growing pseudohelotium ... |

258. Chlorosplenium.—Fries,

| 1860 | 1430 | VIII.1311 | C. æruginosum | De Not., Disc. 22 (1864) | ... Verdigris chlorosplenium ... |
| 1861 | 1431 | „ 1313 | C. omnivirens | Cooke, Austr. Fung. 51 (1883) | ... All-green chlorosplenium |

259. Mollisia.—Fries,

| 1862 | 1437 | VIII.1393 | M. cinerea | ... Karst., M.F. I. 189 (1871) | Ash-coloured mollisia ... ... |

260. Tapesia.—Pers.,

| 1863 | 1438 | VIII.1573 | T. epitephra | ... Sacc. Syll. VIII. 381 (1889) | ... Woolly-base tapesia ... ... |

261. Trichopeziza.—Fckl.,

| 1864 | 1439 | X. 4540 | T. Sphærula | Sacc., Hedw. 155 (1890) | ... Sphærula trichopeziza... ... |

262. Dasyscypha.—Fries, S.M. II. 89 (1822).—

1865	1443	VIII.1924	D. Eucalypti	Sacc. Syll. VIII. 462 (1889)	... Eucalypt dasyscypha ... ...
1866	1442	„ 1876	D. glabrescens	... Sacc. Syll. VIII. 451 (1889)	... Smooth dasyscypha ...
1867	1441	„ 1804	D. lachnoderma	Rehm., Asco. No. 303 (1873)	... Downy dasyscypha ... ...
1868	1444	„ 1938	D. lanariceps	... Sacc. Syll. VIII. 465 (1889)	... Woolly-capped dasyscypha ...
1869	1445	„ 1947	D. terrestris	... Sacc. Syll. VIII. (1889)	Terrestrial dasyscypha ...
1870	1440	„ 1801	D. virginea	... Fckl., Symb. Myc. 305 (1869)	Virgin dasyscypha ... ...

263. Phæopeziza.—Sacc.,

| 1871 | 1376 | VIII.1966 | P. apiculata | ... Sacc., Mich. I. 71 (1877) | ... Apiculate phæopeziza ... ... |

264. Belonidium.—Mont.

| 1872 | 1446 | VIII.2064 | B. aranecosum | Sacc., Syll. VIII. 500 (1889) | Web-like belonidium ... ... |

265. Erinella.—Sacc.

| 1873 | 1447 | | E. lutea ... | ... Phil., Grev. XIX. 61 (1891) | Yellow erinella ... ... |

OF AUSTRALIAN FUNGI—*continued.*

Number.	Habitat. W.A.	S.A.	T.	V.	N.S.W.	Q.	B.	Occurrence.	General Characters.
Obs. II. 305 (1818).									
1855	...		...	V.			...	Rotting leaves ...	Pale, stalked, scattered, wine-glass shaped. Stem slender.
1856	...		T.	...			...	Wood	Ochrey, concave. Stem elongated, cylindrical, arising from interwoven radiating threads.
1857	...		T.	...			...	Leaves of Eucalypts	Top shaped, stalked, smooth, pale, horny brown, minute.
Peziza, Lachnea, Helotium, Lachnella, Mollisia.									
1858	...		T.	...		...	B.	Rotten trunks and inside bark	Minute, gregarious, sessile, globose, then expanded, transparent when moist, downy externally.
1859	...	...	...	V.		...	B	Holly branches and lichen, *Myriangium* growing on Holly	Tufted, hemispherical, then expanded, externally dirty white, disc brown, purple, or rosy grey.
S.V. 356 (1849).—Peziza, Helotium.									
1860	...		...	V.	...	Q.	B.	Fallen wood ...	Shortly stalked or sessile, verdigris green, disc becoming whitish. Stem short, stout. Wood stained employed as green oak in manufacture of Tunbridge ware.
1861	...		T.	...		...	...	Rotten wood ...	Verdigris green, shortly stalked, rather top shaped.
S.M. II. 137 (1822).—Peziza.									
1862	...		T.	V.		...	B.	Decaying wood ...	Gregarious or scattered, sessile, soft, minute, saucer-like, ashy, with entire whitish margin.
M.E. I. 220 (1828).—Peziza, Lachnella.									
1863	...	...	T.	...	...	...	...	Leaves	Minute, white, hemispherical or almost globose, concave, arising from crisp interwoven threads.
Symb. Myc. 295 (1869).									
1864	...	S.A.	...	...		...	..	Dead bark of *Casuarina*	Scattered, minute, sessile, globose, bright sulphur yellow, sprinkled with rough hairs.
Peziza, Hymenoscypha, Lachnella, Helotium.									
1865	...	...	T.	V.	...	...	...	Leaves of *Eucalyptus* and *Casuarina*	Pale olive, plane, margin fringed with rigid dark-purple hairs. Stem cylindrical.
1866	...	...	...	V.	...	...	...	On *Rhipogonum*	Scattered, stalked, white, wine-glass shaped, at first shaggy, then naked, smooth.
1867			T.	...		Q.		Dead bark	Almost hemispherical, shortly stalked, externally snowy white and downy, vermilion within.
1868	...	...	...	V.			...	On *Rhipogonum* ...	Scattered, stalked, ochrey brown, top shaped, at length open, shaggy, sprinkled with purple granules.
1869	...	...	...	..		Q.	...	Bare ground ...	Small, stalked, horn colour, lurid, externally shaggy.
1870	...	...	T.	...		...	B	Wood, bark, branches, &c.	Gregarious, stalked, white, hemispherical, with crowded spreading hair. Stem short.
Mich. I. 71 (1877).									
1871	...	...	...	V.		...	...	Bark ...	Sessile, saucer shaped, fleshy, rather tough, black, smooth, disc with margin entire.
and Dur., Fl. Alg. (1846).									
1872	...	...	T.	...	...	...		Wood	At first nearly globose, then hemispherical, externally web-like, arising from creeping threads.
Syll. VIII. (1889).									
1873	...	...	...	V.		...	...	Crevices of bark ..	Gregarious or scattered, shortly stalked, cup shaped, with short whitish hairs, becoming yellow, then yellowish brown.

Number.	Cooke's Number.	Saccardo's Number.	Scientific Name.	Authority for Name.	English Name.

ORDER XXXV.—ASCOBOLACEÆ,

266. Ascobolus.—Pers.,

1874	1452	VIII. 2161	A. Archeri	Berk., Fl. Tasm. II. 276 (1860) ...	Archer's ascobolus ... ...
1875	1449	„ 2149	A. australis ...	Berk., Linn. Journ. XVIII. 398 (1881) ...	Southern ascobolus ... ...
1876.	1451	„ 2150	A. Baileyi ...	Berk. and Br., Linn. Traus. II. 69 (1883)	Bailey's ascobolus ... . .
1877	1448	„ 2143	A. furfuraceus ...	Pers., Tent. Disp. Meth. 25 (1797)	Scurfy ascobolus
1878	1450		A. Phillipsii ...	Berk., in Cooke's Handb. Austr. Fung. 268 (1892)	Phillips' ascobolus

ORDER XXXVI.—DERMATACEÆ,

267. Urnula.—Fries,

| 1879 | 1453 | VIII. 640 and 218 | U. campylospora | *Cooke*, Handb. Austr. Fung. 268 (1892) ... | ... Curve-spored urnula ... ... |
| 1880 | 1454 | VIII. 331 | U. rhytidea | *Cooke*, Austr. Fung. 52 (1883) ... ... | Wrinkled urnula |

268. Cenangium.—Fries,

| 1881 | 1455 | VIII. 2323 | C. lichenoideum ... | Berk. and Br., Linn. Trans. I. 404 (1879) | Lichenoid cenangium ... ... |

269. Tympanis.—Tode,

| 1882 | | X. 4603 | T. Toomausis | Berk. and Br., Linn. Traus. II. 222 (1887) | ... Tooma tympanis ... ... |

ORDER XXXVII.—BULGARIACEÆ,

270. Ombrophila.—Fries,

1883	1459	X. 4612	O. bulgarioides ...	... Sacc., Myc. Austr. 14 (1890)	Bulgaria-like ombrophila ...
1884	1457	VIII. 2542	O. radicata	... Phil., Grev. XVI. 33 (1887)	Rooting ombrophila ...
1885	1458	„ 2553	O. terrestris	... Phil., Grev. XVI. 75 (1888)	Terrestrial ombrophila
1886	1460	...	O. trachycarpa ...	... Phil., Grev. XIX. 61 (1891)	... Rough-spored ombrophila
1887	1456	VIII. 2526	O. violacea ...	*Fries*, S.V.S. 357 (1849)	Violet ombrophila ...
1887A	1456A	„ „	O. violacea. var. australis...	Cooke, Grev. VIII. 61 (1879) ...	Southern ombrophila ...

271. Orbilia.—Fries,

| 1888 | 1461 | VIII. 2572 | O. chrysocoma ... | *Sacc.* Syll. VIII. 621 (1889) | Golden-yellow orbilia.. ... |
| 1889 | 1462 | „ 2568 | O. decipiens ... | *Sacc.* Syll. VIII. 623 (1889) | Deceptive orbilia ... |

272. Coryne.—Tul., Carp. III. 190 (1865).—

| 1890 | 1463 | VIII. 2617 | C. sarcoides | *Tul.*, Carp. III. 190 (1865) | ... ... Flesh-like coryne ... ... |

OF AUSTRALIAN FUNGI—*continued.*

Number.	Habitat.						B.	Occurrence.	General Characters.
	W.A.	S.A.	T.	V.	N.S.W.	Q.			

BOND., MEM. ASCOB. 20 (1869).

in Gmel. Syst. 1461 (1791).—Peziza.

Number	W.A.	S.A.	T.	V.	N.S.W.	Q.	B.	Occurrence	General Characters
1874	...	...	T.	...	...	...		Charcoal	Wavy, sessile, vinous brown.
1875	...	...	..	...	...	Q.		Dung ...	Brown, cup shaped.
1876	...	...	...	...	...	Q.		Dung ...	Concave, at first ochrey, then vinous brown and flattened, slightly granulate externally.
1877	W.A.	...		...			B.	Old cow dung, &c....	Sessile, globose, then expanded: externally greenish yellow, mealy; disc slightly concave at first, yellowish green, turning blackish brown when old.
1878	...					Q.		Cow dung	Concave with elevated margin, externally wax colour, then tawny, disc ash coloured.

FRIES, S.V. 345 (1849).

S.V.S. 11. 364 (1849).—Peziza, Macropodia.

Number	W.A.	S.A.	T.	V.	N.S.W.	Q.	B.	Occurrence	General Characters
1879	...	...	...	...	...	Q.		Rotten wood	Funnel shaped, sooty or ashy, stalked, deeply wrinkled, margin incurved. Stem similarly coloured.
1880	...	...	...	...	N.S.W.			Ground ...	Sooty brown, nearly sessile, hemispherical, cut, undulately wrinkled, flesh olive.

S.M. 11. 177 (1822).

| 1881 | ... | ... | ... | ... | ... | Q. | | Trunks ... | Tufted, ashy, top shaped, stalked, invested with irregular ashy warts, disc red brown. |

Mœkl. I. 23 (1790).

| 1882 | ... | ... | ... | ... | ... | Q. | | Fruit of Bauskia ... | Gregarious, at first mealy, sphæria shaped. |

FRIES, S.V. 345 (1849).

S.V.S. 357 (1849).—Peziza.

Number	W.A.	S.A.	T.	V.	N.S.W.	Q.	Occurrence	General Characters
1883	...	...	...	...	...	Q.	Rotten wood	Clustered, gelatinous, then hard, at first sessile, then shortly stalked, externally bright ochrey yellow, disc reddish brown.
1884	...	...		V.	...		Swampy places ...	Solitary or tufted, stalked, rather gelatinous. Stem elongated, rooting; disc depressed, wrinkled, liver colour.
1885	...	...		V.			Ground ...	Circular, sessile, gelatinous, concave or flattened, umber brown, a little paler externally.
1886	...	...		V.			Sandy ground among mosses	Somewhat gregarious, sessile, concave, externally wrinkled horizontally, firm, dark red brown.
1887	...	...		V.			Trunks ...	Gregarious or scattered, finally distinctly stalked, violet. Stem obconic, short.
1887a	...	...		V.	...		Branches, &c., in swampy places	Stem longer, flexuous, more ash coloured than type.

S.V. 357 (1849).—Peziza, Calloria.

| 1888 | ... | ... | | V. | ... | ... | Wood | Gregarious, almost globose at first, soon flattened, and rather jelly-like, golden yellow, horny when dry. |
| 1889 | ... | ... | ... | | | Q. | Old rope | Gregarious or scattered, sessile; disc pale-flesh colour, orange red, or pale brown; externally granulose. |

Peziza, Ombrophila, Lichen, Bulgaria, Helvella, Tremella.

| 1890 | ... | ... | T. | V. | ... | ... | B. | Trunks and branches of trees | Tufted, sessile or somewhat stalked, firm, fleshy red, veined below, disc hollowed out. |

Number.	Cooke's Number.	Saccardo's Number.	Scientific Name.	Authority for Name.	English Name.

ORDER XXXVIII.—STICTACEÆ,

273. Stictis.—Pers., Obs. II. 73 (1796).—

1891	1465	X. 4635	S. emarginata	Cooke and Mass., Grev. XVIII. 7 (1889)	Emarginate stictis
1892	1464	VIII. 2795	S. radiata	Pers., Obs. II. 73 (1796)	Radiating stictis
1892a			S. radiata, var. brachyspora	Sacc. and Berl., Rev. Myc. (1885)	Short-spored stictis

ORDER XXXIX.—PHACIDIACEÆ,

274. Pseudopeziza.—Fckl., Symb.

1893		VIII. 2976	P. Cerastiorum	Fckl., Symb. Myc. 291 (1869)	Chickweed pseudopeziza
1894	1467	„ 2971	P. Medicaginis	Sacc., Fung. Ard. No. 90 (1888)	Medicago pseudopeziza
1895	1466	„ 2970	P. Trifolii	Fckl., Symb. Myc. 290 (1869)	Trifolium pseudopeziza

275. Fabræa.—Sacc.,

| 1896 | 1468 | X. 4651 | F. rhytismoides | Sacc. Syll. X. 50 (1892) | Rhytisma-like fabræa |

276. Coccomyces.—De Not.,

| 1897 | 1469 | | C. delta | Cooke, Handb. Austr. Fung. 272 (1892) | Deltoid coccomyces |

ORDER XL.—PATELLARIACEÆ,

277. Patinella.—Sacc..

| 1898 | 1470 | VIII. 3162 | P. tasmanica | Sacc. Syll. VIII. 770 (1889) | Tasmanian patinella |
| 1899 | 1471 | „ 3178 | P. Adamsoni | Sacc. Syll. VIII. 772 (1889) | Adamson's patinella |

278. Karschia.—Koerb.,

| 1900 | 1472 | VIII. 3200 | K. lignyata | Sacc. Syll. VIII. 779 (1889) | Wood-growing karschia |

ORDER XLI.—GYMNOASCEÆ,

279. Exoascus.—Fckl., Enum. Fung.

| 1901 | 1473 | VIII. 3341 | E. deformans | Fckl., Symb. Myc. 252 (1869) | Deforming exoascus (Peach-leaf curl) |

OF AUSTRALIAN FUNGI—continued.

Number	W.A.	S.A.	T.	V.	N.S.W.	Q.	D.	Occurrence.	General Characters.

FRIES, PL. HOM. 86 (1825).

Peziza, Lycoperdon, Sphærobolus, Schmitzomia.

Number	W.A.	S.A.	T.	V.	N.S.W.	Q.	D.	Occurrence.	General Characters.
1891	...	...	...	V.	...	...		Leaves of Eucalypts	Very minute, gregarious, immersed, bursting through, pierced at apex.
1892	...		T.	...		Q.	B.	Wood and branches	Gregarious or scattered, flesh coloured or yellowish, deeply immersed; margin four to six rayed, white, scurfy.
1892A	...	...	...			Q.	...	Rotten stem	Margin narrower than type, and disc ashy grey to violet.

FRIES, S.M. II. 317 (1822).

Myc. 290 (1869).—Phacidium, Ascobolus, Peziza, Mollisia,

Number	W.A.	S.A.	T.	V.	N.S.W.	Q.	D.	Occurrence.	General Characters.
1893	...	...			N.S.W.	...	B.	Leaves and more rarely calyx of *Cerastium vulgatum*	Gregarious, sessile, minute, round, smooth, at first white with reddish grey rim, at last buff with dark-brown rim.
1894	...	...	...	V.	...	...		Leaves of *Medicago*	Scattered, minute, soon flattened, ochrey brown, originating on yellowish spots, then girt by three to four toothed skin.
1895		...	...	V.		...	B	Living but languishing clover leaves	Gregarious, sessile, minute, circular, plane, smooth, smoky yellow; margin thin, torn.

Mich. II. 331 (1881).—Phacidium, Pseudopeziza.

1896	...	...	...	V.	...	...		Living leaves of *Cotula*	Cups clustered together, usually six to eight, minute, externally dark brown; disc closing in drying, blackening, and then resembling *Rhytisma*.

Mem. II. 38 (1817).—Phacidium.

1897	...	...	...	V.	...	...		Leaves of Eucalypts	Innate, three-angled, with three elevated joints, opening in three valves, disc brown.

FRIES, S.V. 345 (1849).

Grev. IV. 22 (1875).—Patellaria, Peziza.

Number	W.A.	S.A.	T.	V.	N.S.W.	Q.	D.	Occurrence.	General Characters.
1898	...	...	T.	...	...	...		Dead wood	Small, sessile, cups concave, then plane; disc reddish brown, then black.
1899	...	...	...	V. .	...	...		Branches of Eucalypts	Circular, plane, cups with distinct margin, quite black.

Parerg. 459 (1863).—Patellaria.

1900	...		...	V.	...		B.	Rotten wood	Scattered or slightly gregarious, sessile, saucer shaped, horny when dry, externally dark red; disc concave, quite black.

BAR. BOT. ZEIT. 158 (1872).

Nass. 29 (1860).—Taphrium, Ascomyces.

1901	...	S.A.	...	V.	N.S.W.	Q.	B.	Peach, &c., leaves.	On under surface of leaves causing blisters, and covered with a whitish bloom.

GENERAL CLASSIFICATION OF TUBEROIDES.

GROUP VI.—TUBEROIDES. VITT.

ARRANGEMENT OF ORDERS (3).

42. ELAPHOMYCETACEÆ.—Gleba or spore-bearing tissue traversed by silky filaments.

43. TUBERACEÆ.—Gleba traversed by branched filaments, or with cavities.

44. ENDOGONACEÆ.—Gleba destitute of internal cavities, continuous.

ORDER XLII.—ELAPHOMYCETACEÆ, TUL.

Genus (1)—
   280. Elaphomyces, Nees.

ORDER XLIII.—TUBERACEÆ, FRIES.

Genus (1)—
   281. Stephensia, Tul.

ORDER XLIV.—ENDOGONACEÆ, FRIES

Genus (1)—
   282. Endogone, Link.

SYSTEMATIC ARRANGEMENT

Number.	Cooke's Number.	Saccardo's Number.	Scientific Name.	Authority for Name.	English Name.

GROUP VI.—TUBEROIDES.—

ORDER XLII.—ELAPHOMYCETACEÆ,

280. ELAPHOMYCES.—Nees,

| 1902 | 1319 | VIII.3481 | E. Leveillei | Tul., Ann. Sci. Nat. 2 Ser. XVI. 21 (1841) | ... Leveille's elaphomyces |

ORDER XLIII.—TUBERACEÆ,

281. STEPHENSIA.—Tul., Compt.

| 1903 | ... | | S. arenivaga | Cooke and Mass., Grev. XXI. 38 (1892) ... | ... Desert stephensia |

ORDER XLIV.—ENDOGONACEÆ,

282. ENDOGONE.—Link,

| 1904 | 1350 | VIII.3597 | E. australis | Berk., Fl. Tasm. II. 282 (1860) ... | ... Southern endogone |

OF AUSTRALIAN FUNGI—*continued.*

Number.	Habitat.						B.	Occurrence.	General Characters.
	W.A.	S.A.	T.	V.	N.S.W.	Q.			

VITT. MON. TUBER. 12 (1831).

TUL. FUNG. HYP. 100 (1851).

Syn. Pl Myc. (1820).

| 1902 | ... | | ... | ... | | Q. | | Under trees | Rounded or depressed, hollowed out on both sides, arising from green crustaceous mycelium. |

FRIES, S.V.S. 437 (1849).

Rend. XXI. 1433 (1845).—Genea.

| 1903 | | S.A. | .. | ... | | | | Sandy soil | ... | Nearly globose, irregular, pale, soft, becoming hard, gathering particles of sand which cohere. |

FRIES, S.V.S. 438 (1849).

Obs Pl. III. 33 (1809).

| 1904 | | | T | | | | | Ground | ... | Hemispherical, white. |

GENERAL CLASSIFICATION OF HYPHOMYCETES.

GROUP VII.—HYPHOMYCETES, MARTIUS.

ARRANGEMENT OF ORDERS (4).

45. MUCEDINACEÆ—Finely filamentous, pale or bright coloured (rarely brownish).
46. DEMATIACEÆ—Finely filamentous, brown or black, rather rigid.
47. STILBEACEÆ—Finely filamentous, pale or brown; fertile threads collected in fascicles (stroma).
48. TUBERCULARIACEÆ—Compact, wart-like, globose, disc-like, superficial or erumpent, waxy or somewhat gelatinous

---

ORDER XLV.—MUCEDINACEÆ, LINK.

ARRANGEMENT OF GENERA (13).

Section 1. Amerosporæ, Sacc.—Conidia spherical or shortly cylindrical, continuous, transparent or brightly coloured.

Sub-section 1. Micronemæ, Sacc.—Threads very short, or scarcely distinct from conidia.

Genera (3)—

283. Oospora, Wallr.	284. Monilia, *Pers.*	285. Oidium, Link.

Sub-section 2. Macronemæ, Sacc.—Threads elongated, distinct from conidia.

Genera (9)—

286. Trichoderma, Pers.	289. Rhinotrichum, Corda.	292. Sepedonium, Link.
287. Aspergillus, *Adans.*	290. Sporotrichum, Link.	293. Verticillium, Nees.
288. Penicillium, Link.	291. Botrytis, *Adans.*	294. Nematogonium, Desm.

Section 2. Didymosporæ, Sacc.—Conidia ovoid, oblong or shortly fusoid, one septate, hyaline or brightly coloured.

Genus (1)—
295. Tricothecium, Link.

ORDER XLVI.—DEMATIACEÆ, FRIES.

ARRANGEMENT OF GENERA (19).

Section 1. Amerosporæ, Sacc.—Conidia continuous, globose, ovoid or oblong.

Sub-section 1. Micronemæ, Sacc.—Threads very short, or scarcely distinct from conidia.

Genera (4)—

296. Coniosporium, Link.	298. Hormiscium, Kunze.	299. Heterobotrys, Sacc.
297. Torula, Pers.		

Sub-section 2. Macronemæ, Sacc.—Threads evident and distinct from conidia.

Genera (3)—

300. Periconia, Tode.	301. Monotospora, Corda.	302. Botryotrichum, Sacc and March.

Section 2. Didymosporæ, Sacc.—Conidia ovoid or oblong, typically one septate.

Sub-section 1. Micronemæ, Sacc.—Threads very short or scarcely distinct.

Genus (1)—
303. Bispora, Corda.

Sub-section 2. Macronemæ, Sacc.—Threads evident and distinct from conidia.

Genera (3)—

304. Fusicladium, Bon.	305. Scolecotrichum, Kunze.	306. Cladosporium, Link.

Section 3. Phragmosporæ, Sacc.—Conidia ovoid, oblong, cylindrical or worm-shaped, two or more septate.

Sub-section 2. Macronemæ, Sacc.—Threads evident, and distinct from conidia.

Genera (4)—

307. Helminthosporium, Link.	309. Cercospora, Fres.	310. Heterosporium, Klotzsch.
308. Brachysporium, Sacc.		

Section 4. Dictyosporæ, Sacc.—Conidia globose, transversely and longitudinally septate, brown.

Sub-section 1. Micronemeæ, Sacc.—Threads very short, or scarcely distinct.

Genus (1)—
311. Sporodesmium, Link.

Sub-section 2. Macronemeæ, Sacc.—Threads evident and distinct from conidia.

Genera (3)—
312. Stemphylium, Wallr. | 313. Macrosporium, Fries. | 314. Fumago, Pers.

ORDER XLVII.—STILBEACEÆ, FRIES.

ARRANGEMENT OF GENERA (6).

Series 1. Hyalostilbeæ, Sacc.—Threads and conidia pallid.

Section 1. Amerosporæ. Sacc.—Conidia globular, ellipsoid or oblong, continuous, transparent or pallid.

Genera (3)—
315. Stilbum, Tode. | 316. Isaria, Pers. | 317. Ceratium, Alb. and Schw

Series 2. Phæostilbeæ, Sacc.—Threads and conidia brown, rigid.

Section 1. Amerosporæ, Sacc.—Conidia globose, oblong or elongated, continuous.

Genus (1)—
318. Harpographium, Sacc.

Section 4. Phragmosporæ, Sacc.—Conidia oblong or cylindrical, two to more septate.

Genera (2)—
319. Podosporium, Schw. | 320. Isariopsis, Fries.

ORDER XLVIII.—TUBERCULARIACEÆ, EHRB.

ARRANGEMENT OF GENERA (13).

Series 1. Tuberculariæ mucedineæ, Sacc.—Threads and conidia white or bright coloured.

Section 1. Amerosporæ, Sacc.—Conidia continuous, ovoid, sigmoid, shortly cylindrical or fusoid.

Genera (6)—
321. Tubercularia, Tode. | 323. Ægerita, Pers. | 325. Hymenula, Fries.
322. Illosporium, Mart. | 324. Fusicolla, Bon. | 326. Thozetia, Berk. and F. v. M.

Section 3. Phragmosporæ, Sacc.—Conidia elongated, fusoid or sickle shaped, typically two or more septate.

Genera (3)—
327. Bactridium, Kunze. | 328. Fusarium, Link. | 329. Microcera, Desm.

Series 2. Tuberculariæ dematiæ, Sacc.—Threads olive or sooty black, conidia same colour or hyaline.

Section 1. Amerosporæ, Sacc.—Conidia continuous, globose, ovoid or elongated.

Genera (4)—
330. Epicoccum, Link. | 332. Myrothecium, Tode. | 333. Actinonema, Sacc.
331. Strumella, Sacc.

Number.	Cooke's Number.	Saccardo's Number.	Scientific Name.	Authority for Name.	English Name.

GROUP VII.—HYPHOMYCETES.—MARTIUS,

ORDER XLV.—MUCEDINACEÆ,

283. OOSPORA.—Wallr., Fl.

1905	1916	X. 7062	O. Aphidis	Cooke and Mass., Grev. XVI. 76 (1888) ...	Aphis oospora ... ...
1906	...	...	O. rutilans	Cooke and Mass., Grev. XXI. 39 (1892)...	Red oospora ...
1907	...	IV. 69	O. vinosella	Sacc., Fung. Ital. 874 (1886)	Vinous oospora

284. MONILIA.—Pers., Syn. 691

| 1908 | ... | IV. 157 | M. fructigena | Pers., Syn. 693 (1801) ... | ... Fruit-growing monilia ... (Brown rot) |

285. OIDIUM.—Link, Berl.

1909	...	IV. 199	O. Chrysanthemi	Rabh., Hedw. I. 19 (1852)	... Chrysanthemum oidium ...
1910	1917	„ 189	O. erysiphoides	Fries, S.M. III. 432 (1829)	... Erysiphe-like oidium ...
1911	1918	.. 190	O. leuconconium	Desm., Ann. Sci. Nat. XIII. 102 (1829) ...	White-dust oidium
1912	1920	X. 7093	O. Lycopersicum ...	... Cooke and Mass., Grev. XVI. 114 (1888)...	Tomato oidium
1913		IV. 219	O. monilioides	Link, Sp. Pl. 122 (1824)	Necklace oidium
1914	1919	„ 191	O. Tuckeri	... Berk., Gard. Chron. 779 (1847) ...	Tucker's oidium (Powdery mildew)

286. TRICHODERMA.—Pers..

| 1915 | 1921 | IV. 284 | T. viride ... | ... Pers., Syn. 230 (1801) ... | Green trichoderma ... ... |

287. ASPERGILLUS.—Adans..

1916	1926	IV. 342	A. Cookei	Sacc. Syll. IV. 71 (1886) ...	Cook's aspergillus ... ...
1917	1922	„ 304	A. glaucus	... Link, Sp. Pl. Fung. I. 67 (1824)	Glaucous aspergillus ...
1918	1924	„ 319	A. Muelleri	... Berk. Linn. Journ. XIII. 175 (1873)	... Mueller's aspergillus ...
1919	1925	„ 326	A. roseus ...	... Link, Sp. Pl. Fung. I. 68 (1824)	... Rose-coloured aspergillus

288. PENICILLIUM.—Link, Berl. Mag. III. 16

| 1920 | | IV. 381 | P. candidum | Link, Obs. Myc. I. 15 (1809) | .. White penicillium |
| 1921 | 1927 | „ 373 | P. glaucum | Link, Obs. Myc. I. 15 (1809) | Glaucous penicillium ... (Common blue mould) |

289. RHINOTRICHUM.—Corda,

1922	1929	IV. 448	R. Carteri	Cooke, Fung. Aust. 60 (1883) ...	. Carter's rhinotrichum ...
1923	1928	., 447	R. microsporum ...	... Berk., Fl. Tasm. II. 272 (1860) ...	Small-spored rhinotrichum
1924	1930	„ 460	R. pulchrum ...	Berk., Linn. Journ. XIII. 175 (1873)	Beautiful rhinotrichum ...
1925	1931	., 469	R. ramosissimum ...	Berk. and Curt., Grev. III. (1875)	Much-branched rhinotrichum ...

290. SPOROTRICHUM.—Link,

| 1926 | 1932 | IV. 507 | S. densum | Link, Obs. Myc. I. 11 (1809) | Dense sporotrichum |

OF AUSTRALIAN FUNGI—*continued.*

Number.	W.A.	S.A.	T.	V.	N.S.W.	Q.	B.	Occurrence.	General Characters.

FL. CRYPT. ERLANG. 334 (1817).

LINK, BERL. MAG. III. 10 (1809).

Crypt. II. 182 (1833).—Torula.

1905						Q		Aphides upon Pumpkin leaves	Threads short, continuous, somewhat tufted, transparent. Conidia in little chains.
1906				V.				Dung	Expanded, crustaceous, red or orange red. Conidia in chains.
1907				V.				With moist *Daldinia concentrica*	Tufted, cushion shaped, dirty yellow to wine colour, minute. Conidia in more or less elongated chains.

(1801).—Oidium, Torula.

| 1908 | | | | V. | | | B. | Apples, Pears, &c. | Compact tufts, cushion shaped, usually running together, downy, whitish then fleshy ochre. Common. |

Mag. III. 18 (1809).

1909				V.			B.	Leaves of Chrysanthemum	Expanded, white. Threads creeping, continuous, transparent. Conidia in long chains.
1910				V.	N.S.W.	Q.	B.	Living leaves of various plants	Broadly expanded, indeterminate, white. Tufts conspicuous, rosy white, threads erect, very slender.
1911				V.	N.S.W.	Q.	B.	Rose leaves, &c.	Tufts expanded, white. Threads creeping, with fertile branches short and erect. Conidial stage of *Sphaerotheca pannosa*.
1912				V.				Stems and leaves of Tomato (*Solanum lycopersicum*)	Tufts expanded, indeterminate, white, spiderweb-like. Threads short, branching, erect.
1913				V.	N.S.W.		B.	Living leaves, &c.	Tufts widely spread, ochrey white. Conidia forming chains like a necklace, dirty white. Conidial stage of *Erysiphe graminis*.
1914		S.A.	T.	V.	N.S.W.	Q.	B.	Vine leaves and grapes	Tufts densely clustered, often running together and forming whitish web-like layer. Conidia barrel shaped.

Obs. Myc. I. 99 (1796).

| 1915 | W.A. | S.A. | T. | V. | N.S.W. | Q. | B. | Bark, wood, leaves, and branches | Tufts nearly circular, cushion shaped, compact, then expanded, first white, then bluish green, afterwards yellowish. Conidial stage of *Hypocrea rufa*. |

Fam. II. 2 (1763).—Mucor, Mouilia.

1916	W.A.	S.A.		V		Q.		Dead plants and leaves	Gregarious, white, intricately interwoven. Fertile threads erect, transparent, crowned with large globose vesicle.
1917	W.A.	S.A.	T.	V.	N.S.W.	Q.	B.	Vegetable substances	Creeping threads fluffy, branched, uncoloured. Fertile threads erect, simple, transparent or glaucous, swelling into spherical vesicle. Conidial stage of *Eurotium herbariorum*.
1918	W.A.			V.				*Lepiota bubalina*, &c.	Snowy white, creeping. Fertile threads erect, rather flexuous.
1919				V.			B.	Soil, damp paper, linen, &c.	Thin, creeping. Fertile threads simple. Conidia globular, rose colour.

(1809).—Aspergillus, Botrytis, Mucor.

| 1920 | | | | V. | | | B. | Leaves and decaying substances | Tufts running together, white. Sterile threads creeping, interwoven white, fertile threads ascending or erect. |
| 1921 | | | | V. | N.S.W. | Q. | B. | Decaying vegetables | Expanded, creeping, white. Sterile threads creeping, intricate. Fertile threads erect, branched at top in a pencil-like manner. Branches erect, once or twice forked. Conidia verdigris green. |

Ic. Fung. I. 17 (1837).

1922					N.S.W.			Wood	White, peziza shaped, rather compact. Threads branched, club shaped, somewhat knotted.
1923			T.					Ground	Threads stuck together, forming nearly cylindrical clubs, with spicules.
1924				V.	N.S.W.			Rotten wood	Forming thin saffron-coloured layer. Threads globosely clavate at top.
1925		S.A.			N.S.W.		B.	Rotten wood	Pale fawn or tan colour. Threads very much branched, ultimate joints elongated and toothed.

Berl. Mag. III. 12 (1809).

| 1926 | | | | | N.S.W. | Q. | | Dead insects, &c. | Threads sparingly branched, transparent, white, densely crowded in a rather thick layer. Allied to *Botrytis Bassiana*, which causes the disease known as "Muscardine" in silkworms. |

Number.	Cooke's Number.	Saccardo's Number.	Scientific Name.	Authority for Name.	English Name.
				291. BOTRYTIS.—_Adans._, Fam. II. 3 (1763).—	
1927	1933	IV. 664	B. vulgaris	... Fries, S.M. III. 398 (1829)	Common botrytis ...
				292. SEPEDONIUM.—Link,	
1928	1935	X. 7206	S. aureo-fulvum	Cooke and Mass., Grev. XVI. 76 (1888) ...	Golden-yellow sepedonium ...
1929	1931	IV. 754	S. chrysospermum...	... _Fries_, S.M. III. 438 (1829)	Golden-spored sepedonium
				293. VERTICILLIUM.—Nees,	
1930	1936	IV. 792	V. eximium	Berk., Linn. Journ. XIII. 475 (1873) ...	Excellent verticillium ...
1931	1938	„ 808	V. lateritium	Berk., in Cooke's Handb. Brit. Fung. 635 (1871) ...	Brick-red verticillium
1932	1937	„ 797	V. niveum	Berk., Fl. Tasm. II. 271 (1860)	Snow-white verticillium
				294. NEMATOGONIUM.—Desm., Ann.	
1933	1939	IV. 867	N. aurantiacum	... Desm., Ann. Sci. Nat. II. 70 (1834)	Orange nematogonium ...
1934	1940	„ 868	N. aureum	... _Sacc._ Syll. IV. 8 (1886)	Golden nematogonium
				295. TRICHOTHECIUM.—Link,	
1935	1941	IV. 881	T. roseum	_Link_, Obs. Myc. I. 16 (1809)	Rosy trichothecium ...

ORDER XLVI.—DEMATIACEÆ, FRIES,

Number.	Cooke's Number.	Saccardo's Number.	Scientific Name.	Authority for Name.	English Name.
				296. CONIOSPORIUM.—Link.	
1936	1942	IV. 1152	C. inquinans	... Dur. and Mont., Fl. Alg. I. 327 (1846)	Black coniosporium
1937	1943	X. 7334	C. pterospermum ...	Cooke and Mass., Grev. XIX. 90 (1891) ...	Wing-spored coniosporium
				297. TORULA.—Pers.,	
1938	1944	IV. 1230	T. herbarum	Link, Sp. Pl. Fung. I. 128 (1824)	Herb-growing torula ... ...
193	1945	X. 7363	T. mycetophila ...	... Cooke and Mass., Grev. XVI. 3 (1887) ...	Fungus-loving torula ...
				298. HORMISCIUM.—Kunze, Myk.	
1940	1947	IV. 1286	H. pithyophilum ...	_Sacc._ Syll. IV. 265 (1886)	Pine-loving hormiscium ...
1941	1946	„ 1283	H. stilbosporum ...	_Sacc._ Syll. IV. 264 (1886)	Stilbum-spored hormiscium ...
				299. HETEROBOTRYS.—Sacc.,	
1942	1948	IV. 1296	H. paradoxa	_Sacc._, Mich. II. 124 (1880)	Paradoxical heterobotrys ...
				300. PERICONIA.—Tode,	
1943	1919	IV. 1329	P. nigrella	_Sacc._ Syll. IV. 274 (1886)	... Black periconia ...

OF AUSTRALIAN FUNGI—*continued*.

Number	Habitat.						B.	Occurrence.	General Characters.
	W.A.	S.A.	T.	V.	N.S.W.	Q.			

Polyactus, Monilia, Peziza.

| 1927 | ... | S.A. | T. | V. | N.S.W. | Q. | B. | Herbs, fruit, flowers, leaves, and branches in decay | Tufts olive grey. Threads fluffy, erect, olive, branched above. Branches shortened, spreading, and branchlets bearing the conglomerate conidia. Mould-like *Botrytis* is only the conidial form of *Peziza*. |

Berl. Mag. III. 18 (1809).—Mucor, Uredo.

| 1928 | ... | ... | ... | V. | ... | ... | | *Polyporus* | Threads creeping, branched. Conidia profuse, globose, forming a golden-tawny powder within decaying *Polypori*. |
| 1929 | W.A. | S.A. | ... | V. | N.S.W. | Q. | B. | *Boletus*, &c. | Threads spread out, then interwoven, rather thick, almost transparent, variously forked, conidia yellow or golden yellow. |

Syst. Pilz. 56 (1816).

1930	...	...	...	...	N.S.W.	...	...	*Clavaria* ...	Threads branched, branches short, thickened at ends, with radiating acute conidia-bearing processes. A beautiful species.
1931	...	...	...	V.		...	B.	Maize ...	Threads elegantly and many times branched in whorls, collected in brick-red velvety or woolly tufts.
1932	...	...	T.	...		...	...	Dead Agarics	White, branched. Branches rather short, thickened at base.

Sci. Nat. II. 69 (1834).—Aspergillus.

| 1933 | ... | ... | ... | ... | | ... | B. | Bark and wood | Tufts velvety, orange tawny, expanded. Sterile threads creeping, thin. Fertile threads erect, swollen at each end. |
| 1934 | ... | ... | ... | ... | | Q. | B. | Bark | Fertile threads erect, short, club shaped, with about four joints. Conidia golden yellow. |

Berl. Mag. III. 18 (1809).—Trichoderma, Puccinia.

| 1935 | ... | ... | ... | V. | | ... | B. | Rotting fruit, branches, leaves, paper, cheese, &c. | Tufts cushion shaped, velvety, rather large ; at first white, then rosy. |

SYST. MYC. III. 335 (1832).

Berl. Mag III. 8 (1809).

| 1936 | W.A. | ... | ... | ... | | ... | | Stems of *Arundo*... | Spread out, very black. Tufts rounded or oblong, run together and irregular. |
| 1937 | ... | ... | ... | V. | | ... | ... | *Lepidosperma* ... | Pustules gregarious, small, bursting through, blackish. Spore body globose, with membranous expansion. |

Ust. Ann. IX. 25 (1795).—Monilia.

| 1938 | ... | ... | ... | ... | | Q. | B. | Rotting herb stems | Tufts expanded, olive to ochrey, then becoming black, somewhat velvety. Sterile threads creeping, sooty. Fertile threads erect, olive, then black. |
| 1939 | ... | ... | ... | V. | | ... | ... | Pileus or cap of *Polyporus cinnabarinus* | Tufts minute, very thin, scattered, black. Threads sparingly branched, nearly straight. |

Hoft. I. 12 (1817).—Antennaria, Torula.

| 1940 | ... | S.A. | ... | ... | N.S.W. | ... | B. | Branches and leaves of *Coniferæ* and *Eucalyptus* | Expanded, thick, superficial, quite black. Chains of conidia indistinctly branched, branches tapering towards apex and slightly curved. |
| 1941 | ... | ... | ... | ... | | Q. | B. | Branches | Tufts bursting through, powdery, run together, quite black. Chains unequal, branched, or simple. |

Mich. II. 21 (1880).

| 1942 | ... | S.A. | ... | ... | | ... | ... | Upper surface of leaves of *Bertya rotundifolia* | Tufted, gregarious, black. Threads straggling, creeping, pale, sooty. Conidia brown. |

Fung. Meck. I. 2 (1790).—Sporocybe.

| 1943 | ... | ... | ... | ... | | Q. | B. | Leaves of *Andropogon* | Very minute, black. Fertile threads simple, thin. Head globose or ellipsoid. |

Number.	Cooke's Number.	Saccardo's Number.	Scientific Name.	Authority for Name.	English Name.
					301. Monotospora.—Corda,
1914			M. fasciculata	Cooke and Mass., Grev. XXI. 39 (1892)	Fasciculate monotospora
					302. Botryotrichum.—Sacc. and
1945	...		B. Lachnella	Sacc., Hedw. 58 (1893)	Lachnella botryotrichum
					303. Bispora.—Corda,
1946	1950	IV. 1632	B. monilioides	Corda, Ic. Fung. I. 9 (1837)	Necklace bispora
					304. Fusicladium —Bon. Handb. 80
1947	1951	IV. 1642	F. dendriticum	Fckl., S.M. 357 (1875)...	Tree-like fusicladium (Black spot of Apple)
1948	1952	„ 1643	F. pyrinum	Fckl., S.M. 357 (1875)...	Pear fusicladium
					305. Scolecotrichum.—Kunze and
1949	1953	X. 7478	S. atrichum	Cooke and Mass., Grev. XVI. 3 (1887)	Black scolecotrichum
1950	...	IV. 1656	S. graminis, var. Avenæ	Erikss., Zeit. Pflkrk. I. 28 (1891)	Oat scolecotrichum
					306. Cladosporium.—Link,
1951	1957	IV. 1698	C. Asteroma	Fckl., S.M. 355 (1875)...	Asteroma cladosporium
1951A	„	...	C. Asteroma, var. minor	Cooke, Handb. Austr. Fung. 376 (1892)	Lesser cladosporium
1952	1959	IV. 1718	C. epiphyllum	Mart., Erl. 351 (1817)...	Epiphyllous cladosporium
1953	1954	IV. 1665	C. herbarum	Link, Obs. Myc. II. 37 (1809)	Herb-growing cladosporium
1953A	„	...	C. herbarum, var. epixylinum	Corda	Wood-growing cladosporium
1954	1958	IV. 1714	C. hypophyllum	Fckl., S.M. 356 (1875)...	Hypophyllous cladosporium
1955	1955	„ 1669	C. oligocarpum	Corda, Ic. Fung. I. 14 (1837)	Few-spored cladosporium
1956	1961	„ 1774	C. papyricolum	Berk. and Br., Linn. Trans. II. 68 (1883)...	Paper-growing cladosporium
1957	1956	„ 1670	C. stenosporum	Berk. and Curt., Grev. III. (1875)	Slender-spored cladosporium
1958	1960	„ 1750	C. Typharum	Desm., Exs. 304	Typha cladosporium
					307. Helminthosporium.—Link,
1959	1966	IV. 1969	H. inconspicuum	Cooke and Ell., Grev. VI. 88 (1878)	Inconspicuous helminthosporium
1960	1963	„ 1973	H. macrocarpum	Grev., Scot. III. 148 (1825)	Large-spored helminthosporium
1961	1965	„ 1966	H. puccinioides	Sacc. and Berl., Rev. Myc. (1885)	Puccinia-like helminthosporium
1962	1962	„ 1971	H. Ravenelii	Curt., in Sill. Journ. 352 (1848)	Ravenel's helminthosporium
1963	1964	„ 2010	H. rhabdiferum	Berk. and Br., Ann. Nat. Hist. XV. 402 (1865)	Rod-bearing helminthosporium
					308. Brachysporium.—Sacc.,
1964	1967	IV. 2039	B. oligocarpum	Sacc. Syll. IV. 424 (1886)	Few-spored brachysporium

OF AUSTRALIAN FUNGI—*continued.*

Number.	Habitat.						B.	Occurrence.	General Characters.
	W.A.	S.A.	T.	V.	N.S.W.	Q.			

Ic. Fung. I. 11 (1837).

| 1944 | ... | ... | ... | V. | | ... | ... | Bark | Tufts bursting through, gregarious, black. Fertile threads erect, densely fasciculate in awl-shaped tufts. |

March., Champ. Copr. Belg. 34 (1885).

| 1945 | ... | ... | ... | V. | ... | ... | ... | Branches and spines of *Bursaria spinosa*, not yet dead | Gregarious, umber to dusky. Tufts of threads in circular bundles, like *Lachnella*. Sterile threads erect. Conidia spherical, rather transparent, granular. |

Ic. Fung. I. 9 (1837).

| 1946 | ... | ... | ... | ... | N.S.W. | Q. | B. | Wood | Expanded, dark brown, powdery. Threads short, rather conical. Conidia sooty brown. Conidial stage of *Bispurella monilifera.* |

(1851).—Cladosporium, Helminthosporium.

| 1947 | ... | S.A. | T. | V. | N.S.W. | Q. | B. | Leaves and fruit of Apple and Pear | Expanded, velvety, olive, often tree-like on leaves. Threads filiform, erect, in bundles. Conidia olive. |
| 1948 | ... | S.A. | T. | V. | N.S.W. | Q. | B. | Leaves and fruit of Pear | Expanded, velvety, olive. Threads short, tapering, toothed at apex. Conidia olive. |

Schm., Myk. Heft. I. 10 (1817).

| 1949 | ... | ... | ... | ... | ... | Q. | ... | Twigs of *Passiflora* | Tufts spread out, run together, black. Threads erect, simple, brown. Conidia dark brown. |
| 1950 | ... | ... | ... | V. | ... | ... | Oats | ... | Spots on leaves elongated, ochrey, drying up. Threads in bundles, point-like, densely clustered. Conidia olive to brown. |

Berl. Mag. VII. 37 (1816).—Dematium.

1951	...	...	...	...	...	...	...	...
1951A	...	...	...	...	...	Q.	...	Foliage of Grape Vine	Tufts in centre of brown spot, disposed in a tree like manner, minute, yellow, becoming greenish.
1952	...	...	...	V.	...	...	B.	Leaves of Oak, &c.	Tufts arranged in a circle, olive, then blackish, large, thick. Threads at first erect, then declining.
1953	W.A.	S.A.	T.	V.	N.S.W.	Q.	B.	Stems, leaves, &c. ...	Tufts densely clustered, run together, forming a velvety yellow-olive then dark-olive layer.
1953A	...	...	...	...		Q.		Wood.	
1954	...	...	...	...		Q.		Lower surface of leaves of *Serjania*	Tufts spread out, then greyish green. Threads branched, flexuous, yellow.
1955	...	...	...	...		Q.	...	Wood	...	Tufts minute, solitary, black. Threads tufted, erect, long, slender, slightly branched, olive.
1956	...	...	...	...		Q.	...	On paper forming dark-grey layer	Threads irregularly branched, transparent above, brown below. Conidia pale brown.
1957	...	...	...	...		Q.	...	Leaves of Pear, &c.	Threads simple, thin, divided, arising from a creeping mycelium, greyish brown below.
1958	...	...	...	V.		...	...	Leaves of *Typha* (Bulrush)	Tufts elongated or oblong, scattered, turning black, seated at first on distinct greyish spot.

Berl. Mag. III. 10 (1809).—Macrosporium.

1959	...	...	...	...		Q.	B.	Failing leaves of Maize	Thin cloud-like stain. Threads elongated, septate knotted, pale brown.
1960	...	...	...	...		Q.	B.	Trunks and branches	Expanded, velvety, dark olive or sooty brown. Threads clustered, lax, simple or sparingly branched.
1961	...					Q.	...	Failing or dead leaves of *Tristania laurina*	Tufts on both surfaces, very black, loosely gregarious, disc shaped, compact, resembling *Puccinia.* Conidia pale sooty brown.
1962	...					Q.	...	Inflorescence of grasses (*Sporobolus indicus*)	Spongy. Threads flaccid, flexuous, knotted, branched. Conidia brown.
1963		S.A.				...	B.	Ripe peaches	...	Expanded, internally black. Threads erect, sparingly branched. Conidia straight, at first oblong, then elongated and somewhat linear, dark brown.

Mich II. 28 (1880).—Helminthosporium.

| 1964 | ... | ... | ... | ... | | Q. | ... | Wood | Tufts minute, linear, nearly parallel. Threads flexuous, simple, in bundles, dark brown. Conidia yellow brown. |

Number.	Cooke's Number.	Saccardo's Number.	Scientific Name.	Authority for Name.	English Name
					309. CERCOSPORA.—Fres., Beitr. 90
1965	1971	X. 7685	C. Daviesiæ	Cooke and Mass., Grev. XVIII. 7 (1889)	... Daviesia cercospora ... ...
1966	1968	„ 7696	C. epicoccoides	Cooke and Mass., Grev. XIX. 91 (1891) ...	... Epicoccum-like cercospora ...
1967	1973	„ 7697	C. Eucalypti	... Cooke and Mass., Grev. XVIII. 7 (1889) ...	... Eucalyptus cercospora
1968	...	...	C. Glycines	... Cooke and Mass., Grev. XXI. 39 (1892) ...	... Glycine cercospora
1969	1969	X. 7678	C. Kennedyæ	... Cooke and Mass., Grev. XIX. 90 (1891) ...	... Kennedya cercospora ...
1970	1972	IV. 2161	C. Solanacea	... Sacc. and Berl., Rev. Myc. (1885)	... Solanum cercospora ...
1971	1970	„ 2200	C. viticola	... Sacc. Syll. IV. 458 (1886)	... Vine cercospora ... ... (Tufted leaf blight)
					310. HETEROSPORIUM.—Klotzsch.
1972	1974	X. 7769	H. epimyces	... Cooke and Mass., Grev. XVI. 80 (1888) ...	... Fungus heterosporium ...
					311. SPORODESMIUM.—Link, Berl
1973	1975	IV. 2391	S. atrofuscum	... Cooke, Grev. XII. 12 (1863)	... Blackish-brown sporodesmium ...
1974	1976	„ 2356	S. melanopodum	... Berk. and Br., Ann. Nat. Hist. V. 459 (1850)	... Black-stalked sporodesmium ...
					312. STEMPHYLIUM.—Wallr., Fl.
1975	1977	IV. 2487	S. pulchrum	... Sacc. Syll. IV. 521 (1886)	Beautiful stemphylium ...
					313. MACROSPORIUM.—Fries,
1976	1980	X. 7837	M. Camelliæ	... Cooke and Mass., Grev. XVII. 42 (1888) ...	... Camellia macrosporium ...
1977	1978	IV. 2501	M. cladosporioides	... Desm., Pl. Crypt. 3 (1857)	... Cladosporium-like macrosporium
1978	1979	„ 2499	M. commune	... Rabh., Fung. Eur. Exs. 1360 ...	... Common macrosporium
1979	...	X. 7853	M. graminum	... Cooke, Grev. XVII. 66 (1889) ...	... Grass macrosporium ...
1980	1982	IV. 2519	M. peponicolum	... Rabh., in Sitz. 101 (1867)	... Gourd-growing macrosporium ...
1981	1981	X. 7841	M. Readeri	... Winter. Rev. Myc. 212 (1886) ...	Reader's macrosporium ...
1982	1980 bis.	IV. 2552	M. Tomato	Cooke, Grev. XII. 32 (1883)	... Tomato macrosporium
					314. FUMAGO.—Pers., Myc.
1983	1983	IV. 2618	F. vagans	Pers., Myc. Eur. I. 9 (1822)	... Creeping fumago ...
					ORDER XLVII.—STILBEACEÆ,
					315. STILBUM.—Tode,
1984		IV. 2714	S. aurantiacum	Bab., Linn. Trans. (1839)	Orange stilbum ... ...
1985	1986	X. 7894	S. caninum	... Cooke and Mass., Grev. XX. 36 (1891) ...	... Dog's dung stilbum
1986	1988	IV. 2705	S. cinnabarinum	... Mont., Fl. Cub. 308 (1842) ...	Vermilion stilbum ...
1987	1987	X. 7879	S. corallinum	... Cooke and Mass., Grev. XIX. 91 (1891) ...	... Coralline stilbum
1988	1984	IV. 2680	S. erythrocephalum	... Ditm., Sturm. Fl. III. (1817) ...	... Red-headed stilbum ...
1989	1985	X. 7893	S. Formicarum	Cooke and Mass., Grev. XVIII. 8 (1889) ...	... Ant stilbum ...

OF AUSTRALIAN FUNGI—*continued.*

Number.	Habitat.						B.	Occurrence.	General Characters.
	W.A.	S.A.	T.	V.	N.S.W.	Q.			

(1850).—Cladosporium, Helminthosporium.

1965	...	...	...	V.	...	...	...		Fading leaves of *Davesia latifolia*	Spots brown, irregular, angular. Threads in bundles, shortened. Conidia bent like a bow, pale brown.
1966	...			V.		...	...		*Eucalyptus* leaves ...	Spots small or run together, purple. Tufts gregarious, rather compact, black. Conidia pale olive.
1967	...		...	V.		...	...		Fading *Eucalyptus* leaves	Spots rather circular, or run together, pale, with rosy margin. Threads shortened. Conidia curved, pale.
1968	...		...	V.		...	...		Living leaves of *Glycine clandestina*	On both surfaces, but chiefly upper, spots definite, irregular, angular, amber. Tufts gregarious on the spots, point-like, black. Conidia almost transparent.
1969	...	...		V.	...	...	...		Leaves of *Kennedya prostrata*	Spots cinnamon brown, irregular and run together. Tufts scattered, black, point-like. Conidia clear olive.
1970	...	...		...	...	Q.	...		Leaves of *Solanum verbascifolium*	Spots nearly circular, brown. Tufts point-like, sooty olive. Conidia rod-like, curved, almost hyaline.
1971	...	...		...	N.S.W.	Q.	...		Vine leaves	Spots on both surfaces, somewhat circular or irregular, when dry ochrey. Threads often on under surface, here and there densely tufted. Conidia ochrey olive.

Herb. Myc. I. 67 (1832).

| 1972 | ... | ... | ... | V. | ... | ... | B. | Decayed Agarics ... | Tufts olive, spread out in more or less dense velvety patches. Threads sparingly forked, pale brown. Conidia minutely warted, pale olive. |

Mag. III. 41 (1809).—Spiloma.

| 1973 | ... | ... | ... | V. | ... | ... | ... | Wood | Expanded, velvety black. Conidia elongated, club shaped, divided in all directions into quadrate cells, dark brown. |
| 1974 | ... | ... | ... | V. | ... | ... | B. | Bark | Tufts broad, black. Conidia nearly globose, opaque, seated on base of variable size. |

Crypt. II. 300 (1833).—Mystrosporium.

| 1975 | W.A. | ... | ... | ... | ... | ... | ... | Rotten wood | Expanded, olive. Threads white, wrinkled, forked or trifid. Conidia a little rough, blackish. |

S.M. III. 373 (1832).

1976	...	...	...	V.	...	...	B.	Living leaves of *Camellia*	Spots circular or run together, pale, with broad brown margin. Threads tufted, pale olive.
1977	...		...	V.	...	...	B.	Leaves and stems of herbs	Spots large, tawny, irregular. Tufts numerous, minute, rounded, velvety, dark olive.
1978	...		...		...	Q.	B.	Stems, leaves, &c. ...	Tufts densely clustered, numerous, brown. Threads in bundles, ascending, brown. Conidia olive. Considered to be conidial condition of *Pleospora herbarum*.
1979	...		...	...	N.S.W.	...	...	Wheat, Sugar cane, &c.	Expanded, very thin. Threads creeping, at length with erect branches, greyish brown. Conidia same colour.
1980	...		...	...		Q.	...	Papaw fruit	Spots large, circular, black. Sterile threads, slender, creeping. Fertile threads, short, erect. Conidia amber brown.
1981	...		...	V.			...	Dry stems of Artichoke	Tufts forming expanded black layer. Threads in minute bundles, erect, brown. Conidia brown.
1982	...		...	V.		Q.	B.	Ripe Tomatoes ...	Circular, black. Threads short, robust, flexuose. Conidia brown.

Eur. I. 9 (1822).—Cladosporium, Torula.

| 1983 | ... | | ... | ... | | Q. | B. | Living leaves of Vine, &c. | Threads creeping, branched in a straggling manner, olive or sooty brown, forming a thin membranous black layer. Conidia in short chains. |

FRIES, MICH. II. 31 (1880).

Mœckl. I. 10 (1790).—Sphærostilbe.

1984	...	...	...		...	Q.	B.	Dead branches ...	Somewhat fasciculate, orange coloured. Stem even, darker downwards. Head somewhat club shaped.
1985	...		...	V.		...	...	Dog's dung	Gregarious, flesh coloured. Head darker, continuous with smooth stem, which is sometimes forked. Conidia transparent.
1986	...		...			Q.	...	Bark	Gregarious, flesh coloured. Head convex to hemispherical. Stem short, mealy.
1987	...		...	V.		...	...	Bark	Tufted, flesh coloured. Stems tapering upwards, shortly branched, mealy. Head orange red.
1988	W.A.		...	V.		...	B.	Dung	Gregarious or somewhat scattered. Stem rather thick, terminated by rosy or red, globose, mealy head.
1989	...	...	...	V.		...	...	Dead ants (*Formica*)	Stems elongated, slender, black, flexuous, slightly thickened below. Head inversely egg shaped, rosy.

M

Number.	Cooke's Number.	Saccardo's Number.	Scientific Name.	Authority for Name.	English Name.
					316. ISARIA.—Pers., Tent.
1090	1997	IV. 2851	I. arbuscula ...	Bres. and Roum., Rev. Myc. 38 (1890) ...	... Dendritic isaria ... ...
1991	1991	„ 2841	I. Cicadæ ...	Miq., Ann. Sci. Nat. X. 379 (1838) ...	... Cicada isaria ... ... (Brazilian Cicada clubs)
1992	1993	„ 2642 and 2439	I. graminiperda (including I. fuciformis, Berk.)	Berk. and F. v. M.. Gard. Chron. 596 (1873)	Grass-destroying isaria
1993	1992	IV. 2828	I. radians ...	Berk., Fl. Tasm. II. 271 (1860) ...	... Radiating isaria
1994	1995	X. 7921	I. suffruticosa ..	Cooke and Mass., Grev. XVIII. 45 (1890) ...	... Shrubby isaria
1995	1991	IV. 2807	I. umbrina ...	Pers., Syn. 689 (1801)	... Umber isaria ...
					317. CERATIUM.—Alb. and Schw., Consp.
1996	1996	IV. 2845	C. hydnoides	.Alb. and Schw., Consp. Fung. Lus. 358 (1805) ...	... Hydnum-like ceratium ...
					318. HARPOGRAPHIUM.—
1997	1998	X. 7919	H. corynelioides ...	... Cooke and Mass., Grev. XVI. 76 (1888) ...	... Corynelia-like harpographium ...
1998	1999	„ 7918	H. quaternarium ...	Cooke and Mass., Grev. XVI. 3 (1887) ...	... Quaternate harpographium
					319. PODOSPORIUM.—Schw.,
1999	2000	IV. 2982	P. grande	Cooke, Grev. XII. 11 (1883)	... Large podosporium ... ...
					320. ISARIOPSIS.—Fries, in Sacc.
2000	2001	IV. 2998	I. clavispora	... Sacc. Syll. IV. 631 (1886)	... Clavate-spored isariopsis ...

ORDER XLVIII.—TUBERCULARIACEÆ,

Number.	Cooke's Number.	Saccardo's Number.	Scientific Name.	Authority for Name.	English Name.
					321. TUBERCULARIA.—Tode, Fung.
2001	2002	X. 7990	T. leguminum	... Cooke and Mass., Grev. XVI. 33 (1887) ...	... Legume-growing tubercularia ...
2002	...	IV. 3002	T. vulgaris	... Tode, Fung. Meck. I. 18 (1790)	... Common tubercularia ...
					322. ILLOSPORIUM.—Mart.,
2003	2004	...	I. flavellum	... Berk. and Br., Linn. Trans. II. 68 (1853)	... Yellow illosporium ... ...
2004	2006	IV. 3106	I. flaveolum	... Sacc., Mich. II. 297 (1880)	... Yellowish illosporium ... ...
2005	2005	X. 8019	I. obscurum	... Cooke and Mass., Grev. XVI. 113 (1888)	... Obscure illosporium ...
					323. ÆGERITA.—Pers., Tent.
2006	2007	IV. 3124	A. candida	Pers., Syn. 684 (1801) ...	... White ægerita ...
					324. FUSICOLLA.—Bon.,
2007	2008	X. 8024	F. incarnata	... Cooke and Mass., Grev. XVII. 8 (1888) ...	... Flesh-coloured fusicolla ...
					325. HYMENULA.—Fries.
2008			H. Eucalypti	Cooke and Mass., Grev. XXI. 39 (1892) ...	Eucalypt hymenula ...
					326. THOZETIA.—Berk. and F. v. M.,
2009	2009	IV. 3213	T. nivea ...	... Berk., Linn. Journ. XVIII., 388 (1881) ...	... Snow-white thozetia ... ...

OF AUSTRALIAN FUNGI—*continued.*

Number.	W.A.	S.A.	T.	V.	N.S.W.	Q.	B.	Occurrence.	General Characters.

Disp. 41 (1797).—Ceratium.

1990	...	...	...	...	...	Q.	...	Rotten wood	Snowy white. Conidia-bearing layer tree-like, branched in tufts from simple base.
1991	...	...	V.		...	...		Head of Cicada ...	Within and growing through joints of dead *Cicada*. Conidia-bearing layer hard and compact, with shortened stem.
1992	S.A.	...	V.	N.S.W.	Q	B.		Grasses (*Lolium*) and germinating cereals	Bright orange, gelatinous, slender, sparingly branched. Conidia minute, globose.
1993	...	T.	...	...	...	...		Bark	Greyish fawn, circular, branched, covered with whorled or forked woolly tufts.
1994	...	...	N.S.W.		...	...		Hairy caterpillar ...	Tufted, white. Stem smooth or slightly mealy, with slender branches interwoven with lateral branchlets. Conidia minute, ellipsoid.
1995	...	...	V.		...	B.		Wood, and *Hypoxylon coccineum*, of which it is conidial form	Clubs without stem, fawn colour, in radiating tufts. Conidia inversely egg shaped, umber.

Fung. Lus. 358 (1805).—Isaria, Tremella.

| 1996 | ... | ... | ... | ... | N.S.W. | Q. | B. | Rotten wood ... | Conidia-bearing layer tapering, simple or sparingly branched, white or yellowish. Conidia ovoid or globose, transparent. |

Sacc., Mich. II. 33 (1880).

| 1997 | ... | ... | ... | ... | V. | | ... | Branches of *Leptospermum scoparium* | Tufted, bursting through, black. Stems composite, radiating, club shaped above, simple or forked. Conidia, curved, transparent. |
| 1998 | ... | ... | ... | ... | ... | Q. | ... | Dead twigs of *Passiflora* | Tufts black, minute. Stems composite, club shaped above. Conidia spindle shaped, transparent, for the most part quarternate. |

Trans. Amer. Phil. Soc. IV. (1832).

| 1999 | ... | ... | ... | V. | ... | | ... | Stems of *Aster argophylla* | Large, black, woolly, forming dense tufts. Threads erect, crowded together, dark brown. |

Mich. II. 33 (1880).—Graphium.

| 2000 | ... | ... | ... | ... | ... | Q. | ... | Vine leaves ... | Minute, olive, arising from circular brown spots. Threads relaxed above and flexuous. |

EHRB. SYLV. MYC. 12 (1818).

Meek. I. 18 (1719).—Tremella, Sphæria.

| 2001 | ... | ... | ... | ... | ... | Q. | ... | Legumes of *Cassia* | Minute, bursting through, flesh colour. Conidia bearers short, straight. |
| 2002 | ... | ... | ... | ... | ... | | B. | Branches... ... | Gregarious, bursting through, vermilion coloured, globular to depressed, more or less shortly stalked. |

Fl. Crypt. Erl. 325 (1817).

2003	...	...	...	...	...	Q.	...	Lichens	Stalked, yellow. Stem short. Conidia globose.
2004	...	...	...	...	...	Q	...	Rotten wood ...	Very minute, gregarious, yellow. Threads branched in a forked manner, twisted.
2005	...	...	...	V.	...		...	Leaves of *Eucalyptus globulus*	Somewhat gregarious, circular, bursting through, minute, sooty brown. Threads branched in a forked manner.

Disp. 40 (1797).—Tubercularia, Sclerotinia.

| 2006 | ... | ... | ... | V. | | ... | B. | Wood and bark ... | Crowded, granule-like, globose to hemispherical, size of poppy or turnip seed, white but yellowish when dry. |

Handb. 150 (1851).

| 2007 | ... | ... | ... | ... | | Q | ... | Dead coriaceous leaves | Pustules small, gregarious, seated on paler spots, rosy flesh colour, somewhat gelatinous or scattered over leaf, stalks, and midribs. |

Pl. Homon. 94 (1825).

| 2008 | ... | ... | ... | V. | | ... | ... | *Eucalyptus* leaves... | On both surfaces; pustules bursting through, disc-like, brownish. Conidia bearers simple, rather thick. |

Linn. Journ. XVIII. 388 (1881).

| 2009 | ... | ... | ... | ... | | Q. | ... | Rotten wood ... | White. Conidia oblong, transparent, acute at each extremity, and terminated by long bristle. |

M 2

Number.	Conic's Number.	Saccardo's Number.	Scientific Name.	Authority for Name.	English Name.
					327. BACTRIDIUM.—Kunze,
2010	2010	IV. 3268	B. flavum	... Kunze and Schw., Myk. Heft. I. 5 (1817)	... Orange bactridium ... ...
2011	2011	„ 3273	B. magnum	... Cooke, Grev. VIII 60 (1879) ...	Great bactridium
					328. FUSARIUM.—Link,
2012	2016	X. 8105	F. hypocreoideum	... Cooke and Mass., Grev. XVI. 76 (1888) ...	... Hypocrea-like fusarium ...
2013	2012	IV. 3283	F. lateritium	... Nees. Syst. 31 (1816)	... Brick-red fusarium ...
2014	2015	X. 8074	F. longisporum	... Cooke and Mass.. Grev. XVI. 4 (1887) ...	... Long-spored fusarium
2015	2013		F. rubicolor	... Berk. and Br., Linn. Trans. 11. 68 (1883)...	... Ruby-coloured fusarium
					329. MICROCERA.—Desm.,
2016'	2017	IV. 3473	M. coccophila	... Desm., Ann. Sci. Nat. X. 359 (1848)	... Coccus-loving microcera ...
2017	2018	X. 8119	M. rectispora	... Cooke and Mass., Grev. XVI. 4 (1887) ...	... Straight-spored microcera
					330. EPICOCCUM.—Link,
2018	2019	IV. 3491	E. scabrum	... Corda, Ic. Fung. 1. 5 (1837)	... Rough epicoccum ... ... (False potato disease)
					331. STRUMELLA.—Sacc.,
2019	2020	X. 8127	S. hysterioides	... Cooke and Mass., Grev. XVIII. 69 (1889)	... Hysterium-like strumella ...
2020	2022	„ 8128	S. patelloidea	Cooke and Mass., Grev. XX. 7 (1891)	... Patelloid strumella
2021	2021	„ 8130	S. Sacchari	Cooke, Grev. XIX. 45 (1890) ...	... Sugar-cane strumella ... (Cane spume)
					332. MYROTHECIUM.—Tode,
2022	2025	IV. 3552	M. inundatum	Tode, Meck. 1. 25 (1790)	inundated myrothecium ...
2023	2024	„ 3550	M. roridum	Tode, Meck. 1. 25 (1790)	... Bedewed myrothecium ...
					333. ACTINOMMA.—Sacc.,
2024	2023	IV. 3564	A. Gastonis	Sacc., Misc. Myc. 1. 28 (1884)	... Gaston's actinomma ... ...

OF AUSTRALIAN FUNGI—*continued*.

Number.	Habitat.						B.	Occurrence.	General Characters.
	W.A.	S.A.	T.	V.	N.S.W.	Q.			

Myk. Hoft. I. 5 (1817).—Tremella.

| 2010 | ... | | ... | ... | | Q. | D. | Rotten wood | Tubercles globose to hemispherical, beautiful orange, rather large. Conidia honey coloured. |
| 2011 | ... | ... | ... | V. | | | | Bare wood | Cushion shaped, somewhat hemispherical or irregular, pale. Conidia club shaped, transparent, large, long. |

Berl. Mag. III. 10 (1809).—Fusisporium, Selenosporium.

2012	...	...	...	...	...	Q.	...	Fading leaves of *Ficus aspera*	Convex, cushion shaped, resembling *Hypocrea*, somewhat disc shaped, orange.
2013	W.A.						B.	Branches	Pustules various, bursting through, brick red. Conidia bow shaped, tapering to each end.
2014	...					Q.	...	Twigs of *Passiflora*	Tufts bursting through, convex, rosy, then whitish. Threads repeatedly forked, transparent.
2015	...	...	...			Q.	...	*Eucalyptus* leaves, spreading over galls	Expanded, greyish flesh colour. Conidia elongated. They spread over the leaves and colour the veins with a tint like that of raspberry cream.

Ann. Sci. Nat. X. 359 (1848).

| 2016 | ... | | | | | Q. | B. | *Cocci* attached to branches | Small, rather tufted, horn-like to conical, rosy, girt at base by thin whitish membrane. Conidial stage of *Sphærostilbe*. |
| 2017 | | | | | | Q. | ... | Coccus of Orange—scale insect (*Chionaspis citri*) | Tufts rather spherical, almost sessile, reddish at first, then pale. Conidia elongated, spindle shape. |

Berl. Mag. VII. 32 (1816).

| 2018 | | ... | ... | ... | | Q. | ... | Leaves and stems of Potato | Gregarious, no spots. Conidia-bearing layer somewhat globose, fleshy, brownish. Conidia brown, rough. |

Mich. II. 36 (1880).

2019	...		...			Q.	...	Decorticated branches	Pustules gregarious, bursting through, rather prominent, resembling *Hysterium*, black. Conidia olive.
2020	...	T.						Naked wood	Conidia-bearing layer circular, plate-like, scattered, superficial, black. Conidia dark olive.
2021	...				N.S.W.	Q.	...	Sugar cane, stalk and leaf	Pustules gregarious, bursting through, black, with short stem-like base. Conidia continuous, dusky.

Meck. I. 25 (1790).

| 2022 | ... | ... | | | | Q. | B. | Putrid Agarics, &c. | Pustules disc shaped or variable, dark olive with a white margin. Conidia olive. |
| 2023 | | | | | | Q. | B. | Old twine | Pustules disc shaped, then run together and distorted, black with a white margin. Conidia pale olive. |

Misc. Myc. I. 28 (1884).

| 2024 | ... | ... | | | | Q. | ... | Phyllodes of fading *Acacia* | Gregarious, superficial, black, flattened, contracted when dry, star shaped. Conidia pale brown. |

GENERAL CLASSIFICATION OF SPHÆROPSIDES.

GROUP VIII.—SPHÆROPSIDES, LEV.

ARRANGEMENT OF ORDERS (5).

49. SPHÆRIOIDACEÆ—Receptacles black, never fleshy nor brightly coloured, entire.
50. NECTRIOIDACEÆ—Receptacles brightly coloured, fleshy or waxy.
51. LEPTOSTROMACEÆ—Receptacles more or less distinctly semicircular.
52. EXCIPULACEÆ—Receptacles cup shaped, saucer shaped, or Hysterium-like.
53. MELANCONIACEÆ—Receptacles absent.

ORDER XLIX.—SPHÆRIOIDACEÆ, Sacc.

ARRANGEMENT OF GENERA (24).

Section 1. Hyalosporæ, Sacc.—Spores hyaline.

Genera (7)—

334. Phyllosticta, Pers.	337. Asteromella, Pass. an l Thuem.	339. Dothiorella, Sacc.
335. Phoma, Fries.	338. Chætophoma, Cooke.	340. Cytospora, Ehr.
336. Aposphæria, Berk.		

Section 2. Phæosporæ, Sacc.—Spores olive or sooty brown.

Genera (4)—

341. Sphæropsis, Lev.	343. Capnodiastrum, Speg.	344. Chætomella, Fckl.
342. Coniothyrium, Corda.		

Section 3. Phæodidymæ, Sacc.—Spores uni-septate, brown.

Genus (1)—
345. Diplodia, Fries.

Section 4. Hyalodidymæ, Sacc.—Spores uni-septate, hyaline or green.

Genera (5)—

346. Ascochyta, Lib.	348. Actinonema, Fries.	350. Diplodina, West.
347. Robillarda, Sacc.	349. Darluca, Cast.	

Section 5. Phragmosporæ, Sacc.—Spores two or many septate, brown.

Genera (2)—

351. Hendersonia, Berk.	352. Stagonospora, Sacc.

Section 6. Dictyosporæ, Sacc.—Spores two or many septate, wall-like, coloured.

Genus (1)—
353. Camarosporium, Schulz.

Section 7. Scolecosporæ, Sacc.—Spores rod shaped, thread-like or elongated, spindle shape, continuous or septate, hyaline or green.

Genera (4)—

354. Septoria, Fries.	356. Phlyctæna. Mont. and Desm.	357. Gamospora, Sacc.
355. Phleospora, Wallr.		

167

ORDER L.—NECTRIOIDACEÆ, Sacc.

ARRANGEMENT OF GENERA (3).

Sub-division 1. Zythieæ, Sacc.—Receptacles nearly globose, Sphæria-like.

Section 1. Hyalosporæ, Sacc.—Spores globose, ovoid or oblong, continuous, hyaline.
Genera (2)—
358. Sphæronæmella, Karst. | 359. Aschersonia, Mont.

Section 2. Scolecosporæ, Sacc.—Spores thread-like or rod shaped, continuous or many septate, hyaline.
Genus (1)—
360. Martinella, Cooke and Mass.

ORDER LI.—LEPTOSTROMACEÆ, Rchb.

ARRANGEMENT OF GENERA (7).

Section 1. Hyalosporæ, Sacc.—Spores globose, ellipsoid or oblong, continuous, hyaline.
Genera (5)—
361. Leptothyrium, Kunze. | 363. Melasmia, Lev. | 365. Sacidium, Nees.
362. Piggotia, Berk. and Br. | 364. Actinothecium, Ces.

Section 2. Scolecosporæ, Sacc.—Spores thread-like or rod shaped, continuous or septate, hyaline.
Genera (2)—
366. Melophia, Sacc. | 367. Leptostromella, Sacc.

ORDER LII.—EXCIPULACEÆ, Corda.

ARRANGEMENT OF GENERA (2).

Section 1. Hyalosporæ, Sacc.—Spores globose, ellipsoid or oblong.
Genus (1)—
368. Dinemasporium, Lev.

Section 2. Scolecosporæ, Sacc.—Spores filiform, elongated.
Genus (1)—
369. Protostegia, Cooke.

ORDER LIII.—MELANCONIACEÆ, Corda.

ARRANGEMENT OF GENERA (7).

Section 1. Hyalosporæ, Sacc.—Conidia globose, ovoid or oblong, continuous.
Genera (2)—
370. Glœosporium, Desm. and Mont. | 371. Pestalozziella, Sacc. and Ell.

Section 2. Didymosporæ, Sacc.—Conidia ovoid or oblong, uniseptate.
Genus (1)—
372. Marsonia, Fisch.

Section 3. Phragmosporæ, Sacc.—Conidia oblong or cylindrical, two to many septate.
Genera (4)—
373. Stilbospora, Pers. | 375. Hyaloceras, Dur. and Mont. | 376. Pestalozzia, De Not.
374. Coryneum, Nees.

Number.	Cooke's Number.	Saccardo's Number.	Scientific Name.	Authority for Name.	English Name.

GROUP VIII.—SPHÆROPSIDES.—

ORDER XLIX.—SPHÆRIOIDACEÆ,

334. PHYLLOSTICTA.—Pers.,

2025	1802	III. 15	P. circumscissa	Cooke, Grev. XI. 150 (1882) ...	Circular phyllosticta ... (Shot-hole fungus)
2026	1813	X. 5072	P. Cordylines	Sacc. and Berl., Misc. Myc. II. 36 (1885)	Cordyline phyllosticta
2027	1805	III. 33	P. Eucalypti	Thuem., Lusit. 374 (1878)	Eucalyptus phyllosticta
2028	1812	,, 219	P. fragaricola	Desm., Pl. Crypt. III. 656	Strawberry phyllosticta
2029	1807	X. 4881	P. Hardenbergiæ ...	Cooke and Mass., Grev. XVI. 5 (1887) ...	Hardenbergia phyllosticta
2030	1810	,, 4903	P. neurospilea	Sacc. and Berl., Misc. Myc. II. 37 (1885)	Vein-spot phyllosticta
2031		,, 5066	P. palmicola	Cooke. Grev. XIV. 89 (1886)	Palm-growing phyllosticta
2032	1806	., 4886	P. phyllodiorum ...	Sacc.. Hedw. 156 (1890)	Phyllode phyllosticta...
2033	1809		P. Platylobii	Cooke and Mass., Grev. XIX. 61 (1891) ...	Platylobium phyllosticta
2034	...		P. Prostantheræ ...	Cooke, Grev. XXI. 39 (1892)	Prostanthera phyllosticta
2035	1803	III. 31	P. Rosæ ...	Desm., Exs. 687	Rose phyllosticta
2036	1804	,, 30	P. Ruborum	Sacc., Mich. II. 342 (1882)	Bramble phyllosticta ...
2037	1808	X. 4979	P. soriformis	Cooke and Mass., Grev., XIX. 47 (1890)	Sorus-shaped phyllosticta

335. PHOMA.—Fries,

2038	1829	III. 940	P. allicola	Sacc. and Roum., Reliq. Lib. Ser. IV. 79 (1884) ...	Allium phoma
2039	1814	,, 467	P. ampelina = Sphaceloma ampelinum	B. and C., Grev. II. (1873)	Vine phoma (Black spot)
2040	1820		P. australis	Cooke, Grev. XV. 17 (1886)	Southern phoma
2041	1829	III. 965	P. Cordylines	Sacc. Syll. III. 162 (1884)	Lily palm phoma
2042	1822	X. 5084	P. Daviesiæ	Cooke and Mass., Grev. XVIII. 7 (1889)	Daviesia phoma
2043	1823	,, 5201	P. Diploglottidis ...	Cooke and Mass., Grev. XVII. 56 (1889)	Diploglottis phoma
2044	1818	III. 649	P. eucalyptidea ...	Thuem., Lus. 563 (1878)	Eucalyptus phoma
2045		,, 923	P. folliculorum ...	Sacc. Syll. III. 155 (1884)	Follicle phoma
2 46	1825	X. 5510	P. Goodeniarum ...	Cooke and Mass., Grev. XVI. 2 (1887) ...	Goodenia phoma
2047	1830	III. 998	P. graminis	West. in Kickx. Fl. Fland. 1. 441 1867)...	Grass phoma
2048	1826	., 796	P. herbarum	West. Exs. 965	Herb phoma ...
2049	1824	X. 5337	P. Lythri	Cooke and Mass., Grev. XVI. 75 (1888)	Lythrum phoma
2050	1817	III. 650	P. Mölleriana	Sacc. Syll. III. 110 (1884)	Möller's phoma
2051	1831	997	P. nitida...	Rob., in Desm. Exs. ...	Shining phoma
2052	1816	., 556	P. notha ...	Berk., Ann. Nat. Hist. 2 Ser. V. 369 (1850)	Spurious phoma

OF AUSTRALIAN FUNGI—*continued.*

Number	W.A.	S.A.	T.	V.	N.S.W.	Q.	B.	Occurrence.	General Characters.

LEV., ANN. SCI. NAT. 3 SER. III. 61 (1845).

SACC. SYLL. III. 1 (1884).

Champ. Com. 55 (1818).

Number	W.A.	S.A.	T.	V.	N.S.W.	Q.	B.	Occurrence.	General Characters.
2025	...	S.A.	T.	V.	N.S.W.	Q.	...	Leaves and fruit of *Prunus Armeniaca* and *P. Cerasus.* &c.	Both surfaces, spots circular, reddish brown, finally falling out and leaving leaf as if riddled with shot.
2026	...	...		...		Q.	...	Leaves of *Cordyline terminalis*	Spots indistinct, becoming pale. Receptacles on upper surface, point-like, pierced.
2027	...	...		V.		...	...	Leaves of *Eucalyptus globulus*	Spots large, irregular, at first dingy brown then whitish, with narrow purple border.
2028	...	S.A.		...		Q.	...	Strawberry leaves	Spots straggling, becoming bleached, with red margin.
2029	...	...		V.		...	...	Leaves of Harden-bergia	Spots on both surfaces, various, tawny.
2030	...	...		...		Q.		Leaves of *Vitis antarctica*	Spots on upper surface, limited by the veins, hence angular, reddish ochrey.
2031	...	...	...	...		Q.	...	Palm leaves ...	Spots on both surfaces, irregular, whitish to ashy. Margin somewhat elevated, brown.
2032	...	...	...	V.		...	...	Phyllodes of *Acacia*	Spots on both surfaces, somewhat circular, whitish, with brown margin.
2033	...	...	...	V.		...	...	Leaves of *Platylo-bium*	Spots on both surfaces, irregular, pale with narrow brown margin. Receptacles minute, gregarious, black.
2034	...	...	...	V.		...	...	Leaves of *Prostan-thera lasianthos*	Spots somewhat circular, pale amber, with raised dark marginal line.
2035	...	...	...	V.		...	B.	Rose leaves ...	Spots rather circular, greenish, then brownish or greyish, with purple border.
2036	...	...	...	V.		...	...	Fading leaves of *Rubus fruticosus*	Spots minute, whitish, often near the veins.
2037	...	...	...	V.		...	...	Leaves of some *Proteaceæ*	Spots on both surfaces, brown, circular, with darker margin.

Novit. Fl. Succ. V. (1819).—Sphæropsis.

Number	W.A.	S.A.	T.	V.	N.S.W.	Q.	B.	Occurrence.	General Characters.
2038	...	...	...	V.	...	...	...	Scapes of *Allium* ...	Receptacles gregarious, spherical, black, very small, obtuse.
2039	...	S.A.	T.	V.	N.S.W.	Q.	...	Vine twigs ...	Sub-cuticular, Hysterium-like, swollen.
2040	...	...	...	V.	...	...	...	Leaves of *Eucalyptus*	Spots brownish, elliptical, surrounded by brown line. Receptacles black, point-like, half immersed.
2041	...	...	...	...		Q.	...	Old leaves of *Crinum pedunculatum*	Receptacles numerous, thickly clustered, on under surface, pustular, quite black.
2042	...	...	...	V.		...	...	Dead leaves of *Daviesia latifolia*	Chiefly on under surface. Receptacles very minute, covered, black, forming nebulous spots.
2043	...	...	...	...		Q.	...	Fading leaves of *Diploglottis Cun-ninghamii*	On under surface, gregarious. Receptacles half immersed, minute, black, pap-like.
2044	...	...	...	V.		...	...	Living or fading leaves of *Euca-lyptus globulus*	Receptacles on under surface, scattered, conically elevated, black, minute.
2045	...	...	...	...		Q.	...	Follicles of a *Mars-denia*	Bursting through. Receptacles gregarious, immersed, black, conical, girt by whitish spots.
2046	...	...	...	V.		...	...	Fading leaves of *Goodenia ovata*	Receptacles scattered, dot-like, minute, black, membranous.
2047	...	...	...	...		Q.	...	Grass stems (*Poa*)	Receptacles globose or angular, black, arranged in series, and forming elongated pustules, wrinkled, dark grey.
2048	...	...	...	V.		...	B.	Herbaceous stems...	Receptacles gregarious, depressed globose, pap-like, black, every-where.
2049	...	...	...	V.		...	...	Fading leaves of *Lathrum hysaupifolia*	On upper surface. Receptacles scattered or gregarious, globose, covered, at length bursting through.
2050	...	...	...	V.		...	...	Fallen leaves of *Eucalyptus globulus*	Receptacles on both surfaces, large, thickly clustered, turgid, shining, dark chestnut.
2051	...	...	...	V.		...	...	Grass	Scattered, minute, shining. Receptacles hemispherical, white within, covered by epidermis, which splits lengthwise.
2052	...	...	...	...		Q.	B.	Dead branches of *Platanus*	Receptacles spurious, circular, elevated here and there.

Number.	Cooke's Number.	Saccardo's Number.	Scientific Name.	Authority for Name.	English Name.
					335. PHOMA.—Fries,
2053	1827	X. 5349	P. plagia ...	Cooke and Mass., Grev. XVII. 55 (1889)	Defined phoma
2054	1832	,, 5390	P. portentosa	Cooke and Mass., Grev. XVI. 2 (1887) ...	Monstrous phoma
2055	1821	,, 5250	P. purpurea	Cooke and Mass., Grev. XV. 97 (1887) ...	Purple phoma
2056	1815	III. 431	P. Rosarum	Dur. and Mont., Alg. 604 (1849)	Rose phoma ...
2057	...	,, 994	P. Sacchari	Sacc. Syll. III. 166 (1884)	Sugar-cane phoma
2058	...	,, 887	P. uvicola = Laestadia Bidwellii	B. and C., Grev. II. (1873) (not Arcang.)	Grape-growing phoma (Black rot)
2059	1819	X. 5115	P. viminalis	Cooke and Mass.. Grev. XVI. 75 (1888) ...	Viminalis phoma
					336. APOSPHÆRIA.—Berk.,
2060	1833	X. 5466	A. Leptospermi ...	Cooke. Grev. XIX. 91 (1891)	Leptospermum aposphæria
					337. ASTEROMELLA.—Pass.
2061	1834	X. 3489	A. acaciæ	Cooke, Grev. XIX. 5 (1890)	Acacia asteromella
2062	1835		A. epitrema	Cooke, Grev. XX. 6 (1891)	Trema asteromella
2063	1836		A. Homalanthi	Cooke and Mass., Grev. XX. 65 (1892) ...	Homalanthus asteromella
					338. CHÆTOPHOMA.—Cooke,
2064	1837	X. 5510	C. eutricha	Sacc. and Berl., Misc. Myc. II. 8 (1885) ...	Well-haired chætophoma
					339. DOTHIORELLA.—Sacc.,
2065	1838	X. 5578	D. Amygdali	Cooke and Mass.. Grev. XIX. 91 (1891) ...	Almond dothiorella
2066	...	,, 5579	D. Eucalypti	Sacc. Syll. X. 229 (1892)	Eucalyptus dothiorella
2067	1839	,, 5599	D. pericarpica	Sacc., Pug. Austr. 15 (1890)	Pericarp dothiorella ...
					340. CYTOSPORA.—Ehr.,
2068	1841	X. 5677	C. verrucula	Sacc. and Berl., Misc. Myc. II. 8 (1885) ...	Warty cytospora
2069	1840	III. 1531	C. xanthosperma ...	Fries, S.M. II. 543 (1823)	Yellow-spored cytospora
					341. SPHÆROPSIS.—Lev.,
2070	1844	...	S. numerosa	Cooke and Mass., Grev. XX. 65 (1892) ...	Numerous sphæropsis
2071	1845	X. 5711	S. phomatoidea	Cooke and Mass., Grev. XVIII. 49 (1890)	Phoma-like sphæropsis
2072	1843	III. 1640	S. Rosarum	Cooke and Ellis, Grev. VI. 2 (1877)	Rose sphæropsis
2073	1842	,, 1720	S. Tricorynes	Berk. and Br., Linn. Trans. II. 68 (1885)	Tricoryne sphæropsis ...
2074	1846	X. 5734	S. Tritici	Cooke and Mass., Grev. XVI. 75 (1888) ...	Wheat sphæropsis
					342. CONIOTHYRIUM.—Corda,
2075	1847	X. 4793	C. olivaceum	Bon. in Fckl., Sym. 377 (1875) ...	Olive coniothyrium
2076	1848	,, 5752	C. septorioides	Cooke and Mass., Grev. XX. 36 (1891) ...	Septoria-like coniothyrium

OF AUSTRALIAN FUNGI—*continued.*

Number.	W.A.	S.A.	T.	V.	N.S.W.	Q.	B.	Occurrence.	General Characters.

Novit. Fl. Suec. V. (1819).—Sphæropsis—*continued.*

Number.	W.A.	S.A.	T.	V.	N.S.W.	Q.	B.	Occurrence.	General Characters.
2053	...	...	...	...		Q.	...	Palm leaves	Spots distinctly defined, glaucous, elliptic or confluent. Receptacles very minute, black.
2054	...	...	V.	...		...	...	Cap of *Polyporus portentosus*	Scattered. Receptacles innate, covered by blackened cuticle, pap-like, black, shining.
2055	...	...	...	...		Q.	...	Foliage of Eucalypts and Tristanias	On both surfaces. Spots circular, purple. Receptacles gregarious, half immersed, black, shining.
2056	...	...	...	...		Q.	...	Rose twigs	Receptacles rather minute, covered by the unbroken or stellately-split epidermis, dark brown, white within.
2057	...	...	...		N.S.W.		...	Leaves and stems of Sugar cane	Receptacles nearly spherical, black, bursting through, scattered or gregarious.
2058	...	...	V.			...	...	Grapes ...	Irregular, bursting through, and then surrounded by the narrow cuticle.
2059	...	...	V.			...	...	Leaves of *Eucalyptus viminalis*	On upper surface. Receptacles immersed, bursting through, black, somewhat globose.

Outl. 315 (1860).

Number.	W.A.	S.A.	T.	V.	N.S.W.	Q.	B.	Occurrence.	General Characters.
2060	...	...	...	V.		...	...	Bark of *Leptospermum*	Receptacles scattered, bursting through, then superficial, minute, black, pap-like, white within.

and Thuem., in M.U. 1689 (1877).

Number.	W.A.	S.A.	T.	V.	N.S.W.	Q.	B.	Occurrence.	General Characters.
2061	...	...	...	V.		...	...	Phyllodes of *Acacia*	Receptacles very numerous, densely crowded, and forming blackish spots, minute.
2062	...					Q.	...	Living leaves of *Trema aspera*	Spots on upper surface, black, somewhat circular. Receptacles minute, rather globose, seated on brown mycelium.
2063	...					Q.	...	Leaves of *Homalanthus populifolius*	Spots somewhat circular, on both surfaces, sooty brown, dotted with minute black receptacles.

Grev. III. 25 (1874).

Number.	W.A.	S.A.	T.	V.	N.S.W.	Q.	B.	Occurrence.	General Characters.
2064	...	...	...			Q.	...	Languid leaves of *Castanospermum australe*	Spots black, often running together. Threads of mycelium, sooty brown. Receptacles dot-like, black.

Mich. II. 5 (1882).

Number.	W.A.	S.A.	T.	V.	N.S.W.	Q.	B.	Occurrence.	General Characters.
2065	...	...	...	V.			...	Bark of Peach and Almond	Receptacles innate, clustered, transversely bursting through, black, opaque, somewhat gelatinous when moist.
2066	...		V.				...	Leaves of Eucalypts	Receptacles globose, seated on a sclerotioid body, black, shining. Sclerotia loosely spongy, pale brown within.
2067				...		Q.	...	Pericarp of *Macrozamia Denisonii*	Receptacles in clusters bursting through, black, cushion shaped, tuberculose, globose or angular.

Syl. Berol. 28 (1820).

Number.	W.A.	S.A.	T.	V.	N.S.W.	Q.	B.	Occurrence.	General Characters.
2068	...	...	...	...		Q.	...	Branches	Receptacles few, immersed, seated on layer soon bursting through, globose or depressed, black.
2069		...	...	V.			...	Branches of *Salix Babylonica*	Receptacles none. Spores issuing in golden tendrils.

Ann. Sci. Nat. III. 62 (1846).

Number.	W.A.	S.A.	T.	V.	N.S.W.	Q.	B.	Occurrence.	General Characters.
2070	...	...	...	V.		...	...	Dead bark	Receptacles gregarious, half immersed, globose, black, becoming flattened.
2071			V.					*Eucalyptus* leaves ...	On under surface. Receptacles scattered over irregular brown spots, at first covered, black.
2072	...	...	...	...		Q.	...	Rose branches ...	Gregarious or scattered. Receptacles covered splitting the epidermis.
2073	...	...	...	...		Q.	...	Leaves of *Tricoryne anceps*	Receptacles minute, black, immersed in substance of leaf.
2074			...	V.	...			Dead leaves and sheaths of Wheat	Receptacles very minute, thickly clustered, at first covered, point-like, black.

Icon. IV. 38 (1854).

Number.	W.A.	S.A.	T.	V.	N.S.W.	Q.	B.	Occurrence.	General Characters.
2075	...		V.				...	Involucres of *Leptospermum lævigatum*	Receptacles scattered, at first covered, then bursting through, rather large, pap-like.
2076	...	...	...	V.			...	Leaves of *Prostanthera lasiantha*	Spots circular, tawny, with broad purple margin. Receptacles mostly in circles upon spots, black.

172

SYSTEMATIC ARRANGEMENT

Number.	Cooke's Number.	Saccardo's Number.	Scientific Name.	Authority for Name.	English Name.
					343. CAPNODIASTRUM.—Speg.,
2077	1849	X. 5809	C. orbiculatum	... Cooke and Mass., Grev. XVIII. 49 (1890)	... Orbicular capnodiastrum ...
					344. CHÆTOMELLA.—Fckl.,
2078	1850	III. 1807	C. brachyspora	... Sacc. and Speg., Mich. I. 260 (1879)	... Short-spored chætomella ...
					345. DIPLODIA.—Fries,
2079	1853	X. 5873	D. canthifolia	... Cooke and Mass., Grev. XX. 36 (1891) ...	... Canthium-leaved diplodia ...
2080	1851	,, 5829	D. lichenopsis	... Cooke and Mass., Grev. XVI. 2 (1887) ...	... Lichen-like diplodia ... ...
2081	...		D. Marsdeniæ	Cooke and Mass., Grev. XXI. 75 (1893) ...	... Marsdenia diplodia ... ...
2082	1852	III. 1090	D. phyllodiorum ...	Penz. and Sacc., Fung. Mort. ...	Phyllode diplodia ...
					346. ASCOCHYTA.—
2083	1855	X. 5957	A. apiospora	Cooke and Mass., Grev. XV. 98 (1887) ...	... Pear-shaped spored ascochyta ...
2084	1854	,, 5964	A. brunnea	Cooke and Mass., Grev. XV. 98 (1887) ...	... Brown ascochyta ... ...
					347. ROBILLARDA.—Sacc.,
2085	1856	III. 2253	R. sessilis	... Sacc., Mich. II. 8 (1880)	... Sessile robillarda ... ...
					348. ACTINONEMA.—Fries,
2086	1857	III. 2257	A. Rosæ ...	... Fries, S.V.S. 424 (1849)	... Rose actinonema ... ...
					349. DARLUCA.—Cast., Cat. Mars.
2087		III. 2263	D. filum ...	... Cast., Cat. Mars. Supp. 53 (1851)	... Thread darluca ... ...
					350. DIPLODINA.—West,
2088	1858	X. 6054	D. Dendrobii	... Cooke and Mass., Grev. XVI. 3 (1887) ...	... Dendrobium diplodina... ...
					351. HENDERSONIA.—Berk.,
2089	1859	III. 2320	H. Eucalypti	... Cooke and Hark., Grev. IX. 128 (1881) ...	... Eucalyptus hendersonia ...
					352. STAGONOSPORA.—Sacc.,
2090	1860	X. 6140	S. orbicularis	... Cooke, Grev. XX. 6 (1891)	... Orbicular stagonospora ...
					353. CAMAROSPORIUM.—Schulz,
2091	1861	X. 6191	C. Eucalypti	Wint., Rev. Myc. 212 (1886)	... Eucalyptus camarosporium ...
					354. SEPTORIA.—Fries,
2092	1870	III. 3051	S. Bromi ...	Sacc., Mich. I. 104 (1879) ...	Brome septoria ... ...
2093	1868	X. 6241	S. epiphylloden ... substituted for S. phyllodi- orum	... Cooke. Handb. Aust. Fung. 356 (1892) ... Sacc., Hedw. 156 (1890).	Epiphyllode septoria ...
2094	1866	,, 6242	S. Hardenbergiæ ...	... Sacc., Hedw. 156 (1890)	... Hardenbergia septoria ... ...

OF AUSTRALIAN FUNGI—continued.

Number.	W.A.	S.A.	T.	V.	N.S.W.	Q.	B.	Occurrence.	General Characters.

Guar. I. 145 (1883).

2077 ... ... Q. — Leathery leaves ... Spots circular, of black interwoven mycelium. Receptacles minute, globose, rather membranous.

Sym. 402 (1875).

2078 ... S.A. ... V. ... ... Bark, branches of Grape vine — Receptacles scattered, rather superficial, globose, then depressed, black, clad with stiff brown bristles.

S.V.S. 416 (1849).

2079 ... ... V. ... ... Leaves of *Oanthium latifolium* — Receptacles scattered, immersed, membranous, dark brown piercing cuticle.

2080 ... ... Q. ... Phyllodes of *Acacia complanata* — Spots brick red, determinate, pale at length, or girt with red zone. Receptacles half-internal, point-like, black.

2081 ... ... Q. ... Follicles of a *Marsdenia* — Receptacles gregarious, black, bursting through. In company with *Phoma folliculorum*, Sacc.

2082 ... ... V. ... ... Living or fading phyllodes of *Acacia*, — Receptacles gregarious, minute, under cuticle then bursting through, black, seated on whitish spots.

Lib. Exs. (1837).

2083 ... ... Q. — Leaves of *Myrtus* and *Backhousia* — Spots on upper surface, circular or irregular, tawny, girt by purple zone. Receptacles minute, innate.

2084 ... ... Q. ... Leaves of Brisbane Box (*Tristania conferta*) — Spots on both surfaces, circular or irregular, pale brown or ochrey, girt by darker elevated line. Receptacles minute, point-like, black.

Mich. II. 8 (1880).—Pestalozzia.

2085 ... ... Q. ... Fading leaves of Vine — Spots small, angular, turning whitish, encircled with red. Receptacles on upper surface.

S.V.S. 424 (1849).—Asteroma.

2086 ... ... V. ... Q. B. Rose leaves ... On upper surface. Spots purplish. Receptacle-like tubercles scattered and collapsing, blackish.

Supp. 53 (1851).—Sphæria, Diplodia.

2087 ... ... Q. B. Leaves of *Sorghum* and *Muehlenbeckia* infested with Uredines — Gregarious, very minute. Receptacles globose, black, shining, pierced.

5 Not. 19 (1866).

2088 ... ... Q. — Leaves of *Dendrobium speciosum* — Receptacles gregarious, innate, black, convex, at length splitting cuticle, shining.

Ann. Nat. Hist. VI. 430 (1841).

2089 ... ... V. ... ... Dead branches and leaves of *Euculyptus* — Receptacles in circular spots, immersed.

Mich. II. 8 (1880).

2090 ... ... V. ... ... Dead leaves of *Eucalyptus* — Spots on both surfaces, small, circular, pale, surrounded by brown line. Receptacles few, in centre of spots, black.

Myk. Beitr. 649 (1870).

2091 ... ... V. ... ... Leaves of *Eucalyptus* — Receptacles on irregular spots, which are pale brown or grey, limited by darker line.

S.M. III. 480 (1832).

2092 ... S.A. ... V. ... ... Leaves of *Bromus*, &c. — Spots obsolete, becoming pale, elongated. Receptacles plentiful, globose to flattened, pierced.

2093 ... S.A. ... V. ... Phyllodes of *Acacia* — Spots on both surfaces, circular, whitish, encircled by brown. Receptacles crowded, point-like, becoming black.

2094 ... S.A. ... ... Leaves of *Hardenbergia monophylla* — Spots on both surfaces, broad, pale, brown at margin. Receptacles point-like, ochrey.

Number.	Cooke's Number.	Saccardo's Number.	Scientific Name.	Authority for Name.	English Name.
					354. SEPTORIA.—Fries,
2095	1869	X. 6429	S. Lepidospermi	Cooke and Mass., Grev. XIX. 91 (1891) ...	Lepidosperma septoria ...
2096	1863	„ 6353	S. Martinii	Cooke, Grev. XIX. 5 (1890)	... Martin's septoria ...
2097	1865	„ 6264	S. Myopori	Cooke and Mass., Grev. XVI. 113 (1888)...	... Myoporum septoria
2098	1864	III. 2683	S. oleandrina	Sacc., Fung. Ven. V. 205 (1873-82)	... Oleander septoria
2099	1867	X. 6245	S. phyllodiorum ... =S. Martiniana ...	... Cooke and Mass., Grev. XIX. 47 (1890) .. Sacc., Syll. X. 351 (1892).	Phyllode septoria
2100	...	III. 3042	S. Tritici	... Desm. IX., Not. 17 (1842)	... Wheat septoria
2101	1862	„ 2611	S. Violæ ...	... West, Exs. Fasc. 2, 91	... Violet septoria
					355. PHLLOSPORA.—Wallr., Fl. Crypt. 7 (1833).—
2102	...	III. 3136	P. Mori ...	Sacc. Syll. III. 577 (1884)	... Mulberry phleospora ... ... (Leaf-spot of Mulberry)
					356. PHLYCTÆNA.—Mont. and Desm.
2103	1871	X. 6518	P. Passiflorae	Cooke and Mass., Grev. XVI. 3 (1887) ...	... Passion-flower phlyctæna ...
					357. GAMOSPORA.—Sacc.,
2104	1872	X. 6529	G. eriosporoides ...	Sacc. and Berl., Rev. Myc. (1885)	... Eriospora-like gamospora ...
					ORDER L.—
					358. SPHÆRONÆMELLA.—Karst.,
2105	1873	III. 3308	S. rufa ...	Sacc. Syll. III. 618 (1884)	... Red sphæronæmella ... ...
					359. ASCHERSONIA.—Mont.,
2106	1874	III. 3313	A. tahitensis	Mont., Ann. Sci. Nat. 122 (1848)	... Tahitian aschersonia ...
					360. MARTINELLA.—Cooke and Mass.,
2107	1875	X. 6553	M. Eucalypti	... Cooke and Mass., Grev. XVIII. 7 (1889)	... Eucalyptus martinella ...
					ORDER LI.—
					361. LEPTOTHYRIUM.—Kunze and
2108	1877	X. 6567	L. aristatum	Cooke, Grev. XX. 6 (1891)	... Bristly leptothyrium ... ...
2109	1876	„ 6568	L. Eucalyptorum ...	... Cooke and Mass., Grev. XVIII. 7 (1889)	... Eucalypt leptothyrium
					362. PIGGOTIA.—Berk. and Br.,
2110	1878	X. 6596	P. substellata	... Cooke, Grev. XX. 6 (1891)	... Stellate piggotia ...

OF AUSTRALIAN FUNGI—*continued.*

Number	Habitat						L.	Occurrence.	General Characters.
	W.A.	S.A.	T.	V.	N.S.W.	Q.			

S.M. III. 480 (1832)—*continued.*

2095	...	...		V. !			...	Leaves of *Lepidosperma*	Spots on both surfaces, greyish, then white, oblong, with broad brown margin. Receptacles small, black.
2096	...			V.			...	Leaves of *Senecio Bedfordii*	Spots on upper surface, grey, run together, surrounded by black line. Receptacles point-like, globose.
2097				V.				Leaves of *Myoporum insulare*	Spots on upper surface, circular, whitish, girt by brown line. Receptacles half immersed, point-like, black.
2098						Q.		Leaves of *Nerium Oleander*	Spots on upper surface, rounded or angular, and run together, turning whitish. Receptacles somewhat large, globose.
2099	...			V.	..			Phyllodes of *Acacia longifolia*	Receptacles closely crowded on both surfaces, without definite spots, often occupying entire surface, immersed, black.
2100				V.	N.S W.		... B.	Fading leaves of Wheat, &c., also stem and ear	On both surfaces. Spots linear lengthwise, whitish with dark-purple margin. Receptacles innate, very minute, black.
2101	...			V.			... B.	Fading Violet leaves	Receptacles minute, numerous, brownish yellow, seated on pale zoned circular spots, girt by reddish-brown ring.

Septoria, Sphærella, Fusarium, Fusisporium.

| 2102 | ... | | ... | ... | V. | | ... | B. Leaves of Mulberry | Spots whitish or ochrey, surrounded by brown. Receptacles innate, globose, gregarious. |

Ann. Sci. Nat. 16 (1847).

| 2103 | ... | ! | ... | .. | ... | Q. | ... | Twigs of *Passiflora* growing on stem | Receptacles very thickly clustered, minute, innate, at length bursting through. |

Rev. Myc. (1883).

| 2104 | ... | | ... | | ... | Q. | ... | Languid leathery leaves | Receptacles on upper surface, interspersed on thin spot-like brown mycelium, point-like, globose to depressed. |

NECTRIOIDACEÆ, SACC. SYLL. III. 613 (1884).

Hedw. 17 (1884).—Sphæronæma.

| 2105 | ... | ... | T. | ... | ... | ... | Pine chips | ... Receptacles awl shaped, acute, reddish brown, paler downwards. |

Ann. Sci. Nat. 3 Ser. X. 121 (1848).

| 2106 | ... | ... | ... | ... | ... | Q. ! | ... | Leaves of climber... | Receptacles minute, seated on hemispherical obtuse yellow layer. |

Grev. XVIII. 7 (1889).

| 2107 | ... | | ... | V. | | | ... | Leaves of *Eucalyptus* | On upper surface. Receptacles very minute, immersed, cracked at mouth, seated on rather circular fleshy reddish-brown layer. |

LEPTOSTROMACEÆ, RCHB. NOM. GEN. 6 (1841).

Schm., Myk. Heft. 11. 79 (1823).

| 2108 | ... | | ... | V. | ... | | ... | Dead leaves of *Eucalyptus* | Receptacles scattered, superficial, circular, dark brown. Spores with oblique bristle at one end. |
| 2109 | | | | V. | | | ... | Fallen leaves of *Eucalyptus* | Receptacles scattered over bleached spots, shield shaped, flattened, black, dehiscing in middle with star-like fissure. |

Ann. Nat. Hist. VII. 2 Ser. 95 (1851).

| 2110 | ... | ... | ... | V. | ... | | ... | Leaves of *Eucalyptus* | On under surface, forming small somewhat circular stellate black patches, composed of flattened receptacles run together. |

176

Number.	Cooke's Number.	Saccardo's Number.	Scientific Name.	Authority for Name.	English Name.
					363. MELASMIA.—Lev.,
2111	1879	X. 6602	M. Eucalypti	Cooke and Mass., Grev. XVI. 75 (1888)	Eucalyptus melasmia ... ...
2112			M. Tecomatis	... Cooke and Mass., Grev. XXII. 37 (1893)	Tecoma melasmia
					364. ACTINOTHECIUM.—Ces.,
2113	1879 bis	III. 339s	A. Scortechinii ...	... Sacc. and Berl., in Sacc. Syll. III. 639 (1884)	Scortechini's actinothecium ...
					365. SACIDIUM.—Nees, in Kunze and Schm.,
2114	1881	X. 6615	S. Camelliæ	... Cooke and Mass.. Grev. XVI. 3 (1887) ...	Camellia sacidium ...
2115	1880	„ 6616	S. Eucalypti	... Cooke and Mass., Grev. XVI. 75 (1888) ...	Eucalyptus sacidium ..
					366. MELOPHIA.—Sacc.
2116	1883	X. 6643	M. Leptospermi ... = M. Victoriæ ...	... Cooke, Grev. XX. 65 (1892) .. Sacc. Syll. X. 428 (1892).	Leptospermum melophia ..
2117	1882	III. 3512	M. Woodsiana ...	... Sacc. and Berl., in Sacc. Syll. III. 659 (1884)	Woodsio's melophia ... ...
					367. LEPTOSTROMELLA.—Sacc.,
2118	1884	X. 6652	L. Eucalypti	... Cooke and Mass., Grev. XIX. 91 (1891) ...	Eucalypt leptostromella ...
					ORDER LII.—
					368. DINEMASPORIUM.—Lev.,
2119	1885	III. 3619	D. hispidulum	... Sacc., Mich. II. 281 (1882)	Hispid dinemasporium... ...
					369. PROTOSTEGIA.—
2120	1886	X. 6715	P. Eucalypti	... Cooke and Mass., Grev. XVI. 75 (1888) ...	Eucalypt protostegia ... ...
					ORDER LIII.—
					370. GLŒOSPORIUM.—Desm. and Mont.,
2121	...	...	G. Alphitouiæ	Cooke and Mass., Grev. XXII. 37 (1893)	Alphitonia glœosporium ..
2122	1900	III. 3755	G. ampelophagum	Sacc., Mich. I. 217 (1879)	Grape-destroying glœosporium ...
2123	1887	X. 6737	G. Citri ...	Cooke and Mass., Grev. XIX. 92 (1891) ...	Citrus glœosporium ... ...
2124	1889	„ 6739	G. citricolum	Cooke and Mass., Grev. XVI. 3 (1887)	Citrus-growing glœosporium ...
2125	1895	„ 6756	G. Denisonii	... Sacc. and Berl., Misc. Myc. II. 10 (1885)...	Denison's glœosporium ...
2126	1894	„ 6819	G. epicladii	Cooke and Mass., Grev. XIX. 92 (1891) ...	Cladium glœosporium ... ...
2127	1895	III. 3751	G. fructigenum	.. Berk., Gard. Chron. 245 (1856) ...	Fruit glœosporium (Ripe rot) ...
2128	1891	X. 6787	G. glaucum	... Cooke and Mass., Grev. XVI. 75 (1888)...	Glaucous glœosporium ...
2129	1892	„ 6786	G. Hedycaryæ	... Cooke and Mass., Grev. XVIII. 7 (1889)	Hedycarya glœosporium ...

OF AUSTRALIAN FUNGI—continued.

Number	Habitat W.A.	S.A.	T.	V.	N.S.W.	Q.	B.	Occurrence	General Characters

Ann. Sci. Nat. 276 (1846).

Number	Habitat	Occurrence	General Characters
2111	V. — Q.	Leaves of *Eucalyptus*	Spots circular or confluent, black. Receptacles few, somewhat gregarious, elliptical, opening by a fissure.
2112	Q.	Leaves of *Tecoma jasminoides*	Receptacles on both surfaces, superficial, circular, wrinkled, black, disc brownish.

Hedw. I. (1852).

| 2113 | Q. | Leaves of *Smilax* | Receptacles linear, straight or curved, somewhat superficial, forked or variously branched, furrowed lengthwise. |

Myc. Heft. II. 64 (1823).

| 2114 | V. | Fading leaves of *Camellia* | Receptacles scattered, superficial, hemispherical, black, opaque, mostly minute. |
| 2115 | V. | Dead leaves of *Eucalyptus globulus* | On both surfaces. Receptacles gregarious, bursting through, small, flattened, black. |

Syll. III. 658 (1884).

| 2116 | V. | Leaves of *Leptospermum laevigatum* | Receptacles scattered on both surfaces, convex, flattened at base, black, white within. |
| 2117 | Q. | Phyllodes of *Acacia harpophylla* | Receptacles distantly scattered, inserted in a thin white filamentous spot-like mass. |

Mich. III. 632 (1882).

| 2118 | V. | Fading leaves of *Eucalyptus* | Spots somewhat circular, on both surfaces, reddish brown, then sooty brown. Receptacles scattered over spots, black. |

EXCIPULACEÆ, CORDA, IC. FUNG. V. 35 (1842).

Ann. Sci. Nat. 274 (1846).—Peziza, Polyuema, Excipula.

| 2119 | W.A. | Wood | Receptacles gregarious or scattered, rather large, cup shaped, black long rigid straight hairs. (No. 1740 wrongly entered as this species.) |

Cooke, Grev. IX. 19 (1880).

| 2120 | V. | Dead leaves of *Eucalyptus incrassatus* | Receptacles immersed, cup shaped, gelatinous, orange coloured, covered by epidermis, at length split. |

MELANCONIACEÆ, CORDA, IC. FUNG. V. 33 (1842).

Ann. Sci. Nat. XII. 293 (1849).—Ramularia, Fusarium.

2121	Q.	Leaves of *Alphitonia excelsa*	Spots irregular or confluent, pale. Pustules bursting through, small, gregarious on spots.
2122	V.	Grapes, rarely vine leaves or branches	Spots rather circular, often run together.
2123	V.	Branches of Lemon	Gregarious, bursting through, pale sooty brown. Pustules rather small, often run together.
2124	Q.	Orange leaves	Spots dark brown, small, rather disc-like, often run together. Pustules immersed.
2125	Q.	Leaves of *Encephalartos Denisonii*	Pustules gregarious, minutely pustulate, covered by epidermis hardly broken, yellowish within.
2126	V.	*Gahnia tetraquetra*	Pustules gregarious in centre of irregular spots, caused by blackened cuticle.
2127	Q.	B.	Pears	Pustules concentric, dull rose colour bursting through, with single pore or fringed mouth.
2128	Q.	Living leaves	Spots rather circular on one or both surfaces, becoming glaucous, rather mealy.
2129	V.	Fading leaves of *Hedycarya Cunninghamii*	On upper surface. Spots circular, turning black. Pustules solitary or gregarious.

Number.	Cooke's Number.	Saccardo's Number.	Scientific Name.	Authority for Name.	English Name.
				370. GLOEOSPORIUM.—Desm. and Mont.,	
2130	1888	III. 3675	G. intermedium	Sacc., Mich. II. 118 (1882)	... Intermediate gloeosporium ...
2131	1902	„ 3757	G. lagenarium	Sacc. and Roum., Rev. Myc. 201 (1880) ...	... Lagenaria gloeosporium
2131A	1903	„ 3758	G. lagenarium, var. Cucurbi- tarum	Cooke, Handb. Austr. Fung. 363 (1892) ...	... Gourd gloeosporium ...
2132	1897	„ 3748	G. leguminis	Cooke and Hark., Grev. IX. 7 (1880)	Legume gloeosporium ...
2133	1896	„ 3747	G. Lindemuthianum = Colleto- trichum	Sacc. and Magn., Mich. I. 129 (1878)	... Lindemuth's gloeosporium (Bean spot)
2134	1899	X. 6811	G. Musarum	Cooke and Mass., Grev. XVI. 3 (1887) ...	Musa gloeosporium
2135	1893	„ 6748	G. nigricans	Cooke and Mass., Grev., XIX. 91 (1891)...	Blackening gloeosporium
2136	1901	„ 6733	G. pestiferum	Cooke and Mass., Grev., XIX. 61 (1891) ...	... Pestiferous gloeosporium
2137	1890	„ 6801	G. subglobosum	Cooke and Mass., Grev. XV. 3 (1887) ...	Sub-globose gloeosporium ...
2138		III. 3752	G. versicolor	Berk. and Curt., Grev. III. (1874) ...	... Colour-changing gloeosporium ... (Bitter rot)
				371. PESTALOZZIELLA.—Sacc. and Ellis.	
2139	1904	X. 6858	P. circularis	Cooke and Mass., Grev. XVIII. 80 (1890)	... Circular pestalozziella ...
				372. MARSONIA.—Fisch.,	
2140	1905	X. 6884	M. Acaciæ	... Cooke and Mass., Grev. XIX. 47 (1890) ...	... Acacia marsonia ...
2141	1906		M. deformans	Cooke and Mass., Grev. XIX. 62 (1891) ...	Deforming marsonia ...
				373. STILBOSPORA.—Pers.,	
2142	1907	X. 6904	S. foliorum	... Cooke, Grev. XX. 6 (1891)	Leaf stilbospora ...
				374. CORYNEUM, Nees, Syst.	
2143	1908	X. 6911	C. viminale	Cooke and Mass., Grev. XX. 36 (1891) ...	Viminalis coryneum ... ...
				375. HYALOCERAS.—Dur. and	
2144	1909	X. 6925	H. dilophosporum ...	... Cooke, Grev. XIX. 5 (1890)	Triseptate-spored hyaloceras ...
				376. PESTALOZZIA.—De Not.,	
2145	1911	III. 4110	P. Acaciæ	... Thuem., Lusit. 576 (1878)	Acacia pestalozzia ... ...
2146	1914	X. 6951	P. Casuarinæ	... Cooke and Mass., Grev. XVIII. 114 (1888)	... Sheoak pestalozzia
2147	1913	III. 4135	P. funerea	Desm., Ann. Sci. Nat. XIX. 235 (1843) ..	... Gloomy pestalozzia ...
2148	...	„ 4146	P. Guepini	Desm., Ann. Sci. Nat XIII. 182 (1840) ...	... Camellia-leaf fungus ...
2149	1915	„ 4161	P. monochæta	Desm., Ann. Sci. Nat. 3 Ser. X. 355 (1848)	One-haired pestalozzia..
2150	1910	„ 4198	P. uvicola	Speg., in Thuem. Pilz. Miu. 13 (1878)	Grape pestalozzia ...
2151	1912	„ 4134	P. versicolor	... Speg., in Sacc. Mich. I. 479 (1879)	Parti-coloured pestalozzia

OF AUSTRALIAN FUNGI—*continued.*

Number.	W.A.	S.A.	T	V.	N.S.W.	Q.	B.	Occurrence.	General Characters.

Ann. Sci. Nat. XII. 295 (1819).—Ramularia, Fusarium—*continued.*

2130	...	...	...	...		Q.	...	Leaves of *Hoya australis*	Pustules gregarious, point-like, black, then bursting through.
2131		...	...	...		Q.	...	Epicarp of Melons, Mango fruit. &c.	Pustules beneath cuticle, bursting through, minute, cushion shaped, somewhat rosy.
2131A	...	...	...	...		Q.		On Gourds, Bananas, and Melons	Spots bright orange, depressed. Conidia club-shaped, shortly stalked.
2132		...		V.			...	Legumes of *Acacia melanoxylon*	Scattered, covered by cuticle. Conidia oval, transparent.
2133	...	...		V.	N.S.W.	Q.	B	Legumes of Bean, Pea, &c.	Spots on fruit, rarely on stem or leaves, roundish, bleached, at first with reddish-brown margin.
2134				...		Q.	...	Ripe Bananas, rendering them dry and insipid	Pustules innate, bursting through, gregarious, rather rosy.
2135				V.			...	Leaves of *Eucalyptus pauciflora*	Without distinct spots, on both surfaces. Pustules densely clustered, becoming black, convex.
2136	...			V.		Q.		Twigs, flower stalks, and fruit of Vine	Pustules gregarious, small, discoid, convex, rosy.
2137	...	...		V.		...	...	Fading leaves of *Goodenia ovata*	Pustules scattered, pale, inconspicuous. Conidia sub-globose.
2138	...	...		V.	N.S.W.			Rotting Apples ...	Spots brown, small, circular, running together. Pustules bursting through, arranged in rings.

Mich. II. 575 (1882).

| 2139 | ... | | ... | V. | | | ... ... | Dead leaves of *Eucalyptus pauciflora* | On both surfaces. False receptacles usually arranged in circles, at first brown, then nearly black and shining. |

in Rab. Fl. Eur. No. 1857.

| 2140 | ... | ... | .. | V. | | ... | | Phyllodes of *Acacia* | Spots irregular or run together, pale or whitish, with brown margin. Pustules gregarious on the spots. |
| 2141 | .. | S.A. | | V. | | | | Cultivated Peas, chiefly on leaves, stipules, leaf stalks, &c. | Pustules gregarious, often run together, brown, distorting the foliage, sometimes on large discoloured spots. |

Syn. Fung. 96 (1801).

| 2142 | ... | ... | ... | V. | ... | | ... ... | Dead leaves of *Eucalyptus* | Pustules in circular paler spots, splitting the cuticle with three or four openings. |

Pilz. 34 (1816).

| 2143 | ... | ... | ... | V. | | .. ... | | Leaves of *Eucalyptus viminalis* | Pustules point-like, flattened, scattered, black, not seated on definite spots. |

Mont. Fl. Alg. 587 (1849).

| 2144 | ... | ... | ... | V. | ... | ... ... | | Leaves of *Leptospermum scoparium* | Pustules gregarious, minute, brown, bursting through, splitting irregularly in centre. |

Micr. Ital. II. (1842).—Coryneum.

2145	...	...	...	V.	...	... ...		Living *Acacia* leaves	On under surface. Pustules gregarious or solitary, hemispherical, seated on irregular dirty ochre spots, with broad rusty margin.
2146	...	...		V.				Branches of *Casuarina*	Pustules gregarious, minute, elliptic, encircled by ruptured epidermis.
2147	...	...				Q.	B.	Leaves of *Elæodendron* and *Myrtus*	Pustules scattered, point-like, black, covered by epidermis, then bursting through.
2148	...		...	...		Q.	B	Foliage of *Alphitonia excelsa* (Red Ash)	Pustules minute, point-like, convex, black, covered, then bursting through.
2149	...	...	...	...		Q.		Leaves of *Eucalyptus*	Pustules scattered or gregarious, often on under surface. Spots variable, becoming stained with black.
2150	...	...	...		N.S.W.	Q.		Vine leaves, Grapes, and Mangos	Pustules globose, then lens shaped, black, beneath cuticle, bursting through.
2151	...		..	...		Q.	...	Leaves of *Cupania anacardioides*	Pustules somewhat lens shaped, covered, then bursting through, causing surrounding parts to blacken.

N 2

Number.	Cooke's Number.	Saccardo's Number.	Scientific Name.	Authority for Name.	English Name.

GENERAL CLASSIFICATION

GROUP IX.—

ORDER LIV.—SACCHAROMYCETACEÆ.—Unicellular, multiplying by budding and by ascospores.

GROUP IX.—SACCHAROMYCETES,

ORDER LIV.—SACCHAROMYCETACEÆ,

377. SACCHAROMYCES.—Meyen, in Wieg.

2152	2028	VIII. 3632	S. apiculatus	Rees, Bot. Unt. 84 (1870)	Apiculate yeast
2153	2026	,, 3620	S. Cerevisiæ	Meyen, in Wieg. Archiv. IV. 109 (1883)...	Beer yeast
2154	2027	,, 3621	S. ellipsoideus	Rees, Bot. Unt. 82 (1870)	Elliptic yeast
2155	2029	,, 3625	S. Mycoderma	Rees, Bot. Unt. 83 (1870)	Scum yeast

OF AUSTRALIAN FUNGI—*continued.*

Number.	Habitat.						B.	Occurrence.	General Characters.
	W.A.	S.A.	T.	V.	N.S.W.	Q.			

OF SACCHAROMYCETES.

SACCHAROMYCETES, REESS.

Genus (1)—
377. Saccharomyces, Meyen.

REESS, BOT. UNT. (1870).

REESS, BOT. UNT. (1870).

Arch. IV. 2 (1838).—Mycoderma, Torula.

2152	...	...	...	B.	In fermentation of wine	Cells lemon shaped, shortly apiculate at each end, rarely united in small scarcely-branched colonies.
2153	...			B.	In beer	Cells mostly round or oval, solitary or united in small colonies.
2154	...			B.	Producing spontaneous fermentation in must	Cells elliptical, solitary or united in little branched colonies.
2155	...	...	...	B.	On fermented fluids, &c.	Cells oval, elliptical or cylindrical, united in very much branched colonies.

GENERAL CLASSIFICATION OF USTILAGINES.

GROUP X.—USTILAGINES, TUL.

ORDER LV.—USTILAGINACEÆ—Parasitic. Mycelium soon disappearing. Spores virtually all unicellular.

ARRANGEMENT OF GENERA (11).

Section 1. Amcrosporæ, Sacc. and De Toni—Spores continuous, sub-solitary.

Genera (4)—

378. Ustilago, Pers.	380. Entyloma, De Bary.	381. Sphacelotheca, De Bary.
379. Tilletia, Tul.		

Section 2. Dictyosporæ, Sacc. and De Toni—Spores agglomerated.

Genera (4)—

382. Doassansia, Cornu.	384. Sorosporium, Rud.	385. Urocystis, Rabh.
383. Thecaphora, Fing.		

Exceptional—Genera (3)—

386. Graphiola, Poit.	387. Cerebella, Ces.	388. Schinzia, Næg.

Number.	Cooke's Number.	Saccardo's Number.	Scientific Name.	Authority for Name.	English Name.

GROUP X.—USTILAGINES,

ORDER LV.—USTILAGINACEÆ,

378. Ustilago.—Pers.,

2156	1701	VII. 1657	U. australis	Cooke, Grev. VIII. 34 (1879) ...	Southern ustilago ...
2157	1705	IX. 1172	U. axicola	Berk., Ann. Nat. Hist. 2 Ser. IX. 200 (1852)	Axis-growing ustilago
2158	1710	VII. 1677	U. bromivora	Waldh., Ustil. 215 (1877)	Brome-destroying ustilago ...
2159	1713	,, 1704	U. bullata	Berk., Fl. N. Zeal. II. 196 (1855)	Blistered ustilago ...
2160	...	,, 1728	U. bursa...	Berk., Hook., Journ. 206 (1854)	Purse ustilago ...
2161		...	U. catenata	Ludw., Zeitsch. f. Pflkrk. III. 139 (1893)	Chain ustilago
2162	1716	VII. 1728	U. Cesatii	Waldh., Ustil. 25 (1877)	Cesati's ustilago ...
2163	...		U. comburens	Ludw., Zeitsch. f. Pflkrk. III. 139 (1893)	Burning ustilago ...
2164	1702	...	U. confusa	Mass., Grev. XX. 65 (1892)	Confused ustilago ...
2165	1703	VII. 1645	U. destruens	Schlecht., Berol. 130 (1823) ...	Destructive ustilago ... ...
2166	...	,, 1644	U. Digitariæ	Rabh., Fung. Eur. 1199 ...	Digitaria ustilago
2167	1714	,, 1712	U. emodensis	Berk., Hook., Journ. III. 202 (1851)	Dark-lilac ustilago
2168	1707	,, 1671	U. leucoderma	Berk., Ann. Nat. Hist., 2 Ser. IX. 200 (1852)	White-skinned ustilago
2169	1708	,, 1675	U. marmorata	Berk., Linn. Journ. XIII. 174 (1873)	Marbling ustilago ...
2170	...	,, 1723	U. maydis	Corda, Icon. V. 3 (1854)	Maize ustilago
2171	1704	,, 1664	U. Muelleriana	Thuem., Myc. Univ. 623 (1879)	Mueller's ustilago ...
2172	1706	,, 1665	U. pilulæformis ...	Tul., Ann. Sci. Nat. 93 (1847)	Pill-shaped ustilago ... ...
2173	1709	,, 1676	U. segetum ...	Ditm., in Sturm's Deutsch. Fl. (1817–51)	Corn ustilago ... ...
2173a			U. segetum, var. tritici	Jensen, Journ. Roy. Ag. Soc. Eng. 407 (1888)	Wheat ustilago ...
2173a			U. segetum, var. avenæ	Jensen, Journ. Roy. Ag. Soc. Eng. 407 (1888)	Oat ustilago ...
2173c			U. segetum, var. nuda hordei	Jensen, Journ. Roy. Ag. Soc. Eng. 407 (1888)	Naked Barley ustilago
2174			U. Spinificis ...	Ludw., Zeitsch. f. Pflkrk. III. 138 (1893)	Spinifex ustilago ...
2175	1712		U. Tepperi	Ludw., Bot. Centr. 341 (1889) ...	Tepper's ustilago ...
2176	1717	VII. 1737	U. utriculosa	Tul., Mem. Ust. 102 (1847)	Swelling ustilago

379. Tilletia.—Tul., Ann.

2177	1719	VII. 1785	T. epiphylla	Berk. and Br., Linn. Trans. II. 67 (1883)	Epiphyllous tilletia ... ...
2178	1718	,, 1760	T. tritici	Winter, Die Pilze 110 (1884)	Wheat tilletia ...

380. Entyloma.—De Bary.

2179	1720		E. Eugeniarum	Cooke and Mass., Grev. XIX. 92 (1891)	Eugenia entyloma ... ...

OF AUSTRALIAN FUNGI—continued.

Number.	Habitat.						B.	Occurrence.	General Characters.
	W.A.	S.A.	T.	V.	N.S.W.	Q.			

TUL., ANN. SCI. NAT. BOT. 14 (1847).

TUL., ANN. SCI. NAT. BOT. 14 (1847).

Syn. 224 (1808).—Uredo, Tilletia, Caeoma, Cintractia.

2156	...	...	...	V.	...	...	...	Spikelets of *Eriachne*	Produced within the ovaries. Spores black, somewhat globose or deformed.
2157	...	...	...	V.		Q.	...	Fruits and panicles of *Cyperus* and *Fimbristylis*	Little dusty irregular balls in axis of lower spikelets. Spores rather pellucid.
2158	...	S.A.	...	V.	N S.W.	Q.	B.	*Bromus mollis* and *arenarius, Anthistiria ciliata*	Produced in inflorescence. Pustules dark brown, soon powdery.
2159	...	S.A.	...	V.	N.S.W.	...	...	Inflorescence of *Triticum*	Pustules black. Spores very pale olive brown.
2160	...	...	...	...		Q.	...	Grain of *Anthistiria frondosa*	Pustules greenish. Spores brownish black.
2161	...	S.A.	...	...		...	...	Spikes of *Cyperus lucidus*	Pustules crumb-like, ashy-black spores joined in a chain.
2162	...	...	...	V.	...	Q.	...	*Paspalum scrobiculatum*	Pustules black. Spores dark brown.
2163	...	S.A.	...	...		...	...	Species of *Stipa* ...	Pustules black, powdery, in stems and panicles which are almost totally destroyed.
2164	...	...	...	V.		...	...	*Panicum paradoxum*	Pustules produced in ovary, soon naked. Mass of spores powdery, violet black.
2165	...	...	...	V.		...	...	*Danthonia* ...	Pustules black, powdery, blackening flowers and panicles, and destroying ovaries.
2166	...	...	...	V.		...	...	*Panicum* ...	Pustules black. Spores brown to orange.
2167	...	...	...	...		Q.	...	Stems, &c., of *Polygonum*	Pustules lobate. Spores dark lilac.
2168	...	...	...	V.		Q.	...	Sheaths of *Carex, Danthonia, &c,*	Pustules black, seated on large spots, covered by whitish crust.
2169	...	S.A.	...	V.	...	...	...	Leaves of *Scirpus prolifer*	Compact. Marbling the yet unbroken epidermis.
2170	...	...	...	...	N.S.W.	...	B	Indian Corn (*Zea Mays*)	Brown in mass with tinge of olive. Spores pale brown, warty.
2171	...	S.A.	...	V.	...	...	...	Seeds of *Juncus planifolius*	Spores at length clustered together, brown.
2172	...	...	...	V.	...	...	...	Ovaries of *Juncus*	Compact, black. Spores black.
2173	...	S.A.	...	V.	N.S.W.	...	B.	*Aristida, Danthonia*	Pustules black to olive brown, powdery, covered by soon ruptured epidermis.
2173a	...	S.A.	...	V.	N.S.W.	Q.	B.	Wheat ...	Spores of one variety do not germinate on the host-plant of another variety.
2173b	...	S.A.	...	V.	N.S.W.	Q.	B.	Oats ...	
2173c	...	S.A.	...	V.	N.S.W.	Q.	B.	Barley ...	
2174	...	S.A.	...	...		...	...	Flowers and spikes of *Spinifex hirsutus*	Pustules olive, destroying the ovaries. Spores grey olive.
2175	...	S.A.	...	...		...	...	*Amphipogon strictus, Neuraches alopecuroides,* and *Danthonia penicillata*	Spores powdery, black, destroying flowers and upper portion of stems.
2176	...	S.A.	...	V.	...	...	B.	Ovaries and stems of *Polygonum minus* and *P. gracile*	Pustules dark violet, turning violet brown, powdery, causing blossoms to swell.

Sci. Nat. 112 (1847).—Uredo, Ustilago, Lycoperdon, Caeoma.

2177	...	...	...	...		Q.	...	Leaves of Maize ...	Pustules short. Spores brown.
2178	...	S.A.	T.	V.	N.S.W.	Q.	B.	Grains of Wheat ...	Pustules olive black, odour of stinking fish, always covered by epidermis, soon powdery.

Bot. Zeit. 101 (1874).

2179	...	...	...	...		Q.	...	Leaves of *Eugenia*	Pustules irregular, dark brown, flattened, rounded, or confluent, in large patches.

Number.	Cooke's Number.	Saccardo's Number.	Scientific Name.	Authority for Name.	English Name.
					381. Sphacelotheca.—De Bary,
2180	1721	VII. 1834	S. hydropiperis ...	De Bary, Vergl. Morph. 187 (1884) ...	Hydropiper sphacelotheca ...
2180A	„		S. hydropiperis, var. columel-lifera	Berk., in Cooke Handb. Austr. Fung. 327 (1892)...	Columella bearing sphacelotheca
					382. Doassansia.—Cornu.,
2181	1722	VII. 1847	D. punctiformis ...	Winter, Rev. Myc. 207 (1886) ...	Point-like doassansia ... ...
					383. Thecaphora.—Fing.,
2182	1723	VII. 1861	T. inquinans	Berk. and Br., Linn. Journ. XIV. 94 (1875)	Defiling thecaphora ... ...
2183	1724	„ 1868	T. Leptocarpi	Berk., Linn. Journ. XVIII. 388 (1881) ...	... Leptocarpus thecaphora
					384. Sorosporium.—Rud.,
2184	1725	VII. 1885	S. Eriachnes	Thuem., Symb. Austr. II. 4 (1878)	... Eriachne sorosporium ...
2185	1726	„ 1884	S. Muellerianum ...	Thuem., Symb. Austr. II. 5 (1878)	... Mueller's sorosporium ...
					385. Urocystis.—Rabh., Klotzsch.
2186	...	VII. 1891	U. occulta	Rabh., Klotzsch, Herb. Myc. II. 393 (1860)	... Hidden urocystis ...
2187	1727	„ 1910	U. solida	Waldh., Ustil. 38 (1877)	... Compact urocystis ... ...
					386. Graphiola.—Poit., Ann.
2188	1728	VII. 1915	G. Phœnicis	Poit., Ann. Sci. Nat. 473 (1824)	... Date graphiola
					387. Cerebella.—Ces.,
2189	1730	VII. 1919	C. Andropogonis ...	Ces., Klotzsch, Herb. 1587 (1851) ...	... Andropogon cerebella ...
2190	1729	„ 1920	C. Paspali	Cooke and Mass., Grev. XVI. 20 (1887)...	Paspalum cerebella ...
					388. Schinzia.—Nægeli,
2191			S. Leguminosarum	Frank., Krauk. Pfl. 652 (1881)	Leguminous schinzia ... ...

OF AUSTRALIAN FUNGI—*continued*.

Number	Habitat						D.	Occurrence.	General Characters.
	W.A.	S.A.	T.	V.	N.S.W.	Q.			

Vergl. Morph. Pilze 187 (1884).—Ustilago, Uredo.

| 2180 | ... | ... | ... | ... | ... | Q. | B. | Ovaries of *Polygonum* | Spore masses black, elongated, projecting from flower, opening to allow escape of spores. |
| 2180A | ... | | ... | | | | | Ovaries of *Polygonum* | Differs only in more distinct columella. |

Ann. Sci. Nat. 285 (1883).

| 2181 | ... | S.A | ... | V. | ... | ... | ... | Leaves of *Lythrum hyssopifolium* | Pustules on both sides, globose, point-like, scattered or gregarious, brownish. |

Linn. X. 230 (1835).

| 2182 | ... | | N.S.W. | Q. | ... | | | Inflorescence of Rice grass (*Leersia hexandra*) | Pustules almost globose, nestling in pales. Spores pale brown. |
| 2183 | | | V. | | ... | ... | | Ovaries of *Leptocarpus tenax* | Spore balls composed of about ten globose spores, ultimately falling away into black powder. |

Linn. IV. 116 1829).

| 2184 | ... | | ... | N.S.W. | ... | Q | | Spikes of *Eriachne* | Mature fruit changed into black powdery mass. Spores brown. |
| 2185 | | S.A. | | V. | ... | ... | | Panicles of *Gahnia filum* | Infesting inflorescence, but hardly visible to naked eye. Spores up to 100, in dark-brown balls. |

Herb. Myc. II. 393 (1860).—Polycystis, Ustilago.

| 2186 | ... | S.A. | ... | V. | N.S.W. | ... | B. | Wheat stems, leaves, glumes | Pustules forming long black streaks. Spores dark brown, one to three celled, surrounded by bladder-like sterile cells. |
| 2187 | ... | | | T. | V. | N.S.W. | ... | ... | *Schœnus imberbis* ... | Pustules black, globose, compact. Spore balls of three to eight. |

Sci. Nat. III. 473 (1821).—Phacidium.

| 2188 | ... | ... | ... | ... | | Q | B. | Date palms | Conceptacles bursting through, opening above, outer layer black and horny. Spores yellow in mass. |

Klotzsch, Herb. 1587 (1851).

| 2189 | ... | ... | ... | ... | ... | Q. | ... | *Heteropogon contortus* | Olive brown, at first covered with spores of same colour. Spores stuck together. |
| 2190 | ... | | | ... | | Q. | ... | Glumes of *Paspalum scrobiculatum* | Convex, hemispherical, twisted and folded, dark olive. Spores olive. |

Linn. XVI. 278 (1842).

| 2191 | ... | ... | ... | V. | | ... | B. | Roots of leguminous plants | Tubercles varying in size and form, coloured like root, containing hyphæ and innumerable minute corpuscles. |

GENERAL CLASSIFICATION OF PHYCOMYCETES.

GROUP XI.—PHYCOMYCETES, DE BARY.

ARRANGEMENT OF ORDERS (5).

Hyphæ well developed—

56. MUCORACEÆ—Threads producing spore sacs.
57. PERONOSPORACEÆ—Threads often branched, bearing active or passive conidia.
58. ENTOMOPHTHORACEÆ—Threads bearing conidia mostly on insects.

Hyphæ obsolete—

59. CHYTRIDIACEÆ—Spore sacs alone, without threads.
60. PROTOMYCETACEÆ—Spore sacs thick walled, slender threads soon disappearing.

ORDER LVI.—MUCORACEÆ, DE BARY.

Genera (5)—
389. Pilobolus, Tode.
390. Mucor, Mich.

391. Phycomyces, Kunze.
392. Spinellus, V. Tiegh.

393. Circinella, V. Tiegh.

ORDER LVII.—PERONOSPORACEÆ, DE BARY.

Genera (4)—
394. Cystopus, Lev.
395. Sclerospora, Schr.

396. Plasmopara, Schr.

397. Peronospora, Corda.

ORDER LVIII.—ENTOMOPHTHORACEÆ, NOWAK.

Genus (1)—
398. Empusa, Cohn.

ORDER LIX.—CHYTRIDIACEÆ, DE BARY.

Genus (1)—
399. Synchytrium, De Bary.

ORDER LX.—PROTOMYCETACEÆ, DE BARY.

Genus (1)—
400. Protomyces, Unger.

SYSTEMATIC ARRANGEMENT

Number.	Cooke's Number.	Saccardo's Number.	Scientific Name.	Authority for Name.	English Name.
			GROUP XI.—PHYCOMYCETES, DE BARY,		
			ORDER LVI.—MUCORACEÆ,		
					389. PILOBOLUS.—Tode,
2192	1689	VII. 592	P. crystallinus	*Tode*, Meck. 41 (1790)	Crystalline pilobolus ...
					390. MUCOR.—*Linn.*,
2193	1691	IX. 1412	M. cervinoleucus ...	... Berk., Fl. Tasm. II. 282 (1860) ...	... Fawn-white mucor ... ...
2194	1690	VII. 613	M. mucedo	... Linn., Sp. Pl. II. 1655 (1753) ...	... Mould mucor ... ...
					391. PHYCOMYCES.—Kunze,
2195	1692	VII. 696	P. nitens ...	*Kunze*, Myk. II. 113 (1823)	... Shining phycomyces ... ...
					392. SPINELLUS.—Van Tiegh,
2196	1693	IX. 1414	S. gigasporus	... Cooke and Mass., Grev. XVIII. 26 (1889)	... Large-spored spinellus... ...
					393. CIRCINELLA.—Van Tiegh and Mon.,
2197	1694	VII. 732	C. umbellata	Van Tiegh and Mon., Ann. Sci. Nat. 300 (1873)	Umbellate circinella
					ORDER LVII.—PERONOSPORACEÆ,
					394. CYSTOPUS.—Lev.. Ann. Sci.
2198	1695	VII. 792	C. candidus	*Lev.*. Ann. Sci. Nat. 371 (1847) ...	... White cystopus
					395. SCLEROSPORA.—Schr., in Cohn's
2199	1696	IX. 1434	S. macrospora	Sacc.. Hedw. 155 (1890)	Large-spored sclerospora
					396. PLASMOPARA.—Schr., in Cohn's
2200	...	VII. 806	P. viticola	... *Berl. and De Ton.*. Sacc. Syll. VII. 239 (1888)	... Vine-growing plasmopara ... (Brown rot or downy mildew)
					397. PERONOSPORA.—Corda,
2201	1697	VII. 877	P. Hyoscyami	De Bary, Ann. Sci. Nat. 123 (1863)	... Henbane peronospora ... ...
2202		„ 857	P. Schleideni	Unger, Bot. Ztg. 315 (1847)	... Schleiden's peronospora ... (Onion mildew)
					ORDER LVIII.—ENTOMOPHTHORACEÆ,
					398. EMPUSA.—Cohn,
2203		VII. 968	E. Muscæ	... Cohn, Nov. Act. Acad. XXV. 317 (1855)..	... House-fly empusa ... ...
					ORDER LIX.—CHYTRIDIACEÆ,
					399. SYNCHYTRIUM.—De Bary and Wor.,
2204	1699	VII. 1002	S. Succisæ	De Bary and Wor., Chytr. (1863)	... ... Succisa synchytrium ... ...
2205	1698	„ 999	S. Taraxaci	... De Bary and Wor., Chytr. (1863)	... Dandelion synchytrium
					ORDER LX.—PROTOMYCETEÆ,
					400. PROTOMYCES.—Unger,
2206	1700	VII. 1120	P. macrosporus	... Unger, Exanth. 344 (1833)	... Large-spored protomyces ...

OF AUSTRALIAN FUNGI—*continued*.

Number.	Habitat.						B.	Occurrence.	General Characters.
	W.A.	S.A.	T.	V.	N.S.W.	Q.			

IN FCKL. SYMB. MYC. 66 (1875).

DE BARY.
Meek. 41 (1790).—Mucor.

| 2192 | ... | ... | ... | ... | ... | Q. | B. | Dung | Gregarious, threads slender, pellucid, weeping, yellowish, club shaped at apex. |

Sp. II. 1185 (1753).

| 2193 | ... | ... | T. | ... | ... | ... | ... | Dung | Threads simple, erect, white below, ochrey above. |
| 2194 | W.A. | S.A. | T. | V. | N.S.W. | Q. | B. | Putrid organic substances | Spore-bearing threads simple, erect, brownish. Spore sacs spherical, dark brown when dry. |

Myk. II. 113 (1823).—Ulva, Mucor.

| 2195 | ... | ... | ... | V. | N.S.W. | ... | B. | Fatty substances | Spore-bearing threads bending, shining, brass colour, continuous. Spore sacs globose, turning black. |

Ann. Sci. Nat. 66 (1875).—Mucor.

| 2196 | ... | ... | ... | V. | ... | ... | ... | Decaying Agarics | Spore-bearing threads simple, bending, shining, olive. Spore sacs somewhat globose. |

Ann. Sci. Nat. 300 (1873).—Mucor, Helicostylum.

| 2197 | ... | ... | ... | ... | ... | Q. | ... | Putrid substances | Spore-bearing threads erect, simple or branched, brown. Spore sacs spherical, becoming bluish. |

DE BARY.
Nat. 371 (1847).—Uredo, Æcidium.

| 2198 | ... | S.A. | ... | V. | N.S.W. | ... | B. | Leaves, stems, &c., of *Cruciferæ* | Pustules bursting through, white, variable. Conidia uniform, globose, colourless. Very common. |

Krypt. Fl. Schl. 236 (1888).

| 2199 | ... | ... | ... | V. | ... | ... | ... | Leaves of *Alopecurus* along with *Puccinia rubigo-vera* | Conidial stage unknown. Reproductive organs covered by epidermis, becoming brownish. |

Krypt. Fl. Schl. 236 (1888).—Peronospora, Botrytis.

| 2200 | ... | ... | ... | V. | ... | ... | ... | Leaves of Grape vine | Threads thick, frequently constricted and swollen, with minute suckers. |

Icon. I. 20 (1854).—Botrytis.

| 2201 | ... | ... | ... | V. | N.S.W. | Q. | B. | Tobacco leaves | Conidia-bearing threads thick, tall, forking. Branches spreading, tapering. |
| 2202 | ... | | | V. | N.S.W. | ... | B. | Leaves of Onion, Garlic, &c. | Forming whitish-grey or greyish-lilac tufts, sometimes covering leaves. Conidia-bearing threads large, erect, forked. |

NOWAK.
Hedw. 57 (1885).—Sporendonema, Entomophthora.

| 2203 | ... | ... | ... | V. | ... | ... | B. | Bodies of house flies and dipterous insects | White mould-like growth. Conidia bearers simple, crowded, club shaped. |

DE BARY AND WORON.
Ber. Nat. Ges. III. 22 (1863).

| 2204 | ... | S.A. | ... | V. | ... | ... | ... | Leaves and leaf stalks of Goodeniaceous plants | Cells containing spore sacs orange red. Galls wart-like, solitary or forming brown crust. |
| 2205 | ... | | ... | V. | | | B. | Leaves of *Compositæ* | Spots crust-like, running together, orange red. Galls small, flattened, scarcely projecting. |

DE BARY.
Exanth. 341 (1833).—Physoderma.

| 2206 | ... | ... | ... | ... | ... | Q. | B. | *Hydrocotyle asiatica* | Spores usually collected in scattered bulging spots, which are at first translucent, then brown. |

GENERAL CLASSIFICATION OF MYXOMYCETES.

GROUP XII.—MYXOMYCETES, WALLR.

ARRANGEMENT OF ORDERS (9).

A.— Wall of spore sac not encrusted with lime—

Section 1. Peritrichæ—Capillitium absent or formed from wall of spore sac.

    61. TUBULINACEÆ—Wall of spore sac not perforated.
    62. CRIBRARIACEÆ—Wall of spore sac perforated.

Section 2. Columelliferæ—Capillitium originating from central columella.

    63. STEMONITACEÆ—Arising from every part of elongated columella.
    64. LAMPRODERMACEÆ—Arising from apical portion of columella.

Section 3. Calotrichæ—Capillitium not springing from columella.

    65. ARCYRIACEÆ—Threads attached.
    66. TRICHIACEÆ—Threads free.

B.— Wall of spore sac with external deposit of lime—

Section 4. Lithodermeæ—Capillitium present.

    67. DIDYMIACEÆ—Threads without lime.
    68. PHYSARACEÆ—Threads with lime.

C.— Without special spore sac.

    69. PLASMODIOPHORACEÆ—Plasmodia, or naked motile masses of protoplasm, formed.

ORDER LXI.—TUBULINACEÆ, MASS.
Genus (1)—
    401. Tubulina, Pers.

ORDER LXII.—CRIBRARIACEÆ, MASS.
Genera (2)—
    402. Enteridium, Ehrb.
    403. Clathroptychium, Rost.

ORDER LXIII.—STEMONITACEÆ, ROST.
Genus (1)—
    404. Stemonitis, Gled.

ORDER LXIV.—LAMPRODERMACEÆ, MASS.
Genus (1)—
    405. Lamproderma, Rost.

ORDER LXV.—ARCYRIACEÆ, ROST.
Genera (4)—
    406. Perichæna, Fries.
    407. Lycogala, Pers.
    408. Prototrichia, Rost.
    409. Arcyria, Hill.

ORDER LXVI.—TRICHIACEÆ, ROST.
Genus (1)—
    410. Trichia, Hall.

ORDER LXVII.—DIDYMIACEÆ, ROST.
Genera (4)—
    411. Chondrioderma, Rost.
    412. Didymium, Schrad.
    413. Spumaria, Pers.
    414. Diachæa, Fries.

ORDER LXVIII.—PHYSARACEÆ, ROST.
Genera (6)—
    415. Craterium, Trent.
    416. Physarum, Pers.
    417. Badhamia, Berk.
    418. Tilmadoche, Fries.
    419. Leocarpus, Link.
    420. Fuligo, Hall.

ORDER LXIX.—PLASMODIOPHORACEÆ, ZOPF.
Genus (1)—
    421. Plasmodiophora, Zopf.

SYSTEMATIC ARRANGEMENT

Number.	Cooke's Number.	Saccardo's Number.	Scientific Name.	Authority for Name.	English Name.

GROUP XII.—MYXOMYCETES,

ORDER LXI.—TUBULINACEÆ,

401. TUBULINA.—Pers., Syn. 197 (1808).—

2207	2032	VII. 1391	T. cylindrica ...	... D.C., Fl. Fr. II. 249 (1815) ...	... Cylindrical tubulina .. ...
2207A	,,	,, 1394	T. cylindrica, var. nitidissima	Cooke, Handb. Austr. Fung. 392 (1892) ...	... Shining tubulina ...
2208	2033	X. 4818	T. spumarioidea ...	... Cooke and Mass., Mon. Myx. 42 (1892) ...	... Spumaria-like tubulina

ORDER LXII.—CRIBRARIACEÆ,

402. ENTERIDIUM.—Ehrb. in Link Jahrb.,

| 2209 | 2034 | VII. 1399 | E. olivaceum | ... Ehrb., Sylv. Ber. II, 54 (1818) ... ... | ... Olive enteridium ... |

403. CLATHROPTYCHIUM.—Rost.,

| 2210 | 2035 | VII. 1396 | C. rugulosum | ... Rost., Mon. 225 (1875) | ... Wrinkled clathroptychium ... |

ORDER LXIII.—STEMONITACEÆ,

404. STEMONITIS.—Gled., Meth. 140 (1753).—

2211	2038	VII. 1365	S. ferruginea	... Ehr., Sylv. Berl. 20 (1818)	Ferruginous stemonitis
2212	2037	,, 1356	S. Friesiana	... De Bary, Rabh., Fung. Eur. 568 (1861–81)	... Fries' stemonitis
2213	2036	,, 1362	S. fusca ...	... Roth., Fl. Germ. I. 548 (1802)	... Brown stemonitis

ORDER LXIV.—LAMPRODERMACEÆ,

405. LAMPRODERMA.—Rost., Vers.

| 2214 | 2039 | VII. 1344 | L. echinulatum ... | ... Rost., Mon. App. 25 (1875) | ... Echinulate lamproderma ... |
| 2215 | 2040 | | L. Listeri | ... Mass., Mon. Myx. 97 (1892) | Lister's lamproderma ... ... |

ORDER LXV.—ARCYRIACEÆ,

406. PERICHÆNA.—Fries, Symb.

| 2216 | 2042 | VII. 1515 | P. applanata | ... Cooke and Mass., Mon. Myx. 116 (1892) ... | ... Flattened perichæna ... ... |
| 2217 | 2041 | ,, 1435 | P. corticalis | ... Rost., Mon. 293 (1875) ... | ... Cortical perichæna ... |

407. LYCOGALA.—Pers., Tent. Disp. 7 (1797).—

| 2218 | 2043 | VII. 1484 | L. epidendrum | ... Rost., Mon. 285 (1875) | ... Tree-growing lycogala ... |

408. PROTOTRICHIA.—Rost.,

| 2219 | 2044 | VII. 1492 | P. metallica | Mass., Mon. Myx. 127 (1892) ... | ... Metallic prototrichia ... ... |

OF AUSTRALIAN FUNGI—*continued.*

Number	Habitat						B.	Occurrence.	General Characters.
	W.A.	S.A.	T.	V.	N.S.W.	Q.			

WALLR., FL. CRYPT. II. 333 (1833).
MASS. MON. MYX. (1892).

Tubulifera, Mucor, Licea, &c.

2207	...	...	T.		...		Q.	B.	Rotten wood	... Spore sacs cylindrical, rounded at apex, gregarious, mostly crowded. Mass of spores chestnut.
2207ᴬ	...	...	...		...		Q.	...	*Eucalyptus microtheca*	Spore sacs shining, golden yellow.
2208	...	...	...	V.			...	...	Running over twigs and on ground	Æthalium irregular, ashy. Cortex membranous, netted with branched veins.

MASS. MON. MYX. (1892).

Grev. II. 51 (1873).—Lycoperdon, Licea.

| 2209 | W.A. | ... | ... | ... | ... | | ... | B. | Wood | ... Æthalium very variable in form, flattened or cushion shaped, olive. |

Mon. 225 (1875).—Fuligo, Licea.

| 2210 | W.A. | ... | ... | ... | | | Q. | B. | Dead twigs, &c. | ... Colour of æthalium variable, red brown or ochrey. Spore sacs bell shaped at apex. |

ROST. MON. (1875).

Clathrus, Comatricha, Trichia.

2211	...	...	...	...			Q.	B.	Rotten wood	... Spore sacs cylindrical, gregarious, on violet-black expansion. Spores rusty cinnamon.
2212	...	...	T.	...	...		Q.	B.	Rotten wood	... Spore sacs globose, ovate, erect. Stem black, shining.
2213	W.A.	...	T.	V.	...		Q.	B.	Rotten wood, &c.	... Spore sacs cylindrical, obtuse, on strongly-developed expansion, which is violet black.

MASS. MON. MYX. (1892).

Syst. Myc. 7 (1873).—Stemonitis.

| 2214 | ... | ... | T. | ... | ... | | ... | | Among moss | ... Spore sacs stalked, dark steel blue or blackish, iridescent. Stem short. Spores spiny. |
| 2215 | ... | ... | T. | ... | | | ... | B. | Moss, wood, &c. | ... Spore sacs stalked, globose, blackish purple, iridescent. Stem elongated, blackish brown. |

ROST. MON. (1875).

Gast. 9 (1818).—Hemiarcyria, Lycoperdon.

| 2216 | ... | ... | ... | ... | | | Q. | ... | Rotting *Cycas* | ... Sessile, much depressed, circular or irregular in outline. Spores in mass clear orange yellow. |
| 2217 | W.A. | ... | ... | ... | ... | | ... | B. | Bark and wood | ... Spore sacs spherical to depressed, crowded. Spores in mass pale yellow. |

Fungus, Lycoperdon, Bovista.

| 2218 | W.A. | ... | ... | V. | ... | | Q. | B. | Stumps ... | ... Æthalium gregarious, rounded, size of pea, shining, distinctly warted, rose colour. |

Mon. Appl. 38 (1875).

| 2219 | ... | ... | T. | ... | | | ... | ... | Wood | ... Spore sacs scattered, stalked or sessile, copper colour with metallic tints. Stem very short. |

Number.	Cooke's Number.	Saccardo's Number.	Scientific Name.	Authority for Name.	English Name.

409. Arcyria.—Hall, Hist. 47 (1768).—Clathrus,

2220	2049	VII. 1459	A. cinerea	Schum., Saell. 1480 (1801)	Grey arcyria
2221	2046	„ 1470	A. ferruginea	Rost., Mon. 280 (1875)	Ferruginous arcyria
2222	2052	X. 4857	A. fuliginea	Mass., Mou. Myx. 169 (1892)	Sooty-brown arcyria
2223	2047	VII. 1461	A. incarnata	Pers., Obs. (1796)	Fleshy arcyria
2224	2048	„ 1464	A. nutans	Grev., Fl. Ed. 455 (1824)	Drooping arcyria
2225	2045	„ 1457	A. punicea	Pers., Disp. 10 (1797)	Reddish arcyria
2226	2050	„ 1512	A. rubiformis	Mass., Mon. Myx. 158 (1892)	Lustrous arcyria
2227	2051	„ 1514	A. serpula	Mass., Mon. Myx. 164 (1892)	Creeping arcyria

ORDER LXVI.—TRICHIACEÆ,
410. Trichia.—Hall. Helv. III. 114 (1768).—

2228	2057	VII. 1499	T. affinis	De Bary. in Rost. Mon. 257 (1875)	Allied trichia
2229	2055	„ 1503	T. contorta	Rost., Mon. 259 (1875)	Contorted trichia
2230	2053	„ 1494	T. fragilis	Rost., Mon. 246 (1875)	Fragile trichia
2231	...	X. 4848	T. Kalbreyeri	Mass., Mon. Myx. 191 (1892)	Kalbreyer's trichia
2232	2054	VII. 1497	T. varia	Pers., Tent. Disp. 10 (1797)	Variable trichia
2233	2056	X. 4847	T. verrucosa	Berk., Fl. Tasm. II. 269 (1860)	Warted trichia

ORDER LXVII.—DIDYMIACEÆ,
411. Chondrioderma.—Rost.,

| 2234 | 2058 | VII. 1282 | C. difforme | Rost., Mon. 177 (1875) | Deformed chondrioderma |
| 2235 | ... | „ 1257 | C. Muelleri | Rost., Mon. 15 (1875) | Mueller's chondrioderma |

412. Didymium.—Schrad., Nov. Pl. Gen. I. 22 (1797).—

2236	2064	X. 4803	D. australe	Mass., Grev. XVII. 7 (1888)	Southern didymium
2237	2059	VII. 1309	D. farinaceum	Schrad., Nov. Pl. Gen. I. (1797)	Mealy didymium
2238	2065	„ 1193	D. flavicomum	Mass., Mon. Myx. 242 (1892)	Yellow-haired didymium
2239	2063	„ 1256	D. pezizoideum	Mass., Mon. Myx. 239 (1892)	Peziza-like didymium
2240	2062	„ 1297	D. serpula	Fries, S.M. III. 126 (1832)	Creeping didymium
2241	2061	„ 1269	D. spumarioides	Fries, Symb. Gast. 20 (1818)	Spumaria-like didymium
2242	2060	„ 1301	D. squamulosum	Fries, S.M. III. 118 (1832)	Scaly didymium

413. Spumaria.—Pers.,

| 2243 | 2066 | VII. 1338 | S. alba | D. C., Fl. Fr. II 261 (1815) | White spumaria |

OF AUSTRALIAN FUNGI—*continued.*

Trichia, Stemonitis, Hemiarcyria, Clathroides, Mucor.

Number	W.A.	S.A.	T.	V.	N.S.W.	Q.	B.	Occurrence.	General Characters.
2220	...	...	...	...	...	Q.	B.	Stumps ...	Gregarious, stalked, ovoid or elongated ovoid. Stem erect, straight, long. Spores usually bright grey.
2221	...	...	...	...	...	Q.	B.	Rotten wood, &c. ...	Spore sacs ovate. Stem usually short. Spores in mass usually brick red, now and then rusty.
2222	...	...	...	...	N.S.W.	...	...	Leaves of *Atheros-permum*	Threads forming a net-work, spiny. Spores globose, smooth, in mass sooty brown.
2223	W.A.	...	...	...	...	Q.	B.	Rotten wood	Spore sacs egg shaped, with evanescent short erect stem. Spores in mass usually flesh colour.
2224	W.A.	...	...	V.	...	Q.	B.	Rotten wood	Spore sacs cylindrical, with short evanescent stem. Capillitium drooping.
2225	...	...	...	...	...	Q.	B.	Rotten stumps	Spore sacs more or less egg shaped, of beautiful lustre, usually with erect stem.
2226	...	...	T.	...	...	...	B.	Dead wood	Spore sacs usually tufted, collected in short stem, often of beautiful metallic lustre.
2227	W.A.	...	...	...	...	...	B.	Rotten wood	Vein-like, creeping, forming a net, or somewhat globose, and sessile on broad base, yellow.

ROST. MON. (1875).

Lycogala, Licea, Lycoperdon.

2228	...	...	T.	...	...	Q.	B.	Rotten wood, &c. ...	Spore sacs clustered, circular or elliptical, sessile on broad base, clear yellow.
2229	...	...	T.	...	...	Q.	B.	Rotten wood	Variable in form, sometimes elongated and twisted, sometimes veined, creeping, hay brown.
2230	W.A.	...	T.	...	...	...	B.	Rotten wood, &c. ...	Spore sacs stalked, or tufted on common stem ; wall smooth, blackish or yellowish.
2231	...	...	T.	...	...	...	...	Fragments of rotting plants	Spore sacs crowded, sessile, globose, yellow, nearly the same colour within.
2232	...	...	T.	...	...	Q.	B.	Stumps ...	Spore sacs variously developed, either stalked or sessile.
2233	...	...	T.	...	...	...	...	Wood ...	Spore sacs brown or chestnut, shining, passing down into long slender stem).

ROST. MON. (1875.)

Mon. 167 (1875).—Physarum.

| 2234 | ... | S.A. | ... | V. | ... | Q. | B. | Bark, leaves, twigs, grass, &c. | Spore sacs sessile, roundish, deformed ; outer wall crustaceous, chalky white. |
| 2235 | ... | ... | ... | ... | ... | Q. | ... | ... ... | Spore sacs discoid, curved upwards, snow white, stalked. Stem straight, rigid, with rusty-brown furrows. |

Mucor, Trichia, Physarum, Chondrioderma, Spumaria.

2236	...	...	...	...	...	Q.	...	Old Auricularia ...	Spore sacs globose or slightly compressed, covered with dense white layer of lime.
2237	...	...	...	V.	...	Q.	B.	Dead leaves, twigs, decaying fruit, &c.	Spore sacs hemispherical or a little flattened, greyish white with lime, or black and shining.
2238	W.A.	...	...	...	...	...	...	Rotten wood	Fructification hemispherical, violet or lilac. Stem elongated, slender, yellowish tan or copper colour.
2239	...	...	...	...	...	Q.	...	Dead wood of *Erythrina vespertilia*	Fructification somewhat nodding, ashy white, arising from mealy crusty cracking membrane.
2240	...	...	...	...	...	Q.	B.	Fallen leaves and rotten wood	Fructification either cushion-like, flattened, or vein-like, creeping.
2241	...	...	T.	...	...	Q.	B.	Leaves, moss, clover, &c.	Spore sacs irregular in shape, snow white or greyish, always in clusters.
2242	...	...	T.	...	...	...	B.	Wood, dead leaves, &c.	Spore sacs either hemispherical and flattened or globose. Stem snow white.

Syn. 162 (1808).—Reticularia.

| 2243 | ... | ... | ... | V. | N.S.W. | Q. | B. | Grass ... | ... Æthalium complex, branching, whitish to grey, spongy. |

Number.	Cooke's Number.	Saccardo's Number.	Scientific Name.	Authority for Name.	English Name.
					414. DIACHÆA.—Fries, Syst. Orb.
2241	2067	VII. 1335	D. leucopoda	... Rost., Mon. 191 (1875)...	... White-stalked diachæa ...
					ORDER LXVIII.—PHYSARACEÆ,
					415. CRATERIUM.—Trent.,
2245	2068	VII. 1233	C. confusum	... Mass., Mon. Myx. 263 (1892) ...	... Confused craterium ... ...
					416. PHYSARUM.—Pers., Obs. Myc. 5 (1799).—
2246	2073	VII. 1189	P. cinereum	Pers., Syn. 170 (1808)...	... Ashy physarum ... ...
2247	...	„ 1171	P. didermoides	... Rost., Mon. 97 (1875) ...	... Two-membraned physarum ...
2248	2072	„ 1192	P. leucophæum ...	... Fries, Symb. Gast. 24 (1818)	... Grey physarum ...
2249	2071	„ 1188	P. leucopus	... Rost., Mon. 101 (1875)... ...	... White-stalked physarum ...
2250	2070		P. Readeri	... Mass., Mon. Myx. 282 (1892)	... Reader's physarum ... ...
2251	2069	VII. 1251	P. rufibasis	... Berk. and Br., Linn. Journ. XIV. 85 (1875)	... Red-based physarum ... ...
2252	2074	„ 1189	P. scrobiculatum ...	... Mass., Mon. Myx. 300 (1892) ...	... Pitted physarum ...
					417. BADHAMIA.—Berk.,
2253	2075	VII. 1150	B. varia	... Mass., Mon. Myx. 319 (1892) ...	... Variable badhamia ... ...
					418. TILMADOCHE.—Fries,
2254	2077	VII. 1217	T. mutabilis	... Rost., Mon. 130 (1875)...	Changeable tilmadoche ...
2255	2076	„ 1244	T. nutans	... Rost., Mon. 127 (1875)	... Nodding tilmadoche ... ...
					419. LEOCARPUS.—Link,
2256	2078	VII. 1242	L. fragilis	Rost., Mon. 132 (1875)...	... Fragile leocarpus ... ...
					420. FULIGO.—Hall, Hist.
2257	2079	VII. 1228	F. varians	... Sommf., Fl. Lapp. 231 (1826) ...	... Variable fuligo ... ...
					ORDER LXIX.—PLASMODIOPHORACEÆ,
					421. PLASMODIOPHORA.—Woron.
2258		VII. 1568	P. Brassicæ	... Woron. Pringsh. Jahrb. XI. 548 (1878) ...	... Turnip Plasmodiophora (Club-Root) (Fingers and Toes)

199

OF AUSTRALIAN FUNGI—*continued.*

Number	Habitat						B.	Occurrence.	General Characters.
	W.A.	S.A.	T.	V.	N.S.W.	Q.			

Veg. I. 143 (1825).—Trichia, Stemonitis.

| 2244 | ... | ... | ... | V. | ... | ... | B. | Leaves, &c. | Spore sacs cylindrical, stalked. Stem short, thickened at base, snow white. |

ROST. MON. (1875).

in Roth. Cat. II. 224 (1806).

| 2245 | W.A. | ... | T. | ... | ... | ... | B. | Leaves, &c. | Spore sacs variable in form, stalked or rarely somewhat sessile, bright, brown, ochrey, or white. |

Lycoperdon, Trichia, Didymium, Tilmadocho.

2246	W.A.	...	...	V.	...	...	B.	Bark, wood, leaves, &c.	Spore sacs globose or hemispherical, sessile or gregarious.
2247	...	...	...	...		Q.	B.	Scales of Onions,and bracts of Maize, Grass, &c.	Spore sacs ovoid, ash coloured,with white mealy covering,and separate membranous outer coat.
2248	..	...	...	...	...		B.	Leaves, &c.	Spore sacs somewhat globose, stalked or sessile, wall thin with white lime patches.
2249	...	...	...	...		...	B.	Wood, &c.	Spore sacs globose, stalked or sessile, wall covered with snow-white coat of lime. Stem white.
2250	...	...	...	V.	...	...		Wood	Spore sacs stalked, greyish, covered with flakes of lime. Stem very thick, brown.
2251	...	...	...	...		Q.	...	Moss	Scattered or gregarious, stalked. Stem elongated, slender, expanding into circular bright-brown base. Spore sacs globose, dull yellow or tawny.
2252	W.A.	...	...	...		...	B.	Charred wood	Spore sacs sessile, on broad or narrowed base, seated on thick spreading expansion.

Outl. 308 (1860).—Physarum.

| 2253 | ... | ... | T. | V. | N.S.W. | ... | B. | Rotten wood,&c. | Spore sacs more or less clustered, sessile or stalked, globose, grey or opaque. |

S.V.S. 454 (1849).—Stemonitis.

| 2254 | W.A. | ... | ... | ... | | Q. | B. | Decayed wood | Spore sacs globose or flattened, usually cracked, yellow or rusty orange, stalked, nodding. |
| 2255 | W.A. | ... | T. | ... | ... | Q. | ... | Rotten wood, &c. | Spore sacs lens shaped, usually cracked, greyish white,stalked, nodding. |

Sp. Pl. I. 25 (1824).—Lycoperdon.

| 2256 | ... | ... | T. | ... | ... | ... | B. | Grass, twigs, moss, &c. | Spore sacs somewhat roundish, sessile, or with thin thread-like coloured stem. |

Helv. III. 110 (1768).—Mucor.

| 2257 | W.A. | S.A. | T. | V. | ... | Q. | B. | Wood, tan, soil, &c. | Spore sacs more or less closely interwoven, bark not always developed ; walls of spore sac usually coloured. |

ZOPF. PILZTH. 129 (1885).

Pringsh. Jahrb. XI., 548 (1878).

| 2258 | ... | ... | ... | V. | N.S.W. | ... | B. | Roots of Crucifers—*Brassica,* &c. | Producing the malformation of the roots of cabbages, cauliflowers, &c., which gives them a clubbed appearance, or several misshapen roots like " fingers and toes." |

Number.	Cooke's Number.	Saccardo's Number.	Scientific Name.	Authority for Name.	English Name.

GROUP I.—
ORDER II.—
Genus 422.—Laccocephalum, McAlp.,

| 2259 | ... | | L. basilapiloides ... | ... McAlp. and Tepp., Proc. Roy. Soc. Vic. VII. N.S. 166 Pl. X. (1894) | Stone-like-base laccocephalum ... |

GROUP III.—
ORDER XIII.—

2260	...	VII. 1026	Uromyces Phaseoli	*Winter*, Die Pilze 157 (1884)	Bean uromyces ... ... (Bean rust)
2261	...		Puccinia Correæ ...	... McAlp., Proc. Roy. Soc. Vic. VII. N.S. 215 (1894)	Correa puccinia ...
2262	...		P. Erechtitis	... McAlp., Proc. Roy. Soc. Vic. VII. N.S. 216 (1894)	Erechtites puccinia ... ...
2263			P. Hypochæris	McAlp., Proc. Roy. Soc. Vic. VII. N.S. 217 (1894)	Hypochæris puccinia ... ...
2264	...		P. Plagianthi	... McAlp., Proc. Roy. Soc. Vic. VII. N.S. 218 (1894)	Plagianthus puccinia ...
2265	...		Æcidium eburneum	... McAlp., Proc. Roy. Soc. Vic. VII. N.S. 218 (1894)	Ivory æcidium ...
2266		IX. 1318	A. monocystis	Berk., Flor. N.Z II. 196 (1855)...	Walled æcidium
2267	...		Puccinia Coprosmatis	Morrison, Vict. Nat. XI. No. 6, 90 (1894)	... Coprosma puccinia ...
2267A	...		P. Coprosmatis, var. Opercularæ	Morrison, Vict. Nat. XI. No. 8, 119 (1894)	... Opercularia puccinia ...
2268	...	VII. 2457	P. investita	Schwein, Syn. N. Am. Fungi (1831)	... Invested puccinia ...

GROUP IV.—
ORDER XV.—

2269	...	I. 1166	Xylaria fulvella ...	... Berk. and Curt., Linn. Journ. X. 380 (1869)	... Tawny xylaria ... ...
2270	...	I. 1282	X. ianthino-velutina	... Mont., Syll. Crypt. (1856)	... Violet-haired xylaria ... ...
2271	...	IX. 2282	Kretzschmaria confusa	... Sacc., Syll. IX. 566 (1891) ..	... Confused kretzschmaria
2272	...		Hypoxylon atrosphæricum ...	Cooke and Mass., Grev. XXII. 68 (1894)	... Black-sphered hypoxylon ...

ORDER XVI.—

| 2273 | | II. 5107 | Phyllachora Grevilleæ | Sacc., Syll. II. 597 (1883) | ... Grevillea phyllachora ... ... |

GROUP V.—
ORDER XXXIV.—

| 2274 | | | Peziza Lyonsiæ | ... Cobb., Ag. Gaz. N.S.W. V. 6, 390 (1894) | ... Lyonsia peziza ... ... |
| 2275 | | | Belonidium parasiticum | Cooke and Mass., Grev. XXII. 68 (1894) | ... Parasitic belonidium ... |

AUSTRALIAN FUNGI.

Number	Habitat.						B.	Occurrence.	General Characters.
	W.A.	S.A.	T.	V.	N.S.W.	Q.			

HYMENOMYCETES, FRIES.
POLYPORACEÆ, FRIES.
Proc. Roy. Soc. Vic. VII. N.S. 166 (1894).

| 2259 | ... | S.A. | ... | ... | ... | ... | ... | Sandy soil in Mallee scrub | Solitary. Cap woody, brownish fawn, surface pitted. Stem compressed oval, dirty fawn, hardened like cap. Pores moderately large, crowded, nearly oval. |

UREDINES, BRONGN.
UREDINACEÆ, BRONGN.

2260	...	...	...	...	N.S.W.	...	B.	All parts of Bean plant (*Phaseolus vulgaris*), more especially on leaves	Pustules brown, scattered, bursting through, surrounded by ruptured cuticle.
2261			T.					Under surface of leaves of *Correa Lawrenciana*	Pustules cushion shaped, circular or interruptedly so, dirty brown, scattered, soon naked.
2262	...	...		V.				Stem and leaves of *Erechtites quadridentata*	Cluster cups pale yellow to orange yellow, causing distortion and swelling. Pustules black, crowded together, and forming swelling.
2263				V.			...	Leaves of *Hypochœris radicata*	Cluster cups on greenish-yellow or brownish circular patches. Pustules intermixed with cluster cups, black, and on both surfaces of leaf.
2264			T.	...			...	Leaves and flowers of *Plagianthus sidoides*	Pustules reddish brown, naked, blistered, scattered. Very common.
2265			T.	V.				Stem, leaves, flowers, and legumes of *Bossiæa cinerea*	Cluster cups ivory colour to brownish, clustered together without definite order.
2266			T.	...			...	Leaves of *Abrotanella forsteroides*	Cluster cups near tips of leaves, large, solitary, surrounded by strong wall arising from matrix.
2267				V.			...	Leaves of *Coprosma Billardieri*	Pustules on under surface, seldom on upper, prominent, deep brown, coalescing. Teleutospores compact, brown.
2267A	...	...	...	V.			...	Leaves and petioles of young plants of *Opercularia varia*	Pustules on under surface, deforming leaf and forming concavity on opposite side, reddish-brown. Teleutospores pale yellowish brown.
2268	...	...		V.			...	Leaves and stems of *Gnaphalium purpureum*	Pustules on both surfaces of leaf, bursting through epidermis, and bordered by ruptured cuticle. Uredospores and teleutospores generally mixed.

PYRENOMYCETES, FRIES.
XYLARIACEÆ, COOKE.

2269	...	...	...	...		Q.		At base of dead stump	Club shaped, rust coloured, papillate. Stem cylindrical, pale tawny. Receptacles with black openings.
2270	...	...	...	...		Q.		Old fruit of a *Flindersia*	Simple or branched, cylindrical and tapering, apex compressed, long violet-brown hair all over.
2271	..	...	..	...	...	Q.		Bark of dead log	Gregarious, stalked, simple. Heads depressed, globose, glaucous, at length black.
2272	...	...	...	...		Q.		Bark	Gregarious, sub-globose, black. Receptacles around the circumference ovate, tent-like.

DOTHIDEACEÆ, NITS.

| 2273 | W.A. | ... | ... | ... | ... | ... | ... | Leaves of *Grevillea* | On both surfaces, scattered, circular, shining black. Receptacles globose, immersed. |

DISCOMYCETES, FRIES.
PEZIZACEÆ, FRIES.

| 2274 | ... | ... | ... | ... | N.S.W. | ... | | Leaves of *Lyonsia reticulata* | Cups somewhat gregarious on both sides, on ashy-grey roundish spots, flat, sessile, round. Leaves appear at a little distance as if attacked by scale insect. |
| 2275 | ... | ... | ... | ... | | Q. | | Parasitic on *Asterina*, growing on leaflets of *Tarrietia trifoliolata* | White Cups very minute, hairless, concave, attached by central papilla, hardly visible to naked eye. |

ADDITIONS.—(A.)—NEW

Number.	Cooke's Number.	Saccardo's Number.	Scientific Name.	Authority for Name.	English Name.

GROUP VII.—
ORDER XLV.—

| 2276 | ... | | Oidium Oxalidis ... | ... McAlp., Proc. Roy. Soc. Vic. VII. N.S. 219 (1894) | Wood-sorrel oidium ... ... |

ORDER XLVI.—
Genus 423.—Stachybotrys,

| 2277 | ... | IV. 1304 | Stachybotrys lobulata | ... Berk., Outl. 343 (1860) ... | ... Lobed stachybotrys ... ... |
| 2278 | ... | „ 1675 | Cladosporium carpophilum ... | Thuem., Fung. Pom. 13 (1878)... | ... Fruit-loving cladosporium ... (Peach freckle) |

ORDER XLVII.—

| 2279 | ... | | Isaria Oncopterae ... | ... McAlp., Proc. Roy. Soc. Vic. VII. N.S. 159 (1894) | Oncoptera isaria ... ... |

GROUP VIII.—
ORDER XLIX.—

| 2280 | ... | III. 2796 | Septoria Dianthi .. | ... Desm., 17 Not. 6, p. 20... | ... Carnation septoria ... ... |

GROUP IX.—

2281	...	VIII. 3622	Saccharomyces conglomera- tus	Reess, Bot. Unt. 82 (1870)	... Conglomerate yeast ... ...
2282	...	„ 3623	S. exiguus	... Reess, Bot. Unt. 82 (1870)	... Small yeast ...
2283		„ 3629	S. Marxianus ...	... Hansen in Ann. de Microg. (1888)	... Marx yeast ... ...
2284		„ 3630	S. membranifaciens	... Hansen Bot. Zeil. 772 (1888) ...	... Membrane-forming yeast ...
2285		„ 3635	S. minor	... Engel, Ferm. (1872) ...	... Lesser yeast ... ...
2286	...	„ 3624	S. Pasteurianus ...	Reess, Bot. Unt. 83 (1870) ...	... Pasteur's yeast ...

GROUP X.—
ORDER LV.—

| 2287 | | | Ustilago Allii | ... McAlp., Proc. Roy. Soc. Vic. VII. N.S. 220 (1894) | Onion ustilago ... ... |
| 2288 | | | U. Poarum | ... McAlp., Proc. Roy. Soc. Vic. VII. N.S. 220 (1894) | Poa ustilago ... ... |

Genus 424.—Tolyposporium, Woron.

| 2289 | | | Tolyposporium Anthistiriae | Cobb, Ag. Gaz. N.S.W. III., pt. 12 (1892) ... | Kangaroo-grass tolyposporium ... |

GROUP XI.—
ORDER LVII.—

| 2290A | | | Peronospora parasitica, var. Lepidii | McAlp., Proc. Roy. Soc. Vic. VII. N.S. 221 (1894) | Lepidium peronospora ... ... |

AUSTRALIAN FUNGI—*continued.*

Number.	Habitat.						D.	Occurrence.	General Characters.
	W.A.	S.A.	T.	V.	N.S.W.	Q.			

HYPHOMYCETES, MARTIUS.
MUCEDINACEÆ, LINK.

| 2276 | ... | ... | ... | V. | ... | ... | ... | | Leaves, leaf stalks, stem, and fruit of *Oxalis corniculata* | Mostly on upper surface of leaves, sometimes on lower, spread out, greyish, powdery. |

DEMATIACEÆ, FRIES.
Corda, Anleit. 57 (1842).

| 2277 | ... | ... | ... | ... | ... | Q. | B. | Damp wall-paper ... | Black, sterile hyphæ creeping, fertile branches ascending, simple or branched, pale upwards. |
| 2278 | ... | ... | ... | ... | N.S.W. | ... | ... | Peaches ... ... | Spots circular, small at first, often confluent, dark green, finally causing cracking of peach. |

STILBEACEÆ, FRIES.

| 2279 | ... | ... | ... | ... | V. | ... | ... | ... | Dead larvæ of *Oncoptera intricata.* Walk. | Dirty-brown root colour. Stem branched, velvety, slender, tips of branches fertile. |

SPHÆROPSIDES, LEV.
SPHÆRIOIDACEÆ, SACC.

| 2280 | ... | ... | ... | V. | ... | ... | B. | Leaves of Carnations | Spots yellowish, oblong, round, or irregular. Receptacles globose to depressed, black to brown. |

SACCHAROMYCETES, REESS.

2281	...	...	...	...	...	...	B.	In fermentation and putrefaction of wine	Cells spheroidal, and forming a conglomeration instead of chains or flakes.
2282	...	...	...	...	...	...	B.	In juices of fermented fruits and fermentation of beer and wine	Cells spherical or top-shaped, united into a few branched colonies, and very small.
2283	...	...	...	...	...	...	...	Grape berry ...	Cells somewhat resembling *C. ellipsoideus.* The asci are many-spored.
2284	...	...	...	...	...	...	...	Saccharine liquids...	
2285	...	...	...	...	...	...	B.	In fermentation of bread	Cells spherical, isolated, double or sometimes in threes; like ordinary yeast but smaller.
2286	...	...	...	...	...	...	B.	In fermentation of wine and self-fermenting beer	Cells oval, oblong to club-shaped, varying in size; in branched colonies or flakes.

USTILAGINES, TUL.
USTILAGINACEÆ, TUL.

| 2287 | ... | ... | ... | V. | ... | ... | ... | Scale leaves of stored Union bulbs | Pustules minute, dark coloured, in parallel lines along veins of leaves, at first covered by epidermis; then powdery black. |
| 2288 | ... | ... | ... | V. | ... | ... | ... | *Poa annua,* especially foliage | Distorting, discolouring, and stunting plants, and forming black powdery masses. |

in Schrœt. Pilzfl. Schles. 276 (1882).

| 2289 | ... | ... | ... | ... | N.S.W. | ... | ... | Inflorescence of Kangaroo-grass (*Anthistiria ciliata*) | Fructification black. Spores compound, consisting of from a few dozen to several hundred thick-walled brown spores. |

PHYCOMYCETES, DE BARY.
PERONOSPORACEÆ, DE BARY.

| 2290A | ... | ... | ... | V. | ... | ... | ... | Leaves, stem, and fruit of *Lepidium ruderale* | Lower surface of leaf attacked first, causing it to curl. Dense white mould forming a felt. |

B.—NEW LOCALITIES, HOSTS, ETC.

GROUP I.—HYMENOMYCETES, FRIES.

ORDER IV.—THELEPHORACEÆ, PERS.

Genus 425.—Soppittiella, Mass. Brit. Fung. Fl. 106 (1892).

1045. THELEPHORA CRISTATA = Soppittiella cristata, Mass. B.
1080. STEREUM OCHROLEUCUM—V.

GROUP II.—GASTROMYCETES, WILLD.

ORDER X.—LYCOPERDACEÆ, EHRH.

1343. BATTARREA PHALLOIDES—Q.

GROUP III.—UREDINES, BRONGN.

ORDER XIII.—UREDINACEÆ, BRONGN.

1446. UROMYCES BETÆ—N.S.W.
1455. UROMYCES ORCHIDEARUM—V. Uredospores intermixed with teleutospores.
1461. MELAMPSORA LINI—T.
1477. PUCCINIA HETEROSPORA—On leaves of a native Hibiscus in Queensland.
1478. PUCCINIA HIERACII—V. On leaves of flowering stems of Hypochœris radicata.
1511. ÆCIDIUM RANUNCULACEARUM—On Ranunculus parriflorus in Victoria.

GROUP IV.—PYRENOMYCETES, FRIES.

ORDER XIV.—HYPOCREACEÆ, DE NOT.

1527. CLAVICEPS PURPUREA—T., N.S.W.

ORDER XV.—XYLARIACEÆ, COOKE.

1619. HYPOXYLON ELLIPTICUM—V. Rotten wood.

ORDER XXVIII.—FOLIICOLACEÆ, FRIES.

1712. SPHÆRELLA FRAGARIÆ—W.A., T.

ORDER XXX.—PERISPORIACEÆ, FRIES.

1756. CAPNODIUM CITRI—W.A., Q. Leaves of oranges and lemons.

GROUP V.—DISCOMYCETES, FRIES.

ORDER XXXIII.—HELVELLACEÆ, Link.

1778. MORCHELLA DELICIOSA—Q. Amongst rotten bark, near stem of gum-tree.

ORDER XXXIV.—PEZIZACEÆ, FRIES.

1869. DASYSCYPHA TERRESTRIS—This form should be restored to its original genus, Helotium terrestre, Berk. and Broome, Linn. Trans. II., 69 (1883). Cooke remarks in Grev. XX, 36 (1891), that this species was originally described in error as being externally villous, whereas it is externally smooth and naked; hence it was wrongly transferred by Saccardo to Dasyscypha.

ORDER XXXIX.—PHACIDIACEÆ, FRIES.

1894. PSEUDOPEZIZA MEDICAGINIS—On both surfaces of leaflets of Medicago sativa in Victoria.

ORDER XLI.—GYMNOASCEÆ, BAR.

1901. EXOASCUS DEFORMANS—T., W.A. On leaves of peaches and nectarines in West Australia.

GROUP VII.—HYPHOMYCETES, MARTIUS.

ORDER XLV.—MUCEDINACEÆ, LINK.

1911. OIDIUM LEUCOCONIUM—T.
1914. OIDIUM TUCKERI—W.A.
1921. PENICILLIUM GLAUCUM—T.

ORDER XLVI.—DEMATIACEÆ, FRIES.

1947. FUSICLADIUM DENDRITICUM—W.A.
1948. FUSICLADIUM PYRINUM—W.A.
1962. HELMINTHOSPORIUM RAVENELII—N.S.W.
1982. MACROSPORIUM TOMATO—N.S.W.

GROUP VIII.—SPHÆROPSIDES, LEV.

ORDER XLIX.—SPHÆRIOIDACEÆ, SACC.

2025. PHYLLOSTICTA CIRCUMSCISSA—W.A.   On apricots.
2039. PHOMA AMPELINA—W.A.   On vines.
2058. PHOMA UVICOLA—N.S.W.
2102. PHLEOSPORA MORI—T.

C.—LIST OF AUSTRALIAN EDIBLE FUNGI.

In the body of the work a number of species are marked Edible, but it is thought desirable to show them together, and give as complete a list as possible. Edible Fungi refers to those which may be eaten with impunity, not necessarily to those which may be eaten with relish. Our native species have still to be tested in most cases, but I have mainly given those Australian species which have been found wholesome in Britain or America. The number is 84.

No.	Name.
6.	Amanita ovoidea.
15.	Amanitopsis vaginata.
20.	Lepiota cepæstipes.
25.	Lepiota excoriata.
32.	Lepiota mastoidea.
35.	Lepiota naucina.
38.	Lepiota procera.
39.	Lepiota rhacodes.
47.	Armillaria mellea.
56.	Tricholoma nudum.
63.	Clitocybe cerussata.
65.	Clitocybe expallens.
67.	Clitocybe fumosa.
69.	Clitocybe infundibuliformis.
71.	Clitocybe laccata.
73.	Clitocybe pruinosa.
81.	Collybia esculenta.
82.	Collybia fusipes.
93.	Collybia radicata.
176.	Pleurotus ostreatus.
178.	Pleurotus petaloides.
181.	Pleurotus pulmonarius.
182.	Pleurotus salignus.
195.	Hygrophorus coccineus.
202.	Hygrophorus miniatus.
208.	Hygrophorus virgineus.
209.	Lactarius pallidus.
210.	Lactarius piperatus.
214.	Russula alutacea.
227.	Cantharellus cibarius.
269.	Marasmius scorodonius.
313.	Lentinus tigrinus.
320.	Panus conchatus.
331.	Panus torulosus.
367.	Volvaria bombycina.
400.	Pholiota mutabilis.
403.	Pholiota præcox.
404.	Pholiota pudica.
496.	Agaricus arvensis.
497.	Agaricus campestris.
500.	Agaricus silvaticus.
528.	Coprinus comatus.
601.	Boletus æreus.

No.	Name.
602.	Boletus æstivalis.
603.	Boletus alliceus.
606.	Boletus badius.
611.	Boletus edulis.
612.	Boletus elegans.
616.	Boletus granulatus.
621.	Boletus luteus.
629.	Boletus scaber.
641.	Fistulina hepatica.
685.	Polyporus intybaceus.
690.	Polyporus mylodes.
691.	Polyporus Mylittæ (Sclerotium, known as "Native Bread").
699.	Polyporus picipes.
723.	Polyporus sulphureus.
728.	Polyporus tumulosus (eaten by Aborigines).
993.	Hydnum coralloides.
1004.	Hydnum lævigatum.
1011.	Hydnum repandum.
1037.	Craterellus cornucopioides.
1183.	Sparassis crispa.
1188.	Clavaria aurea.
1189.	Clavaria botrytes.
1194.	Clavaria cristata.
1196.	Clavaria fastigiata.
1197.	Clavaria flava.
1198.	Clavaria formosa.
1223.	Clavaria rugosa.
1233.	Hirneola auricula-judæ.
1237.	Hirneola polytricha.
1245.	Tremella lutescens.
1289.	Clathrus cibarius.
1392.	Lycoperdon Bovista.
1397.	Lycoperdon gemmatum.
1400.	Lycoperdon lilacinum.
1776.	Cyttaria Gunnii.
1777.	Morchella conica.
1778.	Morchella deliciosa.
1779.	Morchella esculenta.
1780.	Morchella semilibera.
1785.	Leotia lubrica.
1797.	Peziza cochleata.

TABLE I.

Number of Orders, Genera, Species, and Varieties in the different Groups and different Colonies, together with those common to Britain.

Groups	No. of Orders	No. of Genera	No. of Species	No. of Varieties	Australian — Orders	Australian — Genera	Australian — Species	Australian — Varieties	W.A. — Orders	W.A. — Genera	W.A. — Species	W.A. — Varieties	S.A. — Orders	S.A. — Genera	S.A. — Species	S.A. — Varieties	T. — Orders	T. — Genera	T. — Species	T. — Varieties	V. — Orders	V. — Genera	V. — Species	V. — Varieties	N.S.W. — Orders	N.S.W. — Genera	N.S.W. — Species	N.S.W. — Varieties	Q. — Orders	Q. — Genera	Q. — Species	Q. — Varieties	B. — Orders	B. — Genera	B. — Species	B. — Varieties
1. Hymenomycetes	6	106	1,366	25	4	14	19		6	57	137		6	61	151	5	5	16	41		6	88	600	5	61	82	347	8	6	86	618	11	6	81	455	5
2. Gastromycetes	6	39	177	2	3	4	5		5	18	44		4	12	55		5	26	72		6	16	42		5	28		30	16		
3. Uredines	1	9	90	2										6			1		25			27	55	6		
4. Pyrenomycetes	18	83	253	1	8	9	9		9	16	19		3	13	12		11	45	89		10	39	157		49	125	3	33	53	
5. Discomycetes	10	43	125	3	2	3	3		3	7	10		2	7	10		8	21	68	2	21	57	3	31	49	
6. Tuberoidea	3	3	3											1											
7. Hyphomycetes	4	32	123	2	1	1	2		4	11	13		3	12	11		4	30	79		4	28	60	3	50	32	4
8. Sphaeropsides	5	43	128		1	1			2	3			2	5	8		5	31	77		4	25	53	12	12	3
9. Saccharomycetes	1	1	10																					10	5	
10. Ustilagines	1	12	39	4	1	1	2		1	6	16		1	5	14		7	7	21	2	6	5	4		16	9	
11. Phycomycetes	5	12	13	1	1	1			6	12			2	3	4		9	11		5	4	9	9	x	5
12. Myxomycetes	9	21	52	1	1	1	2						2	2	2		12	22	13		13	13	4	38	50	3	9
Totals	69	421	2,384	47	22	33	50		37	125	342	2	35	127	262	8	54	161	339	21	54	291	1,070	12	158	406	12	54	277	1,060	24	54	234	739	9

TABLE II.

Number of Australian Species compared with British and total known Species.

Groups.		No. of Australian Species (1894).	No. of British Species (1892).	Total known Species (1892), (Approximate).
1. Hymenomycetes ...	...	1,266	1,902	10,163
2. Gastromycetes ...	...	177	78	718
3. Uredines	...	90	53	1,428
4. Pyrenomycetes ...	...	253		9,247
5. Discomycetes	...	128	1,275	3,944
6. Tuberoides	...	3		145
7. Hyphomycetes ...	...	123	580	5,004
8. Sphæropsides ...	...	128	685	6,745
9. Saccharomycetes	...	10	8	30
10. Ustilagines	...	39	177	329
11. Phycomycetes ...	...	15	145	686
12. Myxomycetes ...	...	52	137	510
Other Groups ...	...	...		714
Totals ...	...	2,284	5,040	39,663

TABLE III.

Number of Species of Fungi found in Victoria on the following Plants of economic importance.

Name of Plant.	No. of Species of Fungi.	Name of Plant.	No. of Species of Fungi.
Acacia species ...	16	Lucerne	2
Almond ...	3	Maize	1
Apple ...	5	Mulberry	1
Apricot ...	2	Oats ...	4
Bean	1	Onion..	2
Beet	1	Orange	1
Cabbage ...	1	Pea ...	2
Cauliflower ...	1	Peach	4
Celery ...	1	Pear ...	3
Cherry ...	1	Plum ...	3
Clover ...	3	Rye-grass	3
Eucalyptus species	54	Strawberry	1
Flax	1	Tomato	2
Garlic ...	1	Vine ...	8
Lemon ...	1	Wheat	10
Lettuce ...	1		

Total number of Orders	...	69
„ „ Genera	...	424
„ „ Species	...	2,284
Species common to Australia and Britain ...	...	739
Species in West Australia	...	242
„ South Australia...	...	262
„ Tasmania	...	339
„ Victoria	...	1,070
„ New South Wales	...	406
„ Queensland	...	1,060
Species not referred to their respective Colonies	...	50

Proportion of Species of Fungi found in the different Colonies:—

West Australia	...	...	...	...	...	...	...	10·6 per cent.
South Australia	...	...	...	...	...	...	...	11·5 „
Tasmania	...	...	...	...	...	...	...	14·8 „
Victoria ...	...	...	...	...	...	...	...	46·8 „
New South Wales...	...	...	...	...	...	...	...	17·8 „
Queensland	...	...	...	...	...	...	...	46·4 „

NOTE.—It is not to be inferred from the relatively high percentage of Fungi in Victoria and Queensland, for instance, as compared with New South Wales, that they are absent from the latter colony, but rather that they still await investigation and determination there.

HOST-INDEX OF AUSTRALIAN FUNGI.

II.—HOST-INDEX OF AUSTRALIAN FUNGI.

Abrotanella forsterioides, J. Hook.
2266. Æcidium monocystis.—Leaves.

Abutilon Avicennæ, Ger.
1477. Puccinia heterospora.—Leaves.

Abutilon crispum, Don.
1477. Puccinia heterospora.—Leaves.

Acacia sp.
984. Merulius pelliculosus. V.—Branches.
1071A. Stereum hirsutum. var. tinellum.—Rotten wood.
1103. Stereum vittæforme.—Bark.
1456. Uromyces phyllodii.—Phyllodes.
1465. Melampsora phyllodiorum.—Phyllodes.
1521. Uredo leguminum.—Pods.
1655. Trabutia parvicapsa. V.—Phyllodes.
1743. Meliola amphitricha. V.—Leaves.
2024. Actinonema Gastonis.—Phyllodes.
2032. Phyllosticta phyllodiorum. V.—Phyllodes.
2061. Asteromella Acaciæ. V.—Phyllodes.
2082. Diplodia phyllodiorum. V.—Phyllodes.
2093. Septoria epiphyllodea. V.—Phyllodes.
2140. Marsonia Acaciæ. V.—Phyllodes.
2145. Pestalozzia Acaciæ. V.—Leaves.

Acacia complanata, Cunn.
2080. Diplodia lichenopsis.—Phyllodes.

Acacia hakeoides, Cunn.
1458. Uromyces Tepperianus. V.—Branches.

Acacia harpophylla, F. v. M.
1580. Xylaria gracilis.—Wood.
2117. Melophia Woodsiana.—Phyllodes.

Acacia longifolia, Willd.
1656. Trabutia phyllodiæ. V.—Phyllodes.
2099. Septoria phyllodiorum. V.—Phyllodes.

Acacia melanoxylon, R. Br. (Blackwood Tree).
2132. Gloeosporium leguminis. V.—Legumes.

Acacia myrtifolia, Willd.
1458. Uromyces Tepperianus. V.—Branches.

Acacia notabilis, F. v. M.
1448. Uromyces digitatus.—Phyllodes.
1522. Uredo notabilis. V.—Phyllodes.

Acacia penninervis, Sieb.
1641. Phyllachora rhytismoides.—Phyllodes.

Acacia salicina, Lind.
1450. Uromyces fusisporus. V.—Phyllodes.
1458. Uromyces Tepperianus. V.—Branches.

Acacia suaveolens, Willd. (Sweet-scented Acacia.)
1706. Læstadia phyllodiæ. V.—Phyllodes.

Acacia verticillata, Willd.
1695. Physalospora labecula.—Leaves.

Acæna sanguisorbæ, Vahl. (Sheep's-burr.)
1501. Phragmidium Potentillæ. V.—Leaves.

Adiantum sp. (Maiden-hair Fern.)
1166. Cyphella filicola. V.—Fronds.

Ægiceras sp.
1542. Sphærostilbe dubia.—Bark.

Agathis robusta, Masters = Dammara. (Queensland Kauri.)
1702. Læstadia Dammaræ.—Leaves.

Agarics (Fungi).
1555. Hypomyces tomentosus.
1559. Ophionectria agaricicola. V.—Putrid Agarics.
1932. Verticillium niveum.—Dead Agarics.
1972. Heterosporium epimyces. V.—Decayed Agarics.
2022. Myrothecium inundatum.—Putrid Agarics.
2196. Spinellus gigasporus. V.—Decaying Agarics.

Allium sp.
2038. Phoma alliicola. V.—Scapes.

Allium cepa, L. (Onion.)
2202. Peronospora Schleideni. V.—Leaves.
2247. Physarum didermoides.—Scales.
2481. Ustilago Allii. V.—Scale leaves of stored Onion bulbs.

Allium sativum, Bauh. (Garlic.)
2202. Peronospora Schleideni. V.—Leaves.

Alopecurus sp. (Fox-tail Grass.)
2199. Scierospora macrospora. V.
1493. Puccinia Rubigo-vera. V.—Leaves

Alphitonia excelsa, Reiss. (Red Ash.)
2121. Gloeosporium Alphitoniæ.—Leaves.
2148. Pestalozzia Guepini.—Leaves.

Alpinia cœrulea, Benth.
1634. Phyllachora Alpiniæ.—Leaves.

Alsophila sp.
1652. Rhytisma filicinum.—Fronds.

Alsophila Rebeccæ, F. v. M.
1735. Asterella Alsophilæ.—Fronds.

Althæa rosea, Cav. (Hollyhock.)
1485. Puccinia Malvacearum. V.—Leaves and stems.

Alyxia buxifolia, R. Br.
1467. Puccinia Alyxiæ. V.—Leaves.
1707. Sphærella Alyxiæ. V.—Leaves.

Amphipogon strictus, R. Br. (Bearded-heads.)
2175. Ustilago Tepperi.

Andropogon sp.
1943. Periconia nigrella.—Leaves.

Andropogon contortus, L. = Heteropogon contortus.
(Bunch Spear-grass.)
2189. Cerebella Andropogonis.

Ant—Formica sp.

Anthistiria ciliata, L. fil. (Kangaroo-grass.)
2158. Ustilago bromivora. V.—Inflorescence.
2283. Tolyposporium Anthistiriæ.—Inflorescence.

Anthistiria frondosa, R. Br.
2160. Ustilago bursa.—Grain.

Aphides. (On Pumpkin Leaves.)
1995. Oospora Aphides.

Apium graveolens, L. (Celery.)
1468. Puccinia Apii. V.

Apple—Pyrus Malus, L.

Apricot—Prunus Armeniaca, L.

Aristida sp.
2173. Ustilago segetum. V.

Artichoke—Cynara Scolymus, L.

Arundo sp. (Reeds.)
1936. Coniosporium inquinans.—Stems.

Arundo Phragmites, L. (Thatch-reed.)
1484. Puccinia Magnusiana.
1488. Puccinia Phragmitis. V.
1497. Puccinia Tepperi.

Aster argophyllus, Labill. = Olearia. (Musk Tree.)
1736. Asterella subcuticulosa. V.—Fading and dead leaves.
1999. Poiosporium grande. V.—Stems.

Asterina sp. Growing on leaflets of *Tarrietia trifoliolata.*
F. v. M.
2275. Belonidium parasiticum.

Atherosperma moschatum, Labill. (Native Sassafras.)
2221. Arcyria fuliginea.—Leaves.

Aucuba japonica, L.
1700. Pleospora Aucubæ. V.—Leaves.

Auricularia sp. (Fungus.)
2236. Didymium australe.

Australian Beech—Fagus Cunninghamii. Hook.

Avena fatua, L. (Wild Oats.)
1475. Puccinia graminis. V.

Avena sativa, L. (Oats.)
1475. Puccinia graminis. V.
1493. Puccinia rubigo-vera. V.
1950. Scolecotrichum graminis, var. Avenæ. V.
2173n. Ustilago segetum, var. Avenæ. V.—Ear.

Backhousia sp.
2083. Ascochyta apiospora.—Leaves.

Bambusa sp. (Bamboo.)
801. Agaricus versipes.—Roots.
1143. Peniophora bambusicola.—Rotting bamboo.

Banana—Musa Cavendishii, Lamb.

Banksia sp. (Native Honeysuckle.)
123. Mycena subcorticalis.—Log.
1099. Stereum umbrinum—Bark.
1692. Didymosphæria Banksiæ. V.—Leaves.
1882. Tympanis Toomansis.—Fruit.

Banksia integrifolia, L. (Beefwood.)
1709. Sphærella Banksiæ. V.—Fading leaves.

Banksia marginata, Cav.
1741. Parodiella Banksiæ.—Languid leaves.

Barley—Hordeum distichon, Bauh.

Bean—Phaseolus vulgaris, L.

Bean Caper—Zygophyllum ammophilum, F. v. M.

Bearded-heads—Amphipogon strictus, R. Br.

Beefwood—Banksia integrifolia, L.

Beet—Beta vulgaris, L.

Beilschmiedia obtusifolia, Benth. = Nesodaphne.
1462. Melampsora Nesodaphnes.—Fruit.

Bellis perennis, L. (Daisy.)
1487. Puccinia obscura, Schrot. V.

Bertya rotundifolia, F. v. M.
1942. Heterobotrys paradoxa.—Leaves.

Beta vulgaris, L. (Beet.)
1446. Uromyces Betæ. V.—Leaves.

Betula alba, L. (Birch.)
650. Polyporus betulinus.

Bidens pilosus, L.
1519. Uredo Cichoracearum. V.

Birch—Betula alba, L.

Bitter Almond—Prunus Amygdalus var. Amara.

Bitter Bark—Tabernæmontana orientalis, R. Br.

Black Ash—Litsea dealbata, Nees.

Blackbutt Tree—Eucalyptus pilularis, Sm.

Blackwood Tree—Acacia Melanoxylon, R. Br.

Bloodwood Tree—Eucalyptus corymbosa, Sm.

Bluegum Tree—Eucalyptus globulus, Labill.

Boletus sp. (Fungus.)
1552. Hypomyces chrysospermus. V.
1929. Sepedonium chrysospermium. V.

Bossiæa cinerea, R. Br.
2265. Æcidium eburneum. V.—Stems, leaves, flowers, and fruit.

Bottle Gourd—Lagenaria vulgaris, Ser.

Bottle Thistle—Lagenophora Billardieri, Cass.

Bougainvillea sp.
240. Marasmius calobates. V.—Putrid leaves.

Box Eucalypt—Eucalyptus hemiphloia, F. v. M.
„   Eucalyptus largiflorens, F. v. M.

Box Thorn—Bursaria spinosa, Cav.

Bramble—Rubus fruticosus, L.

Brassica oleracea, L. (Cabbage and Cauliflower.)
2258. Plasmodiophora Brassicæ. V.—Roots.

Brisbane Box—Tristania conferta, R. Br.

Bromus sp.
2092. Septoria Bromi. V.—Leaves.

Bromus arenarius, Labill.
2092. Septoria Bromi. V.—Leaves.
2158. Ustilago bromivora. V.—Inflorescence.

Bromus mollis, L.
2158. Ustilago bromivora. V.—Inflorescence.

Bromus sterilis, L.
1475. Puccinia graminis. V.

Bulbine bulbosa, Haw.
1447. Uromyces Bulbinis. V.—Leaves.

Bulrush—Typha sp.

Bunch Spear-grass—Andropogon contortus, L.

Burchardia umbellata, R. Br.
1471. Puccinia Burchardiæ. V.—Leaves.

Bursaria spinosa, Cav. (Boxthorn.)
1609. Nummularia pusilla.—Branches.
1759. Capnodium Walteri. V.—Branches and living
   leaves.
1915. Botryotrichum Lachnella. V. — Branches and
   spines.

Cabbage—Brassica oleracea, L.

Callistemon sp.
1751. Meliola polytricha.—Leaves.

Camellia japonica, L. (Camellia.)
1718. Sphærulina Camelliæ. V.—Leaves.
1976. Macrosporium Camelliæ. V.—Leaves.
2114. Sacidium Camelliæ. V.—Leaves.

Canthium latifolium, F. v. M.
2079. Diplodia canthifolia. V.—Leaves.

Capparis sp. (Caper.)
1695. Didymosphæria conoidella.—Dead branches.

Capparis Mitchelli, Lindl.
1550. Calonectria otagensis.—Twigs.

Carex sp. (Sedge.)
1947. Periconia nigrella.—Leaves.
2168. Ustilago leucoderma. V.—Sheaths.

Carica Papaya, L. (Papaw.)
1980. Macrosporium peponicolum.—Fruit.

Carissa ovata, R. Br.
1472. Puccinia Carissæ.—Leaves.

Carnation—Dianthus Caryophyllus, L.

Cassia sp.
1525. Uredo pallidula.—Leaves, twigs, and legumes.
2001. Tubercularia leguminum.—Legumes.

Cassinia aculeata, R. Br.
1674. Gibberidea plagia. V.—Twigs.

Castanospermum australe, Cunn. and Fraser. (Moreton
   Bay Chestnut.)
1735. Asterina platystoma.—Leaves.
2064. Chætophoma entricha.—Leaves.

Castor-oil Plant—Ricinus communis, L.

Casuarina sp. (Sheoak.)
761. Fomes igniarius. V.—Trunks.
936. Hexagona decipiens. V.—Trunks.
1864. Trichopeziza Sphærula.—Dead bark.
1865. Dasyscypha Eucalypti. V.—Leaves.
2146. Pestalozzia Casuarinæ. V.—Branches.

Cat's-ear—Hypochæris glabra, L.

Cauliflower—Brassica oleracea, L.

Celery—Apium graveolens, L.

Cerastium glomeratum, Thuill.
1893. Pseudopeziza Cerastiorum.—Leaves and calyx.

Cherry—Prunus Cerasus, L.

Chiloglottis diphylla, R. Br.
1455. Uromyces orchidearum. V.—Leaves.

Chionaspis Citri. (Coccus of Orange.)
2017. Microcera rectispora.

Chrysanthemum sp.
1909. Oidium Chrysanthemi. V.—Leaves.

Cicada sp.
1991. Isaria Cicadæ. V.

Citrus sp.
1136. Corticium nudum.—Bark.
1756. Capnodium citri. V.—Leaves.

Citrus Aurantium, L. (Orange.)
2124. Gloeosporium citricolum.—Leaves.
1756. Capnodium citri. V.—Leaves.

Citrus Limonium, Risso (Lemon.)
2123. Gloeosporium Citri. V.—Branches.

Cladium = Gahnia.

Clavaria sp. (Fungus.)
1930. Verticillium eximium.

Clematis aristata, R. Br. (Native Supple-jack.)
1520. Uredo Clematidis. V.—Leaves.

Clematis microphylla, D. C.
1520. Uredo Clematidis. V.—Leaves.

Clover—Trifolium sp.

Club Rush—Scirpus nodosus, Rott.

Coccus sp.
2016. Microcera coccophila.

Compositæ.
1506. Æcidium Compositarum. V.
2205. Synchytrium Taraxaci. V.—Leaves.

Coniferæ.
47. Armillaria mellea. V.—Stumps.
724. Polyporus tabulæformis.—Trunks.
735. Fomes annosus.—Trunks.
1755. Capnodium australe.—Branches.
1940. Hormiscium pithyophilum.—Branches and leaves.

Coprosma Billardieri, J. Hook.
2267. Puccinia Coprosmatis. V.—Leaves.

Cordyline australis, Hook.
2041. Phoma Cordylines.

Cordyline terminalis, Kunth.
2026. Phyllosticta Cordylines.—Leaves.

218

Fir.
1161. Aleurodiscus amorphus.—Trunks and branches.
1268. Calocera stricta. V.—Dead leaves.

Flat-leaved Rush—Juncus planifolius, R. Br.

Flat-weed—Hypochæris radicata, L.

Flax—Linum usitatissimum, L.

Flindersia sp.
1743. Meliola amphitricha. V.—Leaves.
2270. Xylaria ianthino-velutina.—Old fruit.

Flindersia australis, R. Br.
1594. Xylaria scopiformis.—Decaying fruit.

Flooded Gum-tree—Eucalyptus tereticornis, Sm.

Fœniculum vulgare, Mill. (Fennel.)
1953. Cladosporium herbarum. V.—Stems.

Fomes gryphæformis, Cooke = Polyporus (Fungus).
1003. Hydnum isidioides.—Hymenium.

Formica sp. (Ant.)
1989. Stilbum Formicarum. V.—Dead ants.

Fox-tail Grass—Alopecurus sp.

Fragaria vesca, L. (Strawberry.)
1712. Sphærella Fragariæ. V.—Leaves.
2024. Phyllosticta fragaricola.—Leaves.

Gahnia filum, F. v. M. = Cladium filum.
2185. Sorosporium Muellerianum. V.—Panicles.

Gahnia tetraquetra, F. v. M.
2126. Glœosporium epicladii.—V.

Garlic—Allium sativum, Banh.

Geranium sp.
1680. Venturia circinans. V.—Leaves.

Glycine clandestina, Wendl.
1968. Cercospora Glycines. V.—Living leaves.

Gnaphalium purpureum, L.
2268. Puccinia investita, Schwein. V.—Leaves and stems.

Goodeniaceæ.
2204. Synchytrium Succisæ. V.—Leaves and leaf-stalks.

Goodenia sp.
1457. Uromyces puccinioides. V.—Leaves and flower-stalks.
1508. Æcidium Goodeniacearum. V.—Leaves.

Goodenia geniculata, R. Br.
1495. Puccinia Saccardoi. V.—Leaves.

Goodenia pinnatifida, Schlecht.
1508. Æcidium Goodeniacearum. V.—Leaves.

Goodenia ovata, Smith.
2016. Phoma Goodeniarum. V.—Fading leaves.
2137. Glœosporium subglobosum. V.—Fading leaves.

Goodia lotifolia, Sal.
1513. Æcidium soleniforme.
1713. Sphærella Goodiæfolia. V.—Leaves.

Goose-grass—Poa annua, L.
Grape-vine—Vitis vinifera, Banh.

Grass.
241. Marasmius calopus. V.—Roots.
451. Nauceoria frustiosa. Roots
467A. Tubaria inquilina, var. Echola.—Roots.
1045. Thelephora cristata. —Running over grass.
1533. Epichloe cinerea.
1637. Phyllachora graminis. V.—Leaves.
1682. Chætomium comatum. V.—Rotting grass.
1714. Sphærella graminicola. V.—Leaves.
1724. Erysiphe graminis. V.—Leaves and stems.
1992. Isaria graminiperda.—V.
2051. Phoma nitida.—V.
2243. Spumaria alba.—V.
2256. Leocarpus fragilis.

Grass-tree—Xanthorrhœa sp.

Grevillea sp. (Silky Oak.)
169. Pleurotus lampas. V.—Languid stems.
1708. Sphærella atra.—Leaves.
2273. Phyllachora Grevilleæ.—Leaves.

Groundsel—Senecio sp.

Hakea sp.
1517. Uredo angiosperma.—Leaves.

Hakea lorea, R. Br.
1728. Asterina Baileyi.—Leaves.

Hardenbergia = Kennedya.

Hedycarya Cunninghami, Tul.
2129. Glœosporium Hedycaryæ. V.—Fading leaves.

Helianthus annuus, L. (Sun Flower.)
1476. Puccinia Helianthi. V.—Leaves.

Helichrysum sp. (Everlastings.)
1480. Puccinia Kalchbrenneri. V.—Leaves.

Hemarthria compressa, R. Br. (Sugar Grass.)
1493. Puccinia Rubigo-vera.

Heteropogon = Andropogon.

Hibiscus sp.
1477. Puccinia heterospora.—Leaves.

Holly—Ilex Aquifolium, L.

Hollyhock—Althæa rosea, Cav.

Hordeum distichon, Banh. (Barley.)
1475. Puccinia graminis. V.
1493. Puccinia Rubigo-vera. V.
2173c. Ustilago segetum, var. Nuda Hordei. V.—Ear.

Kormogyne cotinifolia, D. C.
943. Hexagonia sericea.—Trunks.

House Fly—Musca domestica.

Hovea longifolia, R. Br.
1730. Asterina hoveæfolia.—Leaves.

Hoya australis, R. Br.
2130. Glœosporium intermedium.

Hydrocotyle asiatica, L.
2206. Protomyces macrosporus.

Hydrocotyle hirta, R. Br.
1486. Puccinia inunlta.—Leaves.

Hymenochæto sp. (Fungus.)
986. Merulius tenuissimus.

Hypochæris glabra, L. (Cat's Ear.)
1478. Puccinia Hieracii.

Hypochæris radicata, L. (Flatweed.)
1478. Puccinia Hieracii. V —Leaves of flowering stems.
2263. Puccinia Hypochæris. V.—Leaves.

Hypoxylon sp. (Fungus.)
1558. Dialonectria tephrothele.

Ilex Aquifolium, L. (Holly.)
1859. Pseudohelotium ilicincolum. V.—Branches.

Insects.
1528. Cordyceps entomorrhiza. V.—Larvæ.
1528A. Cordyceps entomorrhiza, var. Menesteridis. V. Larvæ.
1529. Cordyceps Gunnii. V.—Larvæ.
1530. Cordyceps Hawkesii.—Larvæ.
1531. Cordyceps ophioglossoides.
1532. Cordyceps Taylori. V.—Larvæ.
1926. Sporotrichum densum.—Dead insects.
1994. Isaria suffruticosa.—Hairy caterpillar.

Ironbark-tree—Eucalyptus siderophloia, Hook.

Isolepis = Scirpus.

Jacksonia scoparia, R. Br.
1464. Cronartium Asclepiadeum.—Leaves.
1516. Ræstelia polita. V.—Branches.

Jasminum racemosum, F. v. M.
776. Fomes pallus.—Branches.

Juncus sp. (Rush.)
115. Mycena juncicola. V.—Dead rushes.
1479. Puccinia Junciphila. V.
1658. Phyllachora Junci. V.
2172. Ustilago pilulæformis. V.—Ovaries.

Juncus maritimus, Lam.
1451. Uromyces Junci. V.

Juncus pallidus, R. Br. (Sheathed Rush.)
1518. Uredo armillata. V.

Juncus planifolius, R. Br. (Flat-leaved Rush.)
2171. Ustilago Muelleriana. V.—Seeds.

Kangaroo Grape-vine—Vitis antarctica, Benth.

Kangaroo Grass—Anthistiria ciliata, L.

Kennedya sp. = Hardenbergia.
2029. Phyllosticta Hardenbergiæ. V.—Leaves.

Kennedya monophylla, Vent.
2094. Septoria Hardenbergiæ.—Leaves.

Kennedya prostrata, R. Br. (Native Scarlet-runner.)
1969. Cercospora Kennedyæ. V.—Leaves.

Kochia sedifolia, F. v. M.
1481. Puccinia Kochiæ. V.—Leaves.

Lactuca sp. (Lettuce.)
1490. Puccinia Prenanthis. V.

Lagenaria vulgaris, Sér. = Cucurbita lagenaria. (Bottle Gourd.)
1672. Eutypa polyscia—Epicarp.

Lagenophora Billardieri, Cass. (Bottle Thistle.)
1482. Puccinia Lagenophoræ. V.—Leaves.
1737. Dimerosporium Ludwigianum. V.—Fading leaves.

Leersia hexandra, Swartz. (Native Rice-grass.)
2182. Thecaphora inquinans.—Inflorescence.

Leguminosæ.
1540. Polystigma australiense. V. Leaves, rarely stems.
1742. Parodiella Peri-poriobles. V.—Leaves.
2191. Schinzia Leguminosarum. V.—Roots.

Lemon—Citrus Limonium, Risso.

Lepidium ruderale, L.
2284A. Peronospora parasitica, var. Lepidii. V.—Leaves, stem, and fruit.

Lepidosperma sp. (Sword-rush.)
1699. Anthostomella Lepidosperma.—V.
1937. Coniosporium pterospermum. V.
2095. Septoria Lepidospermi. V.—Leaves.

Lepiota bubalina, Berk. (Fungus.)
1918. Aspergillus Muelleri. V.

Leptocarpus tenax, R. Br.
2183. Thecaphora Leptocarpi. V.—Ovaries.

Leptospermum sp. (Tea-tree.)
152. Pleurotus australis.--Roots.
2060. Aposphæria Leptospermi. V.—Bark.

Leptospermum lævigatum, F. v. M. (Sandstay.)
2075. Coniothyrium olivaceum. V.—Involucres.
2116. Melophia Leptospermi. V.—Leaves.

Leptospermum scoparium, Forst.
1997. Harpographium Corymdioides. V.—Branches.
2144. Hyaloceras diliophosporum. V.—Leaves.

Lettuce—Lactuca sp.

Lichens.
2007. Illosporium flavellum.

Limnanthemum indicum, Thwaites.
1509. Æcidium nymphoidis.—Leaves.

Limosella sp.
1453. Uromyces Limosellæ.—Leaves.

Linum marginale, Cuon. (Native Flax.)
1461. Melampsora Lini. V.—Leaves.

Linum usitatissimum, L. (Flax.)
1461. Melampsora Lini. V.—Leaves.

Litsea sp.
1704. Læstadia Litsæ.—Leaves.

Litsea dealbata, Nees. (Black Ash.)
1643. Dothidella apiculata.—Fading leaves.

Lobelia anceps, Thun.
1469. Puccinia aucta. V.—Leaves.

Lobelia pedunculata, R. Br.
1469. Puccinia aucta. V.—Leaves.

Lobelia platycalyx, F. v. M.
1469. Puccinia aucta. V.—Leaves.

Locellinia cyanopotamia, Sacc. (Fungus.)
1434. Arachnion Drummondi. — Attached to above fungus.

Lolium sp. (Rye-grass.)
1992. Isaria graminiperda. V.

Lolium perenne, L. (Rye-grass.)
1493. Puccinia rubigo-vera. V.
1527. Claviceps purpurea. V.—Inflorescence.
1992. Isaria graminiperda. V.

Lolium temulentum, L. (Darnel, Drake.)
1527. Claviceps purpurea. V.—Inflorescence.

Loosestrife—Lythrum hyssopifolia, L.

Lucerne—Medicago sativa, L.

Lyonsia reticulata, F. v. M.
2274. Peziza Lyonsiæ.—Leaves.

Lythrum hyssopifolia, L. (Loosestrife.)
2049. Phoma Lythri. V.—Fading leaves.
2181. Donssansia punctiformis. V.—Leaves.

Macrozamia = Encephalartos.

Maidenhair Fern—Adiantum sp.

Maize—Zea Mays, L.

Malloe—Eucalyptus incrassata, Labill.

Malva rotundifolia, L. (Dwarf Mallow.)
1485. Puccinia Malvacearum. V.—Leaves and stems.

Mangifera indica, L. (Mango.)
2131. Glæosporium Lagenarium.—Fruit.
2150. Pestalozzia uvicola.—Fruit.

Manna Gum-tree—Eucalyptus viminalis. Labill.

Marsdenia sp.
2045. Phoma follienlorum.—Follicles.
2081. Diplodia Marsdeniæ.—Follicles.

Meadow Grass—Poa sp.

Medicago sativa, L. (Lucerne.)
1703. Læstadia destructiva. V.—Leaves.
1894. Pseudopeziza Medicaginis. V.—Leaves.

Melaleuca sp. (Tea-tree.)
1260. Guepinia merulina.—Rotten wood.
1705. Læstadia Melaleucæ.—Leaves.

Melon—Cucumis Melo, L.

Messmate -Stringybark Tree—Eucalyptus obliqua, L'Her.

Microtis porrifolia, R. Br.
1454. Uromyces Microtidis.—Leaves.

Mint Tree—Prostanthera lasiantha. Labill.

Moluccas Bramble—Rubus Moluccanus, L.

Moreton Bay Chestnut—Castanospermum australe, Cunn.

Morus sp. (Mulberry.)
2102. Phleospora Mori. V.—Leaves.

Moss.
232. Cantharellus lobatus. V.
571. Pleurotus cyphellæformis.
1045. Thelephora cristata Suppitiella cristata.
1140. Corticium simulans. V.
1108. Cyphella moseigena. V.
2215. Lamproderma Listeri.
2231. Didymium spumarioides.
2251. Physarum rufibasis.
2256. Leocarpus fragilis.

Mountain Ash—Eucalyptus virgata. Sieb.

Muehlenbeckia sp.
2087. Darluca filum.

Muehlenbeckia adpressa, Meiss.
1494A. Puccinia rumicis-scutati, var. Muehlenbeckia. V.—Leaves.

Muehlenbeckia Cunninghamii, F. v. M.
1516. Restelia polita. V.—Branches.

Mulberry—Morus sp.

Mnsa sp.
1748. Meliola Musæ.

Musa Cavendishii, Lamb. (Banana.)
2131A. Glæosporium lagenarium, var. Cucurbitarum.
2134. Glæosporium Musarum.—Fruit.

Musca domestica, L. (House-fly.)
2203 Empusa Muscæ. V.

Musk-Tree—Aster argophyllus, Labill.

Myoporum insulare, R. Br.
2097. Septoria Myopori. V.—Leaves.

Myoporum platycarpum, R. Br.
1649. Bagnisiella endopyria. V.—Leaves

Myriangium sp. (Lichen.)
1859. Pseudohelotium ilicincolum.—V.

Myrtus sp. (Myrtle.)
2083. Ascochyta apiospora.—Leaves.
2147. Pestalozzia funerea.—Leaves.

Native Beech—Fagus Cunninghamii, Hook.

Native Flax—Linum marginale, Cunn.

Native Honeysuckle—Banksia sp.

Native Pepper-tree—Drimys aromatica, F. v. M.

Native Raspberry—Rubus parvifolius, L.

Native Rice-grass—Leersia hexandra. Swartz.

Native Sassafras -Atherosperma moschatum. Labill.

Native Scarlet-runner—Kennedya prostrata, R. Br.

Native Supple-jack—Clematis aristata, R. Br.

Native Yam—Dioscorea sp.

Nepenthes sp. (Pitcher plant.)
1823. Hamaria Thozetti.

Nerium Oleander, L. (Oleander.)
2098. Septoria oleandrina.—Leaves.

Nesodaphne—Beilschmiedia.

Nottle—Urtica sp.

Neurachne alopecuroides, R. Br.
2175. Ustilago Tepperi.

Nicotiana Tabacum, L. (Tobacco.)
2201. Peronospora Hyoscyami. V.—Leaves.

Norway-Spruce -Pinus picea, Du Roi.

Oak—Quercus Robur, L.

Oat-grass—Danthonia sp.

Oats—Avena sativa, L.

Olea paniculata, R. Br.
  1767. Lembosia graphioides.—Leaves.

Oleander—Nerium Oleander, L.

Omalanthus populifolius, Grah.
  2063. Asteromella Homalanthi.—Leaves.

Oncoptera intricata, Walk. (Insect.)
  2279. Isaria Oncopterae. V —Dead Larvae.

Onion—Allium cepa, L.

Opercularia varia, J. Hook.
  1807. Æcidium cystosciroides.
  2267A. Puccinia Coprosmatis var. Operculariae. V.—
    Leaves and leaf-stalks.

Orange—Citrus Aurantium, L.

Oxalis corniculata, L.
  2276. Oidium Oxalidis. V.—Leaves, stem, and fruit.

Palm.
  164. Pleurotus Gardneri.—Petioles and half-rotten
    fronds.
  2031. Phyllosticta palmicola —Leaves.
  2053. Phoma plagia.—Leaves.

Panicum sp.
  2166. Ustilago Digitariae. V.

Panicum paradoxum, R. Br.
  2164. Ustilago confusa. V.

Papaw—Carica Papaya, L.

Paspalum scrobiculatum L. (Ditch Millet.)
  2162. Ustilago Cesatii. V.
  2190. Cerebella Paspali.—Glumes.

Passiflora sp. (Passion-flower.)
  266. Marasmius rhytleeps.—Twigs.
  1561. Lisiella Passiflorae.—Stems.
  1949. Seolecotrichum atriellum.—Twigs.
  1998. Harpographium quaternarium.—Dead twigs.
  2014. Fusarium longisporum.—Twigs.
  2103. Phlyctaena Passiflorae.—Twigs.

Pea—Pisum sativum, L.

Peach—Prunus Persica, J. Hook.

Pear—Pyrus communis, L.

Pelargonium australe, Willd.
  1474. Puccinia Geranii. V.—Leaves.

Peppermint Gumtree—Eucalyptus amygdalina, Labill.

Phalaris minor, Retz.
  1475. Puccinia graminis. V.

Phaseolus vulgaris, L. (Bean.)
  2133. Glœosporium Lindemuthianum. V.—Legume.
  2260. Uromyces Phaseoli.—All parts, more especially
    leaves.

Phœnix dactylifera, L. (Date Palm.)
  2188. Graphiola Phœnicis.

Phylica sp.
  402. Pholiota phylicigena.—Trunks.

Pinus sp.
  1157. Coniophora olivacea.—V.—Decayed wood.
  1940. Hormiscium pithyophilum.—Branches.
  2105. Sphaeronæmella rufa.—Pine chips.

Pinus contorta, Doug.
  1144. Peniophora carnea.

Pinus picea, Du Roi. (Norway Spruce.)
  1116. Hymenochaete Mougeotii. V.—Trunks.

Piparomia sp.
  1727. Eurotium lateritium.—Leaves.

Pisum sativum, L. (Pea.)
  2133. Glœosporium Lindemuthianum. V.—Legume.
  2141. Marsonia deformans. V.—Leaves, stipules, &c

Pitcher-plant.—Nepenthes sp.

Pittosporum rubiginosum, Cunn.
  1716. Sphaerella rubiginosa.—Leaves.

Plagianthus sidoides, Hook. •
  2261. Puccinia Plagianthi.—Leaves and flowers.

Plane-tree.—Platanus sp.

Plantago sp.
  1510. Æcidium Plantaginis. V.— Leaves.

Plantain.—Plantago sp.

Platanus sp. (Plane-tree.)
  2052. Phoma notha.—Dead branches.

Platylobium sp.
  2039. Phyllosticta Platylobii. V.—Leaves.

Plum.—Prunus domestica, L.

Poa sp. (Meadow-grass.)
  1189. Puccinia Poarum. V.
  2017. Phoma graminis.—Stems.

Poa annua, L.—(Goose-grass.)
  1489. Puccinia Poarum. V.
  1193. Puccinia rubigo-vera. V.
  2284. Ustilago Poarum. V.—Foliage especially.

Polygonum sp.
  2167. Ustilago emidensis.—Stems, &c.
  2180. Sphacelotheca hydropiperis.—Ovaries.
  2180. Sphacelotheca hydropiperis, var. Columellifera.—
    Ovaries.

Polygonum Hydropiper, L. (Water-pepper.)
  2176. Ustilago utriculosa. V.—Stems and ovaries.

Polygonum minus, Huds.
  2176. Ustilago utriculosa. V.—Stems and ovaries.

Polyporus, sp. (Fungus.)
  1551. Hypomyces aurantius.
  1552. Hypomyces chrysospermus. V.
  1553. Hypomyces membranaceus.
  1554. Hypomyces rosellus.
  1928. Sepedonium aureo-fulvum. V.

Polyporus gryphæformis = Fomes gryphæformis.

Polyporus portentosus, Berk. (Fungus.)
  2054. Phoma portentosa. V.—Cap.

Polystictus cinnabarinus, Cooke = Polyporus. (Fungus.)
  1939. Torula mycetophila. V.—Cap.

Potato.—Solanum tuberosum, L.

Prostanthera lasiantha, Labill (Mint-tree.)
  2031. Phyllosticta Prostantheræ. V.—Leaves.
  2076. Coniothyrium septorioides. V.—Leaves.

Senecio odoratus, Horn.
1170. Cyphella polycephala.

Senecio velleioides, Cunn.
1506. Æcidium Compositarum. V.—Leaves.

Serjania.
1954. Cladosporium hypophyllum.--Leaves.

Sheathed Rush—Juncus pallidus, R. Br.

Sheep's Burr—Acæna sanguisorbæ, Vahl.

She Oak—Casuarina sp.

Silky Oak—Grevillea sp.

Smilax sp. (Sarsaparilla.)
1717. Sphærella smilacicola.—Leaves.
1753. Zukalia loganiensis.—Leaves.
2113. Actinothecium Scortechinii.—Leaves.

Solanum Dallachyi, Benth.
1691. Chætomium cymatotrichum.—Leaves.

Solanum Lycopersicum, L. (Tomato.)
1912. Oidium Lycopersicum. V.—Stem and leaves.
1982. Macrosporium Tomato. V.—Ripe tomatoes.

Solanum tuberosum, L. (Potato.)
2018. Epicoccum scabrum.—Leaves and stems.

Solanum verbascifolium, L.
1970. Cercospora Solanacea.—Leaves.

Sorghum sp.
2087. Darluca filum.

Speedwell—Veronica sp.

Sphæria sp. (Fungus.)
1169. Cyphella parasitica.

Spindle-tree—Euonymus sp.

Spinifex hirsutus, Labill.
2174. Ustilago Spinificis.—Flowers and spikes.

Sporobolus sp.
1533. Epichloe cinerea.

Sporobolus indicus, R. Br. (Tussock Grass.)
1962. Helminthosporium Ravenelii.—Inflorescence.

Spotted Gum—Eucalyptus maculata, Hook.

Spyridium = Cryptandra.

Stipa sp. (Spear Grass.)
2163. Ustilago comburens.

Strawberry—Fragaria vesca, L.

Styphelia sp.
1546. Nectria ferruginea. V.—Leaves, bracts, &c.

Styphelia straminea, Spreng = Cyathodes.
1766. Glonium tardum.—Leaves.

Sugar Cane—Saccharum officinarum, L.

Sugar Grass—Hemarthria compressa, R. Br.

Sunflower—Helianthus annuus, L.

Sweet-scented Acacia—Acacia suaveolens, Willd.

Sword Rush—Lepidosperma sp.

Tabernæmontana orientalis, R. Br. (Bitter Bark.)
1504. Æcidium Apocyni.—Leaves.

Tasmannia = Drimys.

Tea-tree—Leptospermum sp., and Melaleuca sp.

Tecoma jasminoides, Lindl.
2112. Melasmia Tecomatis.—Leaves.

Tetracera Wuthiana, F. v. M.
1752. Meliola Tetraceræ.—Leaves.

Thatch-reed—Arundo Phragmites, L.

Tinea. (Insect.)
1528. Cordyceps entomorrhiza. V.—Larva.

Tobacco—Nicotiana Tabacum, L.

Tomato—Solanum Lycopersicum, L.

Trema aspera, Blume.
1732. Asterina pelliculosa.—Leaves.
1738. Dimerosporium parvulum.—Living leaves.
2062. Asteronella epitrema.—Living leaves.

Tremella fuciformis, Berk. (Fungus.)
1684. Rosellinia tremellicola.

Tricoryne anceps, R. Br.
2073. Sphæropsis Tricorynes.—Leaves.

Trifolium sp. (Clover.)
1459. Uromyces Trifolii.—V.
1642. Phyllachora Trifolii. V.—Leaves.
1895. Pseudopeziza Trifolii. V.—Languishing leaves.
2241. Didymium spumarioides.

Tristania sp.
2055. Phoma purpurea.—Leaves.

Tristania conferta, R. Br. (Brisbane Box.)
1749. Meliola octospora.—Leaves.
2084. Ascochyta brunnea.—Leaves.

Tristania laurina, R. Br.
1961. Helminthosporium puccinioides.—Fading or dead leaves.

Triticum vulgare, Vill. (Wheat.)
1475. Puccinia graminis. V.—Leaves, stem, and ear.
1493. Puccinia rubigo-vera. V.—Leaves, stem, and ear.
1527. Claviceps purpurea. V.—Inflorescence.
1953. Cladosporium herbarum. V.—Leaves, stem, and ear.
1979. Macrosporium graminum.—Leaves.
2074. Sphæropsis Tritici. V.—Dead leaves and sheaths.
2100. Septoria Tritici. V.—Stem, leaves, and ear.
2159. Ustilago bullata. V.—Ears.
2173A. Ustilago segetum, var. Tritici. V.—Ear.
217E. Tilletia Tritici. V.—Grains.
2186. Urocystis occulta. V.—Stem, leaves, glumes.

Tussock Grass—Sporobolus indicus, R. Br.

Typha sp. (Bulrush).
1958. Cladosporium Typharum. V.—Leaves.

Urtica sp. (Nettle).
1544. Æcidium Urticæ. V.

Veronica sp. (Speedwell).
1515. Æcidium Veronicæ.—V.

Vine—Vitis vinifera, Bauh.

Viola sp.
1498. Puccinia Violæ.—V.
2101. Septoria Violæ. V.—Fading leaves.

LIST OF WORKS ON AUSTRALIAN FUNGI.

Q

III.—LIST OF WORKS ON AUSTRALIAN FUNGI.

1. ABBOTT (F.)—"Smut in Wheat." Proc. Roy. Soc., Tasm., 1889.

The nature of smut stated, and means for prevention given.

2. AGRICULTURE DEPARTMENT.—New South Wales, Queensland, South Australia, Tasmania, and Victoria.

Reports and Bulletins issued up to date contain various papers on parasitic fungi, the more important of which are noted under authors' names.

3. Agricultural Gazette of New South Wales. Department of Agriculture, Vols. I.–V.—(continued.) Sydney, 1890–94.

Contains numerous articles on fungus diseases, the more important of which are given under authors' names.

4. ANDERSON (H. C. L.)—"Rust in Wheat : Experiments, and their Objects." Ag. Gaz., N.S.W., I., Pt. I., 1890.

A variety of measures suggested for trial in order to minimize the effects of rust.

5. BACKHOUSE (J.)—"A Narrative of a Visit to the Australian Colonies." 8vo. London, 1843.

Reference at p. 119 to Punk and an edible fungus near Emu Bay in Tasmania; also in Appendix D, p. xl, to the common Mushroom and Mylitta australis.

6. BAILEY (F. M.)—"A General Account of the Flora of Tropical Queensland." Proc. Linn. Soc., N.S.W., II., 1878.

Some of the more important fungi referred to.

7. ———— "Medicinal Plants of Queensland." Ibid. V., 1880. Hirneola auricula-judæ, or Jew's Ear, referred to.

8. ———— "A Synopsis of the Queensland Flora." 8vo. Brisbane, 1883.

9. ———— "A Classified Index of the Indigenous and Naturalized Plants of Queensland." 8vo. Brisbane, 1883.

10. ———— "Contributions to the Queensland Flora." Proc. Roy. Soc., Q., I., 1884.

Eighty-two species of fungi recorded.

11. ———— "Contributions to the Queensland Flora." Pt. II., ibid., 1884.

Eleven species of fungi recorded.

12. ———— "Contributions to the Queensland Flora." Part III., ibid., 1884.

Seven species of fungi recorded.

13. ———— "A Synopsis of the Queensland Flora." First Supplement. 8vo. Brisbane, 1886.

14. ———— "Classified Index of the First Supplement to the Indigenous and Naturalized Plants of Queensland." 8vo. Brisbane, 1886.

15. ———— "A Synopsis of the Queensland Flora." Second Supplement, 8vo. Brisbane, 1888.

16. BAILEY (F. M.)—"Classified Index of the Second Supplement to the Indigenous and Naturalized Plants of Queensland." 8vo. Brisbane, 1888.

17. ———— "Supplement to the Report of the Botany of the Bellenden-Ker Expedition—Fungi collected or observed about the Bellenden-Ker Range." Ann. Rep. Dept. Ag., Q., 1890.

Fifty-seven species of fungi are recorded, fourteen of which are new to Australia.

18. ———— "A Synopsis of the Queensland Flora." Third Supplement. 8vo. Brisbane, 1890.

Includes first addendum to third supplement.

19. ———— "Catalogue of the Plants of Queensland." 8vo. Brisbane, 1890.

Includes second addendum to third supplement.

20. ———— "Contributions to the Queensland Flora." Bull. 1, or Bot. Bulletin I. Dept. Ag., Q., 1890.

Stilumella succhari (Cooke) and Peronospora hyoscyami (De Bary) recorded.

21. ———— "Contributions to the Report of the Botany of the Bellenden-Ker Expedition." Ann. Rep. Dept. Ag., Q., 1891.

Aschersonia tahitensis (Mont.), Entyloma Eugeniarum (Cooke and Mass.), Asterina reptans, (Berk. and Cooke) added.

22. ———— "Additional Fungus Blights observed to have injured Plants during the Year." Ibid., 1891.

Five species are mentioned altogether, occurring on Vines, Hollyhocks, Tobacco plant, and native plants.

23. ———— "Contributions to the Queensland Flora." Bull. 7, or Bot. Bull. II. Dept. Ag., Q., 1891.

24. ———— "Contributions to the Queensland Flora." Bull. 9, or Bot. Bull. III. Dept. Ag., Q., 1891.

Glæosporium pestiferum (Cooke and Mass.) briefly defined.

25. ———— "Contributions to the Queensland Flora." Bull. 15, or Bot. Bull. IV. Dept. Ag., Q., 1891.

Fourteen species of fungi described.

26. ———— "Additional Fungus Blights." Report of Colonial Botanist in Ann. Rep. Dept. Ag., Q., 1892.

Twelve species of fungi are recorded.

27. ———— "Contributions to the Queensland Flora." Bull. 18, or Bot. Bull. V. Dept. Ag., Q., 1892.

Ten species of fungi are recorded, and most of them described.

28. ———— "A Review of the Fungus Blights which have been observed to injure Living Vegetation in the Colony of Queensland." Report Aust. Assoc. Adv. Sci., Hobart, IV., 388, 1892.

Blights are classed under epiphytes and parasites, twenty-five species under the former and one hundred and fifteen under the latter, with hosts.

29. BAILEY (F. M.)—"Contributions to the Queensland Flora." Botany Bull. VIII. Dept. Ag., Q., 1893.

Forty-five species of fungi are recorded.

30. —— "Additional Fungus Blights." Report of Colonial Botanist in Ann. Rep. Dept. Ag., Q., 1893.

Six fungi are noted and two new.

31. —— "Companion for the Queensland Student of Plant Life."

32. —— "Botany abridged " Dept. Ag., Q., 1894.

Edible fungi in Queensland recorded

33. —— "Contributions to the Queensland Flora." Botany Bull. IX. Dept. Ag., Q , 1894.

Thirty-six species described and two recorded without description.

34. BAILEY (F. M.) and GORDON (P. R.)—" Plants reputed Poisonous and Injurious to Stock." 8vo. Brisbane, 1887.

A few fungi are added, injurious to fodder plants. Nine altogether, with an illustration.

BAILEY (F. M.)—[See "Tenison-Woods (J. E)"]

35. BANCROFT (J.)—" Experiments with Indian Wheats in Queensland." Proc. Roy. Soc.. Queensland, 1., Pt. 4, 1884.

Indian Wheats of the tall dark-bearded kinds found to be rust-resisting.

36. BANCROFT (T. L.)—" Notes on Bacterial Diseases of the Roots of Leguminosæ." Proc. Linn. Soc., N.S.W., Vol. VIII., Pt. 1., 1893.

Five leguminous plants affected — Mimosa, Sesbania, Desmodium, Medicago, and Crotalaria.

37. BARWICK (J.)—" Smut in Wheat." Proc. Roy. Soc., Tasm., 1889.

Considers that it is grain damaged in threshing which is smutty, and self-sown grain is never smutty.

38. BELL (R.)—" Some Account of Red Rust and its Remedy." Pp. 10. Ballarat, 1893.

The remedy given is to apply a solution of common salt to the growing wheat plant—1 lb. of salt to 1 gallon of water.

39. BENNETT (G.)—"Gatherings of a Naturalist in Australasia." 8vo. Lond., 1860.

Reference to a luminous agaric.

40. BENSON (A. H.)—" Principal Insect and Fungus Pests in New South Wales, and their remedies." Ag. Gaz., N.S.W., III., Pt. 8, 1892.

Notices injurious fungi, with their remedies, on Citrus, Apple, Pear, Apricot, Plum, and Peach trees, and Vines.

41. —— " Apple Culture." Ibid. V., Pt. 6., 1894.

Fungus diseases of Apple described and illustrated.

42. BERKELEY (M. J.)—" Contributions towards the Flora of Van Diemen's Land." Fungi. Ann. Nat. Hist. III., 1839.

Twenty-seven species given, twelve of which are common European fungi.

43. —— " Description of Two New Fungi in the Collection of Sir W. J. Hooker." Hook., Journ.Bot. II., Pl. 1. 1840.

Lentinus fasciatus is described from Tasmania.

44. BERKELEY (M. J.)—" On some Entomogenous Sphæriæ." Hook., Lond. Journ. Bot. II., Pl. 1, 1843.

Seven described altogether, and one (Sphæria Taylori) described and figured from N.S.W.

45. —— "Decades of Fungi." Decade I. Ibid. III, Pl. 2, 1844.

Three new species are described from Australia—Agaricus nidiformis, Polyporus portentosus, and Aseroë rubra.

46. —— "Decades of Fungi." Decades III.-VII. Ibid. IV., Pl. 2, 1845.

Forty-nine new species described, and some figured.

47. —— " Decades of Fungi." Decades VIII.-X. Ibid. IV., Pl. 2, 1845.

Three new species decribed—Sphæria elevata, S. pulvinulus, and S. inspersa.

48. —— "Decades of Fungi." Decade XI. Ibid. V., 1846.

Four new species described for Australia—Marasmius hepaticus, Hexagonia similis, Polyporus brunneo-leucus, and Peziza fusispora.

49. —— "On Cordyceps Gunnii." Hook., Lond. Journ. Bot. VII., 577, Pl. 22, 1848.

First described and figured.

50. —— "On some Entomogenous Sphæriæ." Linn. Journ. 1., Pl. 1, 1856.

Entomogenous species of Cordyceps mentioned for Australia—C. Gunnii and C. Taylori.

51. —— " Introduction to Cryptogamic Botany." 8vo. London, 1857.

Refers to various Australian fungi, and gives drawings of some such as Cyttaria Gunnii.

52. —— " Flora of Tasmania." Fungi. Hooker's Botany of the Antarctic Voyage. 4to. Pt. III., Vol. II., Pl. 4. London, 1860.

Two hundred and seventy-five species are described, only about eight of which are peculiarly Australian.

53. —— " Outlines of British Fungology." 8vo. Lond., 1860.

Tasmanian fungi referred to at pp. 34 and 35.

54. —— " On a Collection of Fungi from Cuba." Journ. Linn. Soc. X., 1868.

Habitats given for species occurring also in Australia.

55. —— " Australian Fungi, received principally from Baron F. von Mueller and Dr. R. Schomburgh." Ibid. XIII., 1873.

Fungi characterized, and where new described; received during a period of nearly twenty years.

56. —— " Enumeration of the Fungi collected during the Expedition of H.M.S. Challenger," 1874-5. (Third notice.) Ibid. XVI., 1877.

Forty-nine species are recorded altogether from the neighbourhood of Sydney, N.S.W., and new species are described.

57. —— " Gardener's Chronicle." 791, Fig. 130, 1878.

Description of Cordyceps Menesteridis—the same as, or a variety of, C. entomorrhiza.

58. —— "Australian Fungi." Part II. Received principally from Baron F. von Mueller. Ibid. XVIII., 1880.

A number of new species described.

59. BERKELEY (M. J.) and BROOME (C. E.)—"On some species of the genus Agaricus from Ceylon." Trans. Linn. Soc. XXVII., 149–152. Pl. 33–31, 1868.

60. —— "The Fungi of Ceylon." Linn. Journ. Bot., Vols. XI., XIV., and XV., 1870–75.

Contain descriptions of fungi common to Ceylon and Australia, between which countries remarkable coincidences occur in the distribution of some species.

61. —— "List of Fungi from Brisbane, Queensland, with Descriptions of New Species." Ibid. I., 2nd Ser., Bot. I., Pl. 2, 1878.

About one hundred and twenty species recorded.

62. —— Ibid. II., 2nd Ser., Pl. 6, 1882.

Fifty-three species recorded, for the most part common European ones.

63. —— "List of Fungi from Queensland and other Parts of Australia, with Descriptions of New Species." Ibid. II., 2nd Ser., Pt. 3, Pl. 1, 1886.

Supplementary to previous lists.

64. BERKELEY (M. J.) and CURTIS (M. A.)—"Fungi Cubenses (Hymenomycetes)." Journ. Linn. Soc. X., 1867.

Habitats given for species occurring also in Australia.

65. BUCHENO (J. E.)—"On the Potato as an Article of National Diet, and the Potato Disease in connexion with Distress in Ireland." Proc. Roy. Soc., Van Diemen's Land. I., Pt. III., 1851.

Sceptical as to the disease originating with the Aphis; that more likely the insect is the effect than the cause.

66. BLIGHT IN WHEAT.—Dept. Ag., Vict., Bull. No. 1. June, 1888.

Disease similar to "Take-all."

67. BRESADOLA (J.) and SACCARDO (P. A.) — "Pugillus mycetum Australiensium." Malpighia, Genoa, 1890.

Eighty-three species are recorded, with plate, five of which are new.

68. BROWN (R.)—"Miscellaneous Botanical Works (Ray Society): General Remarks—Geographical and Systematical—on the Botany of Terra Australis." I., 1866. (Reprinted from the voyage to Terra Australis, by M. Flinders, London, 1814.)

Ten species of fungi are noted.

69. CAMPBELL (F. M.) [now Mrs. Martin].—"Victorian Fungi hitherto unrecorded." Vict. Nat. II., No. 11, 127, 1886.

Twenty-one species are given.

70. —— "Thirty Species of Fungi hitherto unrecorded for Victoria." Vict. Nat. IV., No. 6, 95, 1887.

71. —— "Vegetable Pathology." Vict. Nat. IV., No. 8, 124, 1887.

Reference is made to the importance of the subject, and the great damage done by fungus pests to our forest trees and cultivated plants.

72. —— "Fungus Pests." Vict. Roy. Com. Veg. Prod., 5th Progress Report, 1888.

Various fungus diseases pointed out, and a large and interesting collection exhibited.

73. CLARSON (W.)—"Blights and their Teachings." Bull. No. 5, Dept. Ag. Vict., Sept., 1889.

General reference to various fungus pests of the orchard.

74. CLARSON (W.)—"The Fruit Garden." Pts. I. and II. Melb., 1889.

Reference is made to fungus diseases under the different fruits.

75. CONN (N. A.)—"Peach Rust in our Orchards (Uromyces amygdali)." Ag. Gaz., N S.W., I., Pt. 1, 1890.

Description, with figure, and treatment recommended.

76. —— "Report on Pumpkin Mould." Ibid. 1890.

Description and drawing of mould belonging to the Erysiphex.

77. —— "Contributions to an Economic Knowledge of the Australian Rusts (Uredineæ)." Ag. Gaz., N.S.W., I., Pt. 3, 1890.

Methods of investigation given, and rust occurring around wheat paddocks recorded.

78. —— "Notes on Diseases of Plants." Ibid. II., Pt. 1, 1891.

Anthracnose on Vines, Fusicladium pyrinum, Sphærella fragariæ, and Puccinia malvacearum noted.

79. —— "Pathological Notes." Ibid. II., Pt. 2, 1891.

Sphærella destructiva (B. and Br.) on Lucerne, and red incrustation on fence-rails noted.

80. —— "Notes on Diseases of Plants." Ibid. II., Pt. 3, 1891.

Bitter Rot of Apple (Glæosporium versicolor), Flax Rust (Melampsora lini), and Peach Rust (Puccinia pruni) noted.

81. —— "Pathological Notes." Ibid. II., Pt. 4, 1891.

Maize Rust (Puccinia maydis) and Apple Scab (Fusicladium dendriticum) noted.

82. —— "Notes on Diseases of Plants." Ibid. II., Pt. 5, 1891.

Cystopus candidus or White Rust, Ustilago maydis or Maize Smut, Puccinia maydis or Maize Rust, Sphærella destructiva on Lucerne, and Water Core in Apples noted.

83. —— "Notes on Diseases of Plants." Ibid. II., Pt. 6, 1891.

Mouldy Core in Apples and Glæosporium pestiferum of the Vine noted.

84. —— "Notes on Diseases of Plants." Ibid. II., Pt. 8, 1891.

Apple Scab (Fusicladium dendriticum) and Strawberry Leaf Blight (Sphærella fragariæ) again noted.

85. —— "Notes on Diseases of Plants." Ibid. II., Pt. 10, 1891.

Onion Mildew (Peronospora Schleideniana), Tobacco Mildew (Peronospora hyoscyami), Banana Disease, and Bread Mould on Oranges noted. Potato Blight described, but not found in Australia.

86. —— "Smut." Ibid. II., Pt. 11, 1891.

Oat Smut (Ustilago avenæ), Wheat Smut (Ustilago tritici and Urocystis occulta), Maize Smut (Ustilago maydis), and Stinking Smut of Wheat (Tilletia fœtens) described.

87. —— "Dialogue concerning the manner in which a Poisonous Spray does its work in Preventing or Checking Blight." Ibid. II., Pt. 12, 1891.

With various drawings showing how a fine spray acts upon the spores of a fungus.

88. Conn (N. A.)—"Contributions to an Economic Knowledge of the Australian Rusts (Uredineæ)." Ibid. III., Pt. 1, 1892.

What has been found out in this and other countries concerning Wheat Rust together with the examination of a number of varieties of wheat, and the kind of rust determined.

89. ——— "Contributions to an Economic Knowledge of the Australian Rusts (Uredineæ)." Ibid. III., Pt. 3, 1892.

Stiff flag, tough cuticle, and glaucousness found to be characteristic of rust-resistant Wheats. Various illustrative drawings and tables, among which the rust-devouring Diplosis is given.

90. ——— "Economic Notes on Plant Diseases." Ibid. III., Pt. 4, 1892.

Apple Scab (Fusicladium dendriticum), Powdery Mildew of Apple (Podosphæra Kunzei), Bitter Rot (Glæosporium versicolor), Mouldy Core and Water Core in Apples. Pear Scab (Fusicladium pyrinum), Shot Hole (Phyllosticta circumscissa). Anthracnose (Glæosporium ampelinum). Tufted-leaf Blight (Cercospora viticola). Strawberry-leaf Blight (Sphærella fragariæ), White Rust (Cystopus candidus). Pumpkin-leaf Oidium (Oidium erysiphoides), and Powdery Mildew of Rose (Sphærotheca pannosa) illustrated and described.

91. ——— "Plant Diseases and how to Prevent them." Ibid. III., Pt. 6, 1892.

Pourridie or Mouldy Root of the Vine, Tufted-leaf Blight of Bean, and Apple Canker described, and remedies prescribed.

92. ——— "Plant Diseases and how to Prevent them." Ibid. III., Pt. 12, 1892.

"Take-all" and Dry Blight of Wheat and Oats and Leaf Curl of Peach described and illustrated. Also two new species of fungi—Cyathus dimorphus and Tolyposporium authisticia.

93. ——— "Contributions to an Economic Knowledge of Australian Rusts (Uredineæ)." Ibid. IV., Pt. 6, 1893.

Seventy-one varieties of Wheat described and illustrated.

94. ——— "Contributions to an Economic Knowledge of Australian Rusts (Uredineæ)." Ibid. IV., Pt. 7, 1893.

Artificial crossing of Wheat and improving Wheats by selection fully described and illustrated.

95. ——— "Plant Diseases and their Remedies—Diseases of the Sugar Cane." Ibid. IV., Pt. 10, 1893.

The gumming of Sugar Cane is due to Bacillus vascularum, and the following five species of fungi are described and illustrated :—Cromyces Kuhnii, Strumella sacchari, Macrosporium gramineum, Phoma sacchari, and an undetermined species causing "Red Rot."

96. ——— "Host and Habitat Index of the Australian Fungi." Ag. Dept., N.S.W. (Miscellaneous Publication, No. 16). p. 44, 1893.

Contains all the fungi recorded in Dr. Cooke's "Handbook of Australian Fungi."

97. ——— "Contributions to an Economic Knowledge of Australian Rusts (Uredineæ)." Ibid. V., Pt. 4, 1894.

Improving Wheat by selection.

98. ——— "Notes on Diseases of Plants." Ibid. V., Pt. 6, 1894.

Bean Anthracnose (Colletotrichum Lindemuthianum). Bean Rust (Cromyces Phaseoli). Peach Freckle (Cladosporium carpophilum), Black Rot of Tomato (Macrosporium Tomato), a Mango Blight (Pestalozzia uvicola, Speg.). Disease of Grass (Helminthosporium Ravenelii) illustrated and described.

99. Conn (N. A.)—"A New Australian Fungus." Ibid. V., Pt. 6, 1894.

Peziza Lyonsiæ described and illustrated.

100. Conn (N. A.) and Olliff (A. S.)—"Insect Larva (Cecidomyia, sp.) Eating Rust on Wheat and Flax." Ag. Gaz., N.S.W., II., Pt. 2, 1891. Also Ann. Nat. Hist. VII., 6th Ser. 1891.

Larva observed under the microscope unmistakably feeding on rust spores. The larva not only devours the spores but spreads the rust.

101. Cooke (M. C.)—"The Beech Morels of the Southern Hemisphere." Pharm. Journ. (3), I., 264, 1870.

Cyttaria Gunnii (B.), found on living branches of Fagus Cunninghamii and F. Gunnii in Tasmania, is figured and described.

102. ——— "Jew's Ear (Hirneola auricula-judæ)." Ibid. 681, 1871.

Figured and popularly described.

103. ——— "Fungi: Their Nature, Influence, and Uses." 8vo. London, 1875.

Various Australian species referred to.

104. ——— "Australian Fungi." Grev. VI. 70. 1877.

Twelve species collected in neighbourhood of Melbourne by Mr. Le Fevre, one of which was undescribed, viz., Trametes scrobiculata (Berk.).

105. ——— "New Zealand Fungi." Grev. VIII. 54, 1879.

Thirty-six Australian species recorded.

106. ——— "Australian Fungi." Grev. IX. 142, 1881.

107. ——— "Australian Fungi." Grev. X. 60, 93. and 131, 1882.

108. ——— "Australian Fungi." Grev. XI. 28 and 57, 1882.

109. ——— "On Xylaria and its Allies." Grev. XI. 81, 1883.

Three new Australian species described.

110. ——— "Australian Fungi." Grev. XI. 97 and 115, 1883.

111. ——— "Hypoxylon and its Allies." Grev. XI. 121, 1883.

Three Australian species described.

112. ——— "Fungi." Trans. Roy. Soc., S.A., Vol. XVI., Pt. 2, 1883.

Eleven species recorded and two described.

113. ——— "Fungi Australiani." 8vo., Pt. 4, pp. 72, London and Melbourne, 1883.

Hymenomycetes, 783 species; Gastromycetes, 111 ; Myxomycetes, 33 ; Æcidiomycetes, 47 ; Discomycetes, 84; Pyrenomycetes, 94; Hyphomycetes, 47 ; Phycomycetes 4 ; total 1,203 species.

114. ——— "Some Exotic Fungi." Grev. XII. 65, 1884.

Meliola densa (Cooke) described.

115. ——— "Some Exotic Fungi — Australasia." Grev. XIV. 11, 1885.

Five species described.

116. ——— "Some Exotic Fungi." Grev. XIV. 89, 1886.

Phyllosticta palmicola (Cooke) described.

117. ——— "Exotic Fungi." Grev. XV. 16, 1886.

Six species described.

118. ——— "Some Australian Fungi." Grev. XV. 93, 1887.

Fourteen species described.

119. COOKE (M. C.)—"Some Australian Fungi." Grev. XVI. 97, 1887.
Nineteen species described.

120. —— "New Australian Fungi." Grev. XVI. 1, 1887.
Thirty-five species described.

121. —— "Two Remarkable Fungi." Grev. XVI. 20, 1887.
Cerebella paspali and Hemiarcyria applanata.

122. —— "Australasian Fungi." Grev. XVI. 30, 1887.
Nineteen species described.

123. —— "Australian Fungi." Grev. XVI. 72, 1888.
Twenty-nine species described or referred to.

124. —— "Australasian Fungi." Grev. XVI. 113, 1888.
Eight species described.

125. —— "Australasian Fungi." Grev. XVII. 7, 1888.
Seven species described.

126. —— "Australian Fungi." Grev. XVII. 55, 1889.
Seven species described.

127. —— "Some Brisbane Fungi." Grev. XVII. 69, 1889.
Four species described.

128. —— "New Australian Fungi." Grev. XVIII. 1, 1889.
Forty-two species described.

129. —— "New Australian Fungi." Grev. XVIII. 25, 1889.
Six species described, and one new genus—Seismosarca.

130. —— "Australian Fungi." Grev. XVIII. 1, 1890.
Thirty-six species mostly described.

131. —— "Australian Fungi." Grev. XVIII. 49, 1890.
Two species described—Sphæropsis phomatoidea and Capnodiastrum orbiculatum.

132. —— "Australian Fungi." Grev. XVIII. 80, 1890.
Three species described.

133. —— "On Campbellia—New Genus." Grev. XVIII. 87, 1890.

134. —— "Australian Fungi." Grev. XIX. 5, 1890.
Five species described.

135. —— "Australian Fungi." Grev. XIX. 44, 1890.
Fifteen species and new genus (Chainoderma) described.

136. —— "Australian Fungi." Grev. XIX. 89, 1891.
Twenty-one species described.

137. —— "Additions to Dædalea." Grev. XIX. 93, 1891.
Dædalea Muelleri described.

138. —— "Trametes and its Allies." Grev. XIX. 98, 1891.
Sclerodepsis, Trametes, and Hexagonia noticed.

139. —— "New Sub-genus of Agaricus." Grev. XIX. 104, 1891.
Metraria insignis described.

140. —— "Additions to Merulius." Grev. XIX. 108, 1891.
Merulius pelliculosus described.

141. COOKE (M. C.)—"Australian Fungi." Grev. XX. 4, 1891.
Seventeen species described.

142. —— "Species of Cyphella." Grev. XX. 9, 1891.
Cyphella australiensis described.

143. —— "Notes on Clavariei." Grev. XX. 10, 1891.
Clavaria Muelleri (Berk.) and C. tasmanica (Berk.) are described.

144. —— "Notes on Thelophoræ." Grev. XX. 11, 1891.
Species of Hymenochæte and Corticium referred to.

145. —— "Apple Scab (Fusicladium dendriticum)." Grev. XX. 27, 1891.
Noticing profusion of examples from Australia and methods of checking quoted.

146. —— "Ceylon in Australia." Grev. XX. 29, 1891.
Ceylon species of fungi found also in Australia.

147. —— "Australian Fungi." Grev. XX. 65, 1892.
Four species described.

148. —— "A Mystery Solved." Gardener's Chronicle, 20th Oct., 1892.
Sclerotium of Mylitta with Polyporus ovinus.

149. —— "Vegetable Wasps and Plant Worms—A Popular History of Entomogenous Fungi or Fungi Parasitic upon Insects." S.P.C.K., Lond. 1892.
Seven Australian forms are noted and described.

150. —— "Handbook of Australian Fungi." 8vo., pp. 458. Pl. 36. London, 1892.
Gives a technical description of over two thousand species, and is intended to include all Australian fungi known to date.

151. —— "Australian Fungi—A Supplement to Handbook." Grev. XXI. 35, 1892.
Twenty-four species recorded as additions and corrections to the Handbook.

152. —— "Fungi." Trans. Roy. Soc., S.A., XVI, Pt. II., 1893.
Eleven species recorded, and two described—Stephensia araneiraga (Cooke and Mass.) and Diploderma Subulosum (Cooke and Mass.).

153. —— "Exotic Fungi." Grev. XXI. 75, 1893.
A new species (Diplodia Marsdeniæ, C. and M.) described.

154. —— "Australian Fungi." Grev. XXII. 36, 1893.
Nine new species described, and seven old ones recorded.

155. —— "Australian Fungi." Grev. XXII. 68, 1893.
Two new species described—Hypoxylon atrosphæricum (Cooke and Mass.) and Belonidium parasiticum (Cooke and Mass.).

COOKE (M. C.)—[See "Kalchbrenner (K.)"]

156. COOKE (M. C.) and MASSEE (G.)—"Gloeosporium pestiferum." Gardener's Chronicle, London, 1891.

157. CORDA (A. C. J.)—"Icones Fungorum hucusque cognitorum." 6 vols. Folio. 1837-54.

158. COUESLAND (F.)—"Disease of the Vine (or Oidium Tuckeri), and its Remedy." Pamphlet. Melb., 1876.

159. CRAWFORD (F. S.)—"The Apricot Disease." Proc. Roy. Soc., S.A., 1884.
Phyllosticta circumscissa and Helminthosporium rhabdiferum.

160. CRAWFORD (F. S.)—"Report en the Fusicladiums, the Codlin Moth, and certain other Fungus and Insect Pests attacking Apple and Pear Trees in South Australia." Pp. 70. Pl. 7. Adelaide, 1886.

Fusicladium dendriticum and *F. pyrinum* fully described and illustrated, and their history given, together with several remedies suggested.

161. —— "Insect and Fungus Pests." Vict. Roy. Com. Veg. Prod., 4th Progress Report and App. No. 5, 1887.

Refers to Fusicladiums and other fungus pests, and suggests a central department in Victoria for Vegetable Pathology.

162. —— "Insect and Fungus Pests." Proc. of First Congress of Ag. Bureau of S.A. Pt. 6. Adelaide, 1890.

Describes and illustrates rust in Wheat.

163. CRICHTON (D. A.)—"The Australasian Fruit Culturist." 8vo. Melbourne, 1893-4.

Under the heading of the various fruits gives the principal fungus diseases.

164. DESPEISSIS (J. A.)—"Anthracnosis, or Black Spot of the Grape." Ag. Gaz., N.S.W., II., Pt. 7, 1891.

Drawings of *Sphaceloma ampelinum* and methods of treatment and remedies.

165. —— "Oidium in Grapes." Ibid. V., Pt. 10, 1894.

Experiments with different preparations in its treatment.

166. DRUMMOND (J.)—"Fungi of Swan River, West Australia." Hook., Lond. Journ. Bot. I., 1842.

A general account of some luminous fungi, in letter dated 1841.

167. ELLIS (Smith)—"Preventive for Rust in Wheat." 1p. 15, Melbourne. 1890.

Recommends that the seed-wheat be reaped when perfectly ripe, then to be kept perfectly dry, and finally sown in a wet seed bed, as the "Rust Smoke" perishes in water.

168. ENDLICHER (S.)—"Iconographia Generum Plantarum." Folio Vindob. 1838.

Ascra pentactina figured = *A. rubra, var. pentactina*.

169. FISCHER (Ed.)—"Versuch einer Syst. Uebersicht uber die bisher bekannaten Phalloideen." 8vo. Berlin, 1886.

170. —— "Unters. z. vergl. Entwicklungsgeschichte und Systematik der Phalloideen, in Neue Denkschr. d. Allgem.—Schweiz. gesellsch. f. d. ges. Naturwiss." Bd. XXXII., p. 4, Zürich, 1889.

171. —— "Beiträge zur Kenntniss exotischer Pilze." Hedwigia, 2, 1891.

Refers to *Mylitta australis* (Berk.) and *Lentinus Cynthus* (Berk. and Br.).

172. FRIES (E. M.)—"Plantæ Preissianæ in Australasia Coll. Fungi." 8vo. Hamburg, 1846-7.

Forty-one species enumerated.

173. FULTON (T. W.)—"The Dispersion of the Spores of Fungi by the Agency of Insects, with Special Reference to the Phalloidei." Ann. Bot. III., No. X. 1889.

Ten Australian species referred to.

174. GALLOWAY (T. B.)—"Rust of Flax." Journ. Myc. V., No. 4, 1889.

Melampsora lini (D. C.) Tul., received from S.A.

GORDON (P. R.)—[See " Bailey (F. M.)"]

175. GRAY (G. R.)—"Notices of Insects that are known to form the bases of Fungoid Parasites." London, 1858 (privately printed).

Cordyceps Hawkesii (Gray) described and figured.

176. "Grevillea: A Monthly, now Quarterly, Record of Cryptogamic Botany and its Literature." Edited by Dr. M. C. Cooke, and now by G. Massee. 8vo. Illustrated. Vols. I.-XXII. London. 1872-94—(continued).

177. HAMLET (W. M.)—"Anthrax in Australia, with some Account of Pasteur's Method of Vaccination." Trans. Int. Med. Cong. Anstr., 522, 1889.

178. HAVILAND (E.)—"On a Microscopic Fungus (Oidium monilioides) Parasitic on Cucurbitaceæ." Proc. Linn. Soc., N.S.W., I., 2nd Ser., 1886.

179. HEDWIGIA.— " Ein Notizblatt für Cryptogamische Studien." 8vo. Illustrated. Vols. I.-XXXII.—(continued). Dresden, 1852-94.

HILL (W. H. F.)—[See " McAlpine (D.)"]

180. JENSEN (F. L.)—"The Strawberry (Fragaria): Its History and Cultivation." Ag. Gaz., N.S.W, III., Pt. 7, 1892.

Strawberry-leaf Blight referred to.

181. "Journal of the Bureau of Agriculture, South Australia." Monthly. Edited by A. Molineux, F.L.S. Vols. I.-VII. Adelaide, 1889-94—(continued).

Contains numerous useful notes on disease-causing fungi.

182. KALCHBRENNER (K.)—"Phalloidei novi vel minus cogniti." 8vo. Buda-Pest, 1880.

183. —— "Fungi in reg. div. Australiæ et Asiæ a Jul. Remy collecti." 1863-6. 8vo. Toulouse, 1880.

Pholiota prominens described and illustrated (will be given in Supplement).

184. —— "Fungi of Australia—Basidiomycetes." Grev. VIII. 151, 1880.

Twenty species described.

185. —— "Definition of some new Australian Fungi." Proc. Linn. Soc., N.S.W., VII., 1882.

Eight species described.

186. —— "Fungi aliquot Australiæ orientalis." Proc. Linn. Soc., N.S.W., VII., 1882.

Five species described.

187. —— "New Species of Agaricus discovered in Western Australia." Proc. Linn. Soc., N.S.W., VII., 1882.

Five new species described, and seven species recorded.

188. —— "Description of Two New Fungi." Proc. Linn. Soc., N.S.W., VIII., 1883.

Polyporus Pentzkyi (Kalch.) and *Paxillus hirtulus* (F. v. M.) described.

189. —— "Gastromycetes novi vel minus cogniti." 8vo. Buda-Pest, 1883.

Australian species described, with coloured illustrations.

190. KALCHBRENNER (K.) and COOKE (M. C.)—"Australian Fungi." Grev. IX. 1, 1880.

Sixteen species recorded, fifteen of which are described, and a new genus (*Anthurus*) constituted.

191. KYNGDON (F. R.)—"Rust in Wheat." Address at Conference. Ag. Gaz., N.S.W., I., Pt. 1., 1890.

Suggests experiments for dealing with it.

192. LABILLARDIÈRE (J. J.)— " Relation du Voyage à la Recherche de la Pérouse." 2 vols., 4to , with atlas in folio, Paris, An. VIII., 1799-1800 ; or English translation, in 1 vol., 4to., or 2 vols., 8vo., London, 1800.

193. ——— " Novæ Hollandiæ Plantarum Specimen." 2 vols., 4to, Paris 1804-1806.

Genus *Asroë* founded, and *A. rubra* described and figured, from a specimen found in Tasmania.

194. LAMB (S.) — " Tobacco: Its Cultivation in Northern Queensland." Bull. 6, Dept. Ag., Q., 1890.

Tobacco Blight referred to, and remedy recommended.

195. ——— " Tobacco : Its Cultivation in Southern Queensland." Bull. 15, Dept. Ag., Q., 1892.

Blue Mould (*Peronospora hyoscyami*) stated to be very destructive in North Queensland, but not as common in South Queensland.

196. LÉVEILLÉ (J. H.)—"Champignons Exotiques," in Ann Sci. Nat., 3rd Ser., Vols. II.-III., 8vo. Paris, 1844.

Sphæria (Conferva) atra, N. sp. [*Sphærella*], and *S. (Conf.)labecula*, N. sp. [*Physalospora*] from New Holland, described.

197. ——— " Descr. des Champ. de l'Herbier du Mus. de Paris," in Ann. Sci. Nat. (3) V. 111., 249, 1846.

Thelephora (Stereum) Leuchardtiana, Lev., from Moreton Bay, and *Sphæria (Conferta) tephrosia*, Lev., with *Dothidea Grevillii*, Lev., from Swan River, described.

198. LIVERSIDGE (A.)—" Disease in the Sugar Cane, Queensland." Pamphlet, pp. 34 (no date).

Disease known as " Rust " and fungus considered to be the consequence of and not cause of disease.

199. LUDWIG (F.)—"Ueber einige neue Pilze aus Australien " (On some New Fungi from Australia). Bot. Cent., Pt. XLIII., 1890, and Cassel 1889.

200. ——— " Contributions to the Fungal Flora of Australia." Translated and communicated by J. G. O. Tepper. Roy. Soc., S. A , XIV. 55, 1891.

Contains lists of the Australian *Uredineæ*, *Ustilagineæ*, and the parasitic enemies of Eucalypts and Acacias.

201. ——— " Ueber neue Australische Rostkrankheiten " (On New Australian Rust Diseases). Zeitschrift f. Pflanzenkrankheiten. Vol. II., Pt. 3, 130-4, 1892.

Two new species are described—*Puccinia Tepperi* (Ludw.) on *Arundo Phragmites*, and *Puccinia munita* on the underside of the leaves of *Hydrocotyle hirta*. Also *Puccinia Magnusiana* (Körn) on *Arundo Phragmites*.

202. ——— "Ueber einige Rost- und Brand-pilze, Australiens" (On some Rust and Smut Australian Fungi). Zeitschrift f. Pflanzenkrankheiten. Vol. III., Pt. 3, 137-9, 1893.

Five species described, four of which are new— viz., *Puccinia Burchardia*, *Ustilago Spinifleis*, *U. cumbureus*, and *U. catenata*.

203. MAIDEN (J. H.)—"Australian Indigenous Plants providing Human Food and Food Adjuncts." Proc. Linn. Soc., N.S.W., 1888.

Agaricus (Psalliota) campestris and *Mylitta australis* mentioned.

204. ——— " The Useful Native Plants of Australia (including Tasmania)." 8vo. Sydney and London, 1889.

Agaricus campestris and *Mylitta australis* noted with reference to their edible qualities.

205. MAIDEN (J. H.)—"A Bibliography of Australian Economic Botany." Tech. Ed. Series, No. 10. Sydney, 1892.

References given to economic fungi.

206. ——— " Native Bread or Native Truffle (*Polyporus Mylittæ*, C. and M., syn. *Mylitta australis*, Berk.)" Ag. Gaz., N.S.W., IV., Pt. 12, 1893.

Description and drawing given and preliminary chemical examination made.

207. MASSEE (G.)—"Monograph of the Genus Lycoperdon." Journ. Roy. Micr. Soc., Pl. II , 1887.

One hundred and twenty-nine species recorded, eleven of which are found in Australia.

208. ——— "Monograph of the Genus Calostoma. Desv. (*Mitremyces, Nees*)." Ann. Bot. II., No. V., 1888.

Calostoma fuscu and *C. lurida* described.

209. ——— " A Revision of the Trichiaceæ." Journ. Roy Micr. Soc., Pl. 4, 1889.

Fifty-one species recorded, seven of which are found in Australia.

210. ——— "Monograph of the British Gastromycetes." Ann. Bot. IV., No. XIII., 1889.

The habitats are given for Australia.

211. ——— "Monograph of the Genus Podaxis." Journ. Bot., Feb., 1890.

212. ——— "Monograph of the Thelephoreæ." Part I.— Journ. Linn. Soc., Pl. 3, 1890. Part II.—Journ. Linn. Soc., Pl. 3, 1891.

A number recorded for Australia.

213. ——— "Monograph of Myxogastres." 8vo. Coloured plates. Lond., 1892.

Forty-five Australian species are described.

214. ——— " Notes on Exotic Fungi in the Royal Herbarium, Kew." Grev. XXI. 1, 1892.

Cyathus Budeyi (Mass.) from Brisbane described.

215. ——— " Australian Fungi." Grev. XXII. 17, 1893.

Two new species described — *Puccinia Kochia* (Mass.) and *Xylaria Readeri* (F. v. M.); and *Phoma uricola* recorded.

MASSEE (G.)—[See " Cooke (M. C.)"]

216. McALPINE (D.)—" Report of the Vegetable Pathologist." Dept. Ag., Vict , Bull. 12, 1891.

Objects of section of vegetable pathology dealing with fungus pests stated, and reference to reports upon *Cromyces betæ* and *Plasmodiophora Brassicæ*.

217. ——— " The Life-history of the Rust of Wheat." Ibid., Bull. 14, Pl. I.-II., 1891.

A popular account of the different phases of rust, (*Puccinia graminis*).

218. ——— " Rust of Wheat." Ibid., Bull. 14, 1891.

Uredospores of *Puccinia graminis* found during the winter season as well as in summer, and *Puccinia poarum* also observed all the year round.

219. ——— " Report of Wheat Blight." Ibid., Bull. 14, 1891.

Gives an account of *Septoria tritici*.

220. ——— " Report on Club Root of Cauliflowers, Cabbages, Turnips, and other Cruciferous Plants." Ibid., Bull. 14, Pl. III., 1891.

Description of *Plasmodiophora Brassicæ*, with preventive measures.

221. McALPINE (D.)—"Beet-leaf Rust or Blighted Mangel Leaves." Ibid., Bull. 14, 1891.

Description of *Uromyces betæ*, with preventive measures.

222. —— "Root Fungus of Raspberry (Raspberry-root Disease)." Ibid., Bull. 14, 1891.

Mycelium of a fungus found on the roots.

223. —— "Report on Peach and Plum Leaf Rust (Puccinia pruni)." Ibid., Bull. 14, Pl. V.-VI., 1891.

Notes the appearance of this disease not only on the leaves but also on the fruit of the Peach. A full account of it is given, together with preventive and remedial measures.

224. —— "Report on Rust in Wheat as Victorian Delegate to New South Wales." Proc. Rust in Wheat Conf., Sydney, 1891.

225. —— "Ueber die Verwendung geschrumpfter Körner von rostigem Weizen als Saatgut." (On the use of Shrivelled Grains of Rusty Wheat for Seed). Zeitschrift f. Pflanzenkrankheiten, III. Pt. 4, 1892.

Gives the results of experiments with rust-shrivelled and plump grain, showing that 87 per cent. of the former germinated, as against 67 per cent. of the latter.

226. —— "Report on Rust in Wheat as Victorian Delegate to South Australia." Proc. Rust in Wheat Conf., Adelaide, 1892.

227. —— "Native Bread (Polyporus Mylittæ C. and M.)." Aust. Jour. Pharm. VIII. 291, 1893.

Fresh specimen of sclerotium described, along with its edible qualities.

228. —— "The Undescribed Uredospores of Puccinia Burchardiæ, Ludwig." Vict. Nat. X. 192, 1894. (Read Nov. 1893.)

Uredospores on stem and leaves described.

229. —— "Report on Rust in Wheat Experiments, 1892-3." Pp. 66. Illustrated with maps and drawings. Govt. Printer, Melbourne, 1894.

Results obtained from 313 experimental plots.

230. —— "Report on Rust in Wheat as Victorian Delegate to Brisbane." Proc. Rust in Wheat Conf., Brisbane, 1894.

Results obtained from 255 plots at School of Horticulture, Burnley ; also from farmers' wheat-testing experiments.

231. —— "Botanical Nomenclature, with special reference to the Fungi." Report Aust. Assoc. Adv. Sci., Adelaide, V., 414, 1893.

232. —— "Australian Fungi." Proc. Roy. Soc., Vict., VII., N.S., 1895.

Twenty-eight species recorded, eight of which are new to science, in addition to one new variety.

233. —— "Systematic Arrangement of Australian Fungi, together with Host-index and List of Works on the subject." Dept. of Agriculture. Govt. Printing Office, Melbourne, 1894.

Gives a list of all known Australian fungi up to date, numbering 2278 species, with habitats, occurrence, general characters, &c. Also the host-plants with their accompanying fungi, and a numbered list of all works relating to the subject.

234. McALPINE (D.) and HILL (W. H. F.)—"The Entomogenous Fungi of Victoria." Proc. Roy. Soc., Vict. VII., N.S., 159, 1895.

Isaria Oncopteræ (McAlp.) described on dead larvæ of *Oncoptera intricata*.

235. McALPINE (D.) and TEPPER (J. G. O.)—"A New Australian Stone-making Fungus" (*Laccocephalum basila-pilodes*, McAlp. and Tepp.) Proc. Roy. Soc., Vict. VII., N.S. 166. Pl. X., 3 Figs., 1895.

A new genus (*Laccocephalum*, McAlp.) constituted and the stone-forming species fully described.

236. "Michelia, Commentarium Mycologicum Italicum." Edited by P. A. Saccardo. 8vo.—(*continued*.) Padua, 1877-94.

Published at irregular intervals, and devoted exclusively to fungi.

237. MORRISON (A.)—"Notices of Victorian Fungi : New or imperfectly described *Uredineæ*." Vict. Nat. XI., No. 6, 90, 1894.

Puccinia Coprosmatis new and uredospores of *Uromyces Orchidearum* (Cooke and Mass.) first described.

238. —— "Notices of Victorian Fungi : New or imperfectly described *Uredineæ*—(*continued*.)" Vict. Nat. XI., No. 8, 1894.

Puccinia Coprosmatis, var. *Operculariæ*, new, and *Puccinia investita* (Schwein) new to Australia.

239. MÜCKE (C.)—"The 'Take-All' " (*Xenodochus cerealium*, F. v. M.), with five plates. Prize Essay, pp. 19. Melb., 1870.

The cause of this disease supposed to be due to a fungus.

240. MUELLER (F. v.)—" Fragmenta Phytographiæ Australiæ—Supplement." The fungi determined by Berkeley, von Thuemen, Kalchbrenner, and Cooke. 8vo., Vol. XI. Melb. 1880.

One thousand and sixty-nine species of fungi are recorded.

241. —— "Census of the Genera of Plants, hitherto known as Indigenous to Australia." Proc. Roy. Soc., N.S.W., XV., 1881.

The genera of Australian fungi are recorded, with authority for names and year of publication.

242. —— "Additions to Census of the Genera of Plants." Proc. Roy. Soc., N.S.W., XVII., 1883.

Several genera of fungi are given.

243. —— "Notes on Victorian Fungs." Vict. Nat. II., No. 6, 76, 1885.

Gives list of fungi obtained by the Botanic Department of Melbourne from 1882 to 1884 as new for Victoria. Compiled from successive records furnished by Dr. M. C. Cooke, M.A. One hundred and one species are enumerated.

244. —— " Further Additions to Census of the Genera of Plants " Proc. Roy. Soc., N.S.W., XX., 1886.

Several genera of fungi recorded.

245. —— "Notes on Rare Victorian Fungi." Vict. Nat. III., No. 10, 140, 1887.

Cyttaria Gunnii (Berk.) and *Cordyceps Taylori* are given.

246. —— "Notes on some New and Rare Plants." Proc. Roy. Soc., Tasm., 1887.

Two new fungi for Tasmania are given—*Diploderma glauenm* (Cooke and Mass.) and *Castoreum radicatum* (Cooke and Mass.).

247. —— "List of Fungi named by Dr. M. C. Cooke—collected near Lake Bonney by Miss Webb." Proc. Roy. Soc., S.A., XI., 1888.

Thirty-five species are recorded.

248. MUELLER (F. v.)—"Select Extra-tropical Plants, readily Eligible for Industrial Culture or Naturalization." Melb. Successive editions and translations up to 1891.

References to useful and edible fungi.

249. MUSSON (C. T.)—"Notes on Insect and Fungous Pests." Ag. Gaz., N.S.W., V., Pt. 8, 1894.

Notices fungus diseases of Apple, Pear, Peach, Apricot. and Vine, and suggests remedies.

250. —— "Notes on Insect and Fungous Pests." Ibid. V., Pt. 9, 1894.

Suggests measures for dealing with such pests in the spring.

251. —— "Notes on Insect and Fungous Pests." Ibid. V., Pt. 10, 1894.

Suggestions for carrying out quarantine against such pests.

252. "Native Bread (Mylitta australis)." Ag. Gaz., N.S.W., III., Pt. 1, 1892.

Referred to in general notes.

253. "New Victorian Fungs." Vict. Nat. III., No. 6, 80, 1886.

Fifteen species named by Dr. M. C. Cooke, forwarded by Baron von Mueller.

254. PLOWRIGHT (C. B.)—"British Uredineœ and Ustilagineœ." 8vo. London, 1889.

Occasional references to Australian species.

255. RALPH (T. S.)—"On Dry Rot." Proc. Roy. Soc., Vict., VI., 1861.

Remarks chiefly in reference to Pines.

256. —— "On the Æcidium affecting the Senecio vulgaris. or Groundsel." Vict. Nat. VII., No. 2, 18, 1890.

257. READER (F.)—"Notes on some hitherto Unrecorded Victorian Fungi." Vict. Nat. II., 66, 1885.

258. —— "Note on Hirneola polytricha." Vict. Nat. IV., 174, 1887.

259. REICHARDT (H. W.)—"Fungi, in Reise der Oesterreichischen regatte Novara um die Erde in den Jahren 1857, 1858, 1859." Botanischer Theil. 4, Wien., 1870.

Two species—Lycogala lejosperma (Rehdt.) and Hydnum griseo-fuscescens (Rehdt.)—described, from Sydney.

REMY (J.)—[See "Kalchbrenner (K.)"]

260. "Revue Mycologique, recueil trimestriel illustré consacré à l'Etude des Champignons. Edited by C. Roumeguère. 8vo., Vols. I.-XV.—(continued)." Toulouse, 1879-94.

261. "Rust in Wheat—Minutes of Proceedings at a Conference of Delegates from Victoria, South Australia, New South Wales, and Queensland." Melb., March, 1890.

A series of experiments, and the issuing of questions to farmers suggested. Appendix to report on "The Nature of Rust in Wheat," by A. N. Pearson.

262. "Rust in Wheat—Report of the Conference at Sydney." Sydney, 1891.

Delegates from the four colonies as above, and results of experiments, together with replies from farmers given.

263. "Rust in Wheat—Report of the Proceedings of the Conference at Adelaide (Third Session)." Adelaide, 1892.

Delegate from Tasmania in addition, and reports on replies from farmers and on experiments given.

264. "Rust in Wheat." Ag. Gaz., N.S.W., III., Pt. 7, 1892.

Details of mode of conducting experiments.

265. "Rust in Wheat—Report of the Proceedings of the Rust in Wheat Conference at Brisbane (Fourth Session)." Brisbane, 1894.

Reports from the various delegates and descriptions of prominent varieties of Wheat with illustrations.

266. SACCARDO (P. A.)—"Sylloge Fungorum omnium hucusque cognitorum. 8vo, Vols. 1.-X., Padua, 1882-92.

Describes all the known species of fungi up to June, 1892 = 39,663.

267. —— "Notes Mycologiques—Mycetes aliquot australienses." Series I., Hedwigia, 125, 1889. Series II., Bull, Soc. Myc., France, V., 115, 1890. Series III., Hedwigia, 1890. Series IV., Hedwigia, 1893.

Series II. consists of twenty-two species, four of which are new. Series III. consists of eighteen species, nine of which are new. Series IV. consists of twelve species, four of which are new.

268. SACCARDO (P. A.) and BERLESE (A. N.)—"Fungi australienses." Rev. Myc. No. 26, 1885, or Atti del. R. Instit. Veneto di Scienze, &c., Venetiæ, 1885.

Fifty-one species recorded with figures. Three new genera are constituted—Scortechinia, Gibellia, and Gomospora; and eighteen new species described.

SACCARDO (P. A.)—[See "Bresadola (J.)"]

269. SCHLECTENDAL (D. F. L. V.)—"De Ascroës genere Dissertatio." Berlin, 1847.

270. —— "Eine neue Phalloidee, nebst Bemerkungen üb. d. ganze Familie derselben." Linnæa. Bd. XXXI, 101, 1861.

Contains general remarks on the Phalloideœ, which are perhaps better represented in Australia than in any of the other great divisions of the globe.

271. SKUSE (F. A. A.)—"The New Zealand Vegetable Caterpillar." Vict. Nat. VIII., Nos. 2 and 3, 47, 1891.

Criticises paper by T. Steel on the subject, and expresses doubts as to the Sphœria Robertsii being associated with the caterpillar of Hepialus virescens.

272. SOUTHALL (W.)—"Note on a Specimen of Mylitta australis (with figure)." Pharm. Journ. (3) XV., 210, 1884.

273. SPICER (W. W.)—"Ergot." Proc. Roy. Soc., Tasm., 1877. On Lolium perenne.

274. STEELE (T.)—"The New Zealand Vegetable Caterpillar." Vict. Nat. VIII., No. 8, 110, 1890.

Refers to Australian specimens of entomogenous fungi.

275. STEPHEN (T.)—"Smut in Wheat." Proc. Roy. Soc., Tasm., 1889.

Steeps for prevention given.

276. SUTHERLAND (G.)—"The South Australian Vine-growers' Manual." Adelaide, Gov. Printer, 1892.

Oidium, Black Rot, Peronospora viticola, Black Spot or Anthracnose, Pox or Glœosporium ampelophagum referred to at pp. 101-104.

277. TATE (R.)—"A List of the Charas, Mosses, Liverworts, Lichens, Fungs, and Algals of Extra-topical S.A." Proc. Roy. Soc., S.A., Vol. IV., 1881.

Eighty-six species of fungi are recorded.

278. TAYLOR (R.)—" Description of the Bulrush Caterpillar (Sphæria Robertsii)." Tasm. Journ. I., Pl. 1. 1842.
A drawing of *Sphæria innominata* is given from N.S.W.

279. TENISON-WOODS (J. E.)—" Botanical Notes on Queensland." Proc. Linn. Soc., N.S.W., VII., 1882.
Hexagonia crinigera (Fr.) mentioned.

280. TENISON-WOODS (J. E.) and BAILEY (F. M.)—" On some Fungi of New South Wales and Queensland." Proc. Linn. Soc., N.S.W., 1880.
Notes on the genera and more remarkable species, followed by a classified list.

281. TEPPER (J. G. O.)—" Red Rust : Its Nature, Approximate Cause, and Probable Cure." Proc. Roy. Soc., S.A., III., 1879.

282. —— " Botanical Notes Relating to S.A." Proc. Roy. Soc., S.A., VI., 1883.
Additions to the list of Australian fungi—eighteen species (eight unrecorded for Australia and ten new to S.A.).

283. —— " Fungi collected near Clarendon 1882-3." Proc. Roy. Soc., S.A., VIII., 1885.
Eleven species are recorded identified by Dr. Cooke.

284. —— " Additional Lichens and Fungi of S.A. collected from 1880-85." Proc. Roy. Soc., S.A., IX., 1887.
Five species of fungi recorded.

285. —— " Notes on South Australian Fungi." Proc. Roy. Soc., S.A., XII., 1889.
List of fungi new or rare for S.A.—fourteen species—together with Australian *Ustilagineæ*.

286. —— " Additional Species of Australian Fungi." Proc. Roy. Soc., S.A., XIII., 1890.
Records twenty-four species collected by himself in S.A., and described by Winter, Saccardo, and Ludwig in various journals.

287. —— " ' Take-all ' and its Remedies." Ag. Gaz. N.S.W., III., Pt. 1, 1892.
Fungus not considered to be the cause of the disease, but simply starvation of the crop.
TEPPER (J. G. O.)—[See " Ludwig (F. M.)" and also " McAlpine (D.)"].

288. THOMPSON (E. H.)—" A Handbook to the Insect Pests of Farm and Orchard." Bull. No. 1. Dept. Ag. Tasmania.
Treats also of fungus pests, particularly *Puccinia pruni* and *Fusicladium dendriticum.*

289. THUEMEN (F. von.)—" Symbolæ ad Floram Mycologicam Australiæ I." Grev. I., 1875."

290. —— Ibid. II., Flora, 1878.

291. —— " Mycotheca Universalis." Bayreuth 1879.
Ustilago Muelleriana recorded as a new species.

292. TISDALL (H. T.)—" Fungi of Country East of Mount Baw Baw." Vict. Nat. I., No. 15, 169, 1885.
Seven species of *Agaricus* in its extended sense are recorded, determined by Dr. Cooke.

293. —— " Fungi of North Gippsland." Vict. Nat. II., No. 9, 106, 1886.
Eight species described, and found a *Polyporus* developed from *Mylitta australis.*

294. —— " Notes on Fungi in Mines." Proc. Roy. Soc., Vict. XXIV., Parts I.-II., 1887.

295. —— " Victorian Agarics." Vict. Nat. IV., No. 12, 203, 1888.
Forty-three species briefly described, and localities given.

296. TISDALL (H. T.)—" Fungi of the Season." Vict. Nat. VI., No. 7, 107, 1889.
Fungi found in or near Melbourne.

297. —— " A Curious Fungus." Vict. Nat. VI., No. 7, 119, 1889.
Species of *Cordyceps* growing from an ant *Formica corisobrina*, and found by Mr. C. French, Government Entomologist.

298. —— " Victorian Fungs new to Science." Vict. Nat. VII., No. 7, 96, 1890.
Seven new species recorded, and six of these described.

299. —— " On a Species of Isaria." Vict. Nat. X., No. 6, 90, 1893.
Found on a cocoon, supposed to be that of the moth *Darala ocellata.*

300. " Tobacco Industry in the Adelong and Tumut Districts." Ag. Gaz., N.S.W., II., Pt. 1, 1891.
Tobacco Blight (*Peronospora*) referred to and remedies suggested.

301. TRYON (H.)—" Report on Insect and Fungus Pests." Dept. Ag., Queensland. 8vo., pp. 238. Brisbane, 1889.
Records fungus diseases in Apple, Pear, Peach, Almond, Orange, Vine, Pumpkin, Potato, Maize, and Wheat.

302. TURNER (F.)—" Xylostroma giganteum., Fr. (a peculiar fungus)." Ag. Gaz., N.S.W., III., Pt. 6, 1892.
Obtained from the heart-wood of several Eucalypts.

303. TULASNE (L. R.)—" Fungi Hypogæi, p. 199, folio, 1851."
Refers to *Mylitta australis*, &c.

304. " Victorian Royal Commission on Vegetable Products, 1885-94." 8vo. Melbourne.
Ten Progress Reports issued, and fungus pests occasionally referred to.

305. WALLACE (R.)—" The Rural Economy and Agriculture of Australia and New Zealand." 8vo. London, 1891.
References to rust in Wheat, Anthracnose of the Vine, Oidium, Ergot, and *Peronospora* in Australia.

306. WALLIS (A. R.)—" The Vine Disease, Oidium Tuckeri." Two plates. Ann. Rep. Dept. Ag., Vict., 1873.

307. —— " A New Disease among Rye Grass." Two plates. Ibid., 1873.
Isaria graminiperda (Berk. and F. v. M.) as the cause of it, described in *Gardener's Chronicle,* 596, 1873.

308. WEDL (Miss).— " List of Species of Agaricus and Panus, discovered near Lake Bonney." Proc. Roy. Soc., S.A., X., 1887.
Seven species of *Agaricus* recorded, and one of *Panus (P. corbonarius).*

309. WINTER (G.)—" Exotische Pilze II., Hedwigia, 1885."
Four new species described.—*Uromyces vesiculosa, Asterina microthyrioides, Phyllachora nervisequia,* and *Lembosia orbicularis, Meliola cladotricha* (Lev.) also further described.

310. —— " Fungi Australienses." Revue Mycologique, Toulouse, 1886, and Rev. Myc. 1888.

311. WOOLLS (W.)—" A Contribution to the Flora of Australia." 8vo., pp. 255. Sydney, 1867.
Occasional references are made to fungi in the neighbourhood of Sydney.

CORRECTIONS.

P. 13.—No. 180.—S.A. in wrong column.

P. 124.—Order XX. should be XXV.

P. 156.—No. 193 should be 1939.

P. 178.—No. 2133.—*Colletotrichum* should be in brackets as a synonym.

P. 180.—No. 2152.—Add *Carpozyma* as a synonym, since Apiculate yeast is considered by some to belong to this genus.

By Authority: Robt. S. Brain, Government Printer, Melbourne.

www.ingramcontent.com/pod-product-compliance
Lightning Source LLC
Chambersburg PA
CBHW020855270326
41928CB00006B/708